Challenging Problems in Geometry

ALFRED S. POSAMENTIER

Professor of Mathematics Education
The City College of the City University of New York

CHARLES T. SALKIND

Late Professor of Mathematics
Polytechnic University, New York

DOVER PUBLICATIONS, INC.

New York

Copyright

Bibliographical Note

This Dover edition, first published in 1996, is an unabridged, very slightly altered republication of the work first published in 1970 by the Macmillan Company, New York, and again in 1988 by Dale Seymour Publications, Palo Alto, California. For the Dover edition, Professor Posamentier has made two slight alterations in the introductory material.

Library of Congress Cataloging-in-Publication Data

Posamentier, Alfred S.
Challenging problems in geometry / Alfred S. Posamentier, Charles T. Salkind.
p. cm.
Originally published: New York: The Macmillan Company, 1970.
ISBN-13: 978-0-486-69154-1 (pbk.)
ISBN-10: 0-486-69154-3 (pbk.)
1. Geometry—Problems, exercises, etc. I. Salkind, Charles T., 1898– . II. Title.
QA459.P68 1996
516′.0076—dc20 95-52535
CIP

Manufactured in the United States by Courier Corporation
69154309
www.doverpublications.com

CONTENTS

INTRODUCTION

The challenge of well-posed problems transcends national boundaries, ethnic origins, political systems, economic doctrines, and religious beliefs; the appeal is almost universal. Why? You are invited to formulate your own explanation. We simply accept the observation and exploit it here for entertainment and enrichment.

This book is a new, combined edition of two volumes first published in 1970. It contains nearly two hundred problems, many with extensions or variations that we call *challenges*. Supplied with pencil and paper and fortified with a diligent attitude, you can make this material the starting point for exploring unfamiliar or little-known aspects of mathematics. The challenges will spur you on; perhaps you can even supply your own challenges in some cases. A study of these nonroutine problems can provide valuable underpinnings for work in more advanced mathematics.

This book, with slight modifications made, is as appropriate now as it was a quarter century ago when it was first published. The National Council of Teachers of Mathematics (NCTM), in their *Curriculum and Evaluation Standards for High School Mathematics* (1989), lists problem solving as its first standard, stating that "mathematical problem solving in its broadest sense is nearly synonymous with doing mathematics." They go on to say, "[problem solving] is a process by which the fabric of mathematics is identified in later standards as both constructive and reinforced."

This strong emphasis on mathematics is by no means a new agenda item. In 1980, the NCTM published *An Agenda for Action*. There, the NCTM also had problem solving as its first item, stating, "educators should give priority to the identification and analysis of specific problem solving strategies.... [and] should develop and disseminate examples of 'good problems' and strategies." It is our intention to provide secondary mathematics educators with materials to help them implement this very important recommendation.

ABOUT THE BOOK

Challenging Problems in Geometry is organized into three main parts: "Problems," "Solutions," and "Hints." Unlike many contemporary problem-solving resources, this book is arranged *not* by problem-solving technique, but by topic. We feel that announcing the technique to be used stifles creativity and destroys a good part of the fun of problem solving.

The problems themselves are grouped into two sections. Section I, "A New Twist on Familiar Topics," covers five topics that roughly parallel the sequence of the high school geometry course. Section II, "Further Investigations," presents topics not generally covered in the high school geometry course, but certainly within the scope of that audience. These topics lead to some very interesting extensions and enable the reader to investigate numerous fascinating geometric relationships.

Within each topic, the problems are arranged in approximate order of difficulty. For some problems, the basic difficulty may lie in making the distinction between relevant and irrelevant data or between known and unknown information. The sure ability to make these distinctions is part of the process of problem solving, and each devotee must develop this power by him- or herself. It will come with sustained effort.

In the "Solutions" part of the book, each problem is restated and then its solution is given. Answers are also provided for many but not all of the challenges. In the solutions (and later in the hints), you will notice citations such as "(#23)" and "(Formula #5b)." These refer to the definitions, postulates, and theorems listed in Appendix I, and the formulas given in Appendix II.

From time to time we give alternate methods of solution, for there is rarely only one way to solve a problem. The solutions shown are far from exhaustive, and intentionally so, allowing you to try a variety of different approaches. Particularly enlightening is the strategy of using multiple methods, integrating algebra, geometry, and trigonometry. Instances of multiple methods or multiple interpretations appear in the solutions. Our continuing challenge to you, the reader, is to find a different method of solution for every problem.

The third part of the book, "Hints," offers suggestions for each problem and for selected challenges. Without giving away the solution, these hints can help you get back on the track if you run into difficulty.

USING THE BOOK

This book may be used in a variety of ways. It is a valuable supplement to the basic geometry textbook, both for further explorations on specific topics and for practice in developing problem-solving techniques. The book also has a natural place in preparing individuals or student teams for participation in mathematics contests. Mathematics clubs might use this book as a source of independent projects or activities. Whatever the use, experience has shown that these problems motivate people of all ages to pursue more vigorously the study of mathematics.

Very near the completion of the first phase of this project, the passing of Professor Charles T. Salkind grieved the many who knew and respected him. He dedicated much of his life to the study of problem posing and problem solving and to projects aimed at making problem solving meaningful, interesting, and instructive to mathematics students at all levels. His efforts were praised by all. Working closely with this truly great man was a fascinating and pleasurable experience.

Alfred S. Posamentier
1996

PREPARING TO SOLVE A PROBLEM

A strategy for attacking a problem is frequently dictated by the use of analogy. In fact, searching for an analogue appears to be a psychological necessity. However, some analogues are more apparent than real, so analogies should be scrutinized with care. Allied to analogy is structural similarity or pattern. Identifying a pattern in apparently unrelated problems is not a common achievement, but when done successfully it brings immense satisfaction.

Failure to solve a problem is sometimes the result of fixed habits of thought, that is, inflexible approaches. When familiar approaches prove fruitless, be prepared to alter the line of attack. A flexible attitude may help you to avoid needless frustration.

Here are three ways to make a problem yield dividends:

(1) The result of formal manipulation, that is, "the answer," may or may not be meaningful; find out! Investigate the possibility that the answer is not unique. If more than one answer is obtained, decide on the acceptability of each alternative. Where appropriate, estimate the answer in advance of the solution. The habit of estimating in advance should help to prevent crude errors in manipulation.

(2) Check possible restrictions on the data and/or the results. Vary the data in significant ways and study the effect of such variations on the original result.

(3) The insight needed to solve a generalized problem is sometimes gained by first specializing it. Conversely, a specialized problem, difficult when tackled directly, sometimes yields to an easy solution by first generalizing it.

As is often true, there may be more than one way to solve a problem. There is usually what we will refer to as the "peasant's way" in contrast to the "poet's way"—the latter being the more elegant method.

To better understand this distinction, let us consider the following problem:

> If the sum of two numbers is 2, and the product of these same two numbers is 3, find the sum of the reciprocals of these two numbers.

Those attempting to solve the following pair of equations simultaneously are embarking on the "peasant's way" to solve this problem.

$$x + y = 2$$
$$xy = 3$$

Substituting for y in the second equation yields the quadratic equation, $x^2 - 2x + 3 = 0$. Using the quadratic formula we can find $x = 1 \pm i\sqrt{2}$. By adding the reciprocals of these two values of x, the answer $\frac{2}{3}$ appears. This is clearly a rather laborious procedure, not particularly elegant.

The "poet's way" involves working backwards. By considering the desired result

$$\frac{1}{x} + \frac{1}{y}$$

and seeking an expression from which this sum may be derived, one should inspect the algebraic sum:

$$\frac{x + y}{xy}$$

The answer to the original problem is now obvious! That is, since $x + y = 2$ and $xy = 3$, $\frac{x + y}{xy} = \frac{2}{3}$. This is clearly a more elegant solution than the first one.

The "poet's way" solution to this problem points out a very useful and all too often neglected method of solution. A reverse strategy is certainly not new. It was considered by Pappus of Alexandria about 320 A.D. In Book VII of Pappus' *Collection* there is a rather complete description of the methods of "analysis" and "synthesis." T. L. Heath, in his book *A Manual of Greek Mathematics* (Oxford University Press, 1931, pp. 452-53), provides a translation of Pappus' definitions of these terms:

> *Analysis* takes that which is sought as if it were admitted and passes from it through its successive consequences to something which is admitted as the result of synthesis: for in analysis we assume that which is sought as if it were already done, and we inquire what it is from which this results, and again what is the antecedent cause of the latter, and so on, until, by so retracing our steps, we come upon something already known or belonging to the class of first principles, and such a method we call analysis as being solution backward.

> But in *synthesis*, reversing the progress, we take as already done that which was last arrived at in the analysis and, by arranging in their natural order as consequences what before were antecedents, and successively connecting them one with another, we arrive finally at the construction of that which was sought: and this we call *synthesis*.

Unfortunately, this method has not received its due emphasis in the mathematics classroom. We hope that the strategy recalled here will serve you well in solving some of the problems presented in this book.

Naturally, there are numerous other clever problem-solving strategies to pick from. In recent years a plethora of books describing various problem-solving methods have become available. A concise description of these problem-solving strategies can be found in *Teaching Secondary School Mathematics: Techniques and Enrichment Units*, by A. S. Posamentier and J. Stepelman, 4th edition (Columbus, Ohio: Prentice Hall/Merrill, 1995).

Our aim in this book is to strengthen the reader's problem-solving skills through nonroutine motivational examples. We therefore allow the reader the fun of finding the best path to a problem's solution, an achievement generating the most pleasure in mathematics.

PROBLEMS

SECTION I
A New Twist on Familiar Topics

1. Congruence and Parallelism

The problems in this section present applications of several topics that are encountered early in the formal development of plane Euclidean geometry. The major topics are congruence of line segments, angles, and triangles and parallelism in triangles and various types of quadrilaterals.

1-1 In any $\triangle ABC$, E and D are interior points of $\overline{AC}$ and $\overline{BC}$, respectively (Fig. 1-1). $\overline{AF}$ bisects $\angle CAD$, and $\overline{BF}$ bisects $\angle CBE$. Prove $m\angle AEB + m\angle ADB = 2m\angle AFB$.

Challenge 1 Prove that this result holds if E coincides with C.

Challenge 2 Prove that the result holds if E and D are exterior points on extensions of $\overline{AC}$ and $\overline{BC}$ through C.

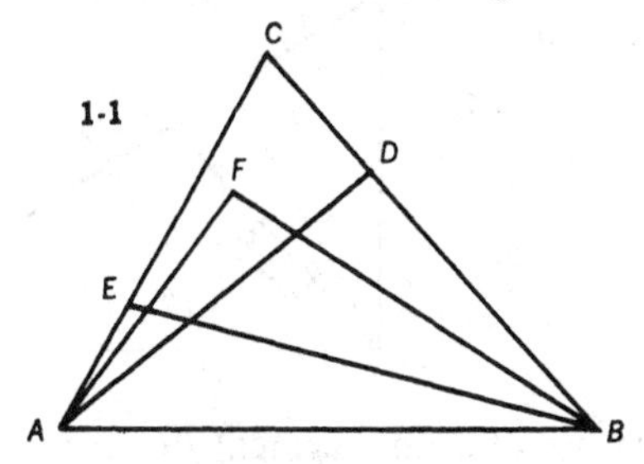

1-2 In $\triangle ABC$, a point D is on $\overline{AC}$ so that $AB = AD$ (Fig. 1-2). $m\angle ABC - m\angle ACB = 30$. Find $m\angle CBD$.

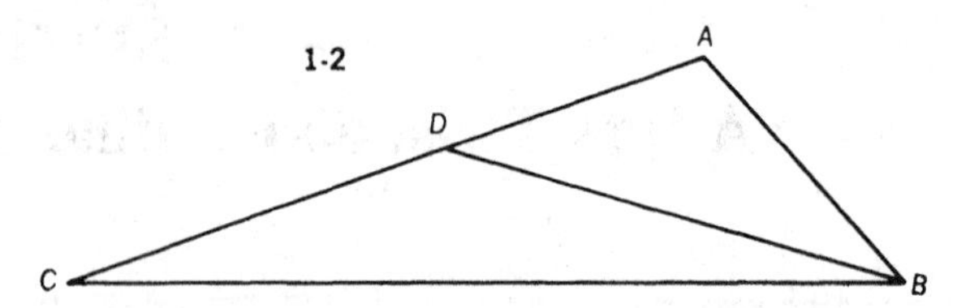

1-3 The interior bisector of $\angle B$, and the exterior bisector of $\angle C$ of $\triangle ABC$ meet at D (Fig. 1-3). Through D, a line parallel to $\overline{CB}$ meets $\overline{AC}$ at L and $\overline{AB}$ at M. If the measures of legs $\overline{LC}$ and $\overline{MB}$ of trapezoid $CLMB$ are 5 and 7, respectively, find the measure of base $\overline{LM}$. Prove your result.

Challenge Find $\overline{LM}$ if $\triangle ABC$ is equilateral.

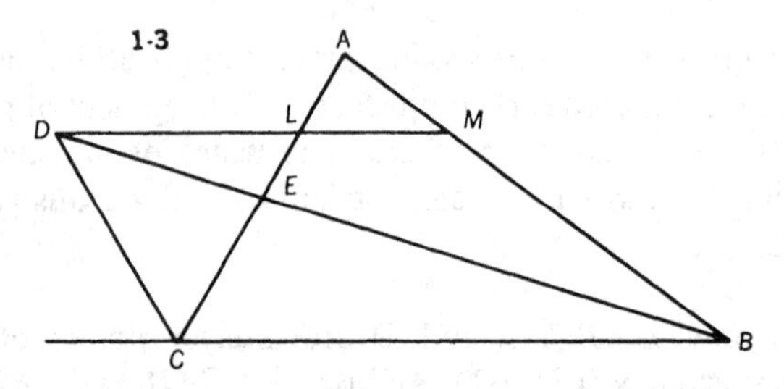

1-4 In right $\triangle ABC$, $\overline{CF}$ is the median to hypotenuse $\overline{AB}$, $\overline{CE}$ is the bisector of $\angle ACB$, and $\overline{CD}$ is the altitude to $\overline{AB}$ (Fig. 1-4). Prove that $\angle DCE \cong \angle ECF$.

Challenge Does this result hold for a non-right triangle?

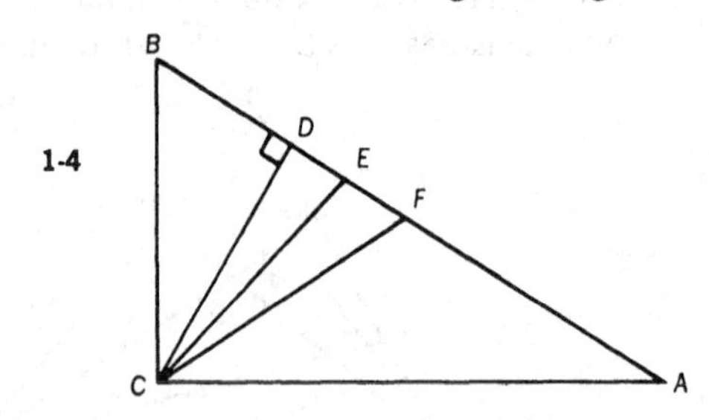

1-5 The measure of a line segment $\overline{PC}$, perpendicular to hypotenuse $\overline{AC}$ of right $\triangle ABC$, is equal to the measure of leg $\overline{BC}$. Show $\overline{BP}$ may be perpendicular or parallel to the bisector of $\angle A$.

1-6 Prove the following: if, in $\triangle ABC$, median $\overline{AM}$ is such that $m\angle BAC$ is divided in the ratio 1:2, and $\overline{AM}$ is extended through M to D so that $\angle DBA$ is a right angle, then $AC = \frac{1}{2} AD$ (Fig. 1-6).

Challenge Find two ways of proving the theorem when $m\angle A = 90$.

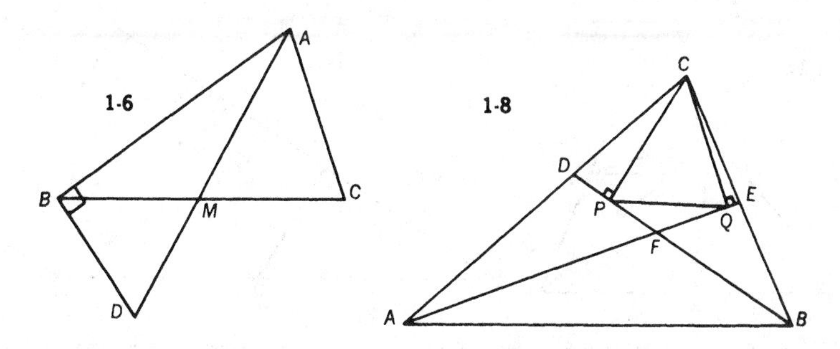

1-7 In square $ABCD$, M is the midpoint of $\overline{AB}$. A line perpendicular to $\overline{MC}$ at M meets $\overline{AD}$ at K. Prove that $\angle BCM \cong \angle KCM$.

Challenge Prove that $\triangle KDC$ is a 3–4–5 right triangle.

1-8 Given any $\triangle ABC$, $\overline{AE}$ bisects $\angle BAC$, $\overline{BD}$ bisects $\angle ABC$, $\overline{CP} \perp \overline{BD}$, and $\overline{CQ} \perp \overline{AE}$ (Fig. 1-8), prove that $\overline{PQ}$ is parallel to $\overline{AB}$.

Challenge Identify the points P and Q when $\triangle ABC$ is equilateral.

1-9 Given that $ABCD$ is a square, $\overline{CF}$ bisects $\angle ACD$, and $\overline{BPQ}$ is perpendicular to CF (Fig. 1-9), prove $DQ = 2PE$.

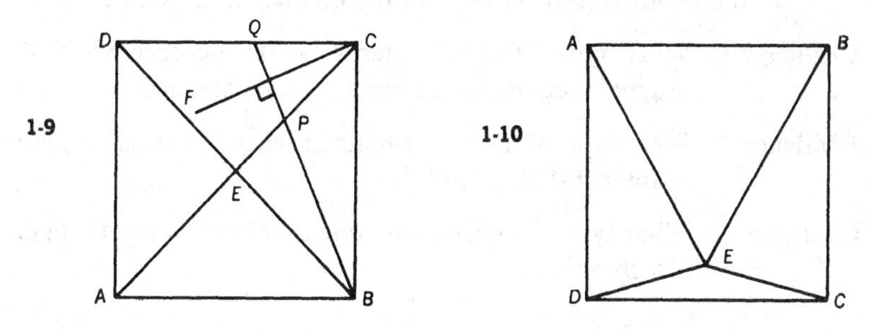

1-10 Given square $ABCD$ with $m\angle EDC = m\angle ECD = 15$, prove $\triangle ABE$ is equilateral (Fig. 1-10).

1-11 In any $\triangle ABC$, D, E, and F are midpoints of the sides $\overline{AC}$, $\overline{AB}$, and $\overline{BC}$, respectively (Fig. 1-11). $\overline{BG}$ is an altitude of $\triangle ABC$. Prove that $\angle EGF \cong \angle EDF$.

Challenge 1 Investigate the case when $\triangle ABC$ is equilateral.

Challenge 2 Investigate the case when $AC = CB$.

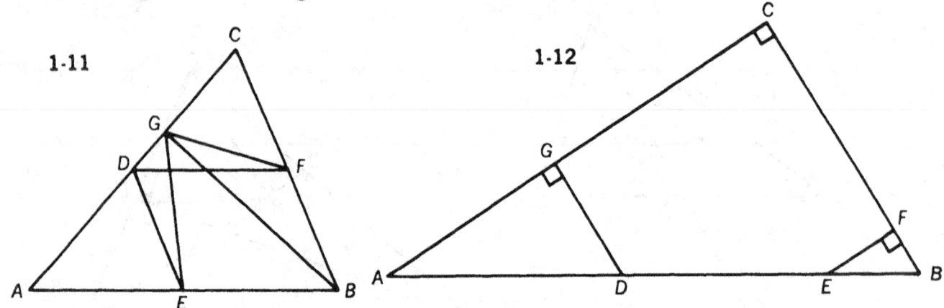

1-12 In right $\triangle ABC$, with right angle at C, $BD = BC$, $AE = AC$, $\overline{EF} \perp \overline{BC}$, and $\overline{DG} \perp \overline{AC}$ (Fig. 1-12). Prove that $DE = EF + DG$.

1-13 Prove that the sum of the measures of the perpendiculars from any point on a side of a rectangle to the diagonals is constant.

Challenge If the point were on the extension of a side of the rectangle, would the result still hold?

1-14 The trisectors of the angles of a rectangle are drawn. For each pair of adjacent angles, those trisectors that are closest to the enclosed side are extended until a point of intersection is established. The line segments connecting those points of intersection form a quadrilateral. Prove that the quadrilateral is a rhombus.

Challenge 1 What type of quadrilateral would be formed if the original rectangle were replaced by a square?

Challenge 2 What type of figure is obtained when the original figure is any parallelogram?

Challenge 3 What type of figure is obtained when the original figure is a rhombus?

1-15 In Fig. 1–15, $\overline{BE}$ and $\overline{AD}$ are altitudes of $\triangle ABC$. F, G, and K are midpoints of $\overline{AH}$, $\overline{AB}$, and $\overline{BC}$, respectively. Prove that $\angle FGK$ is a right angle.

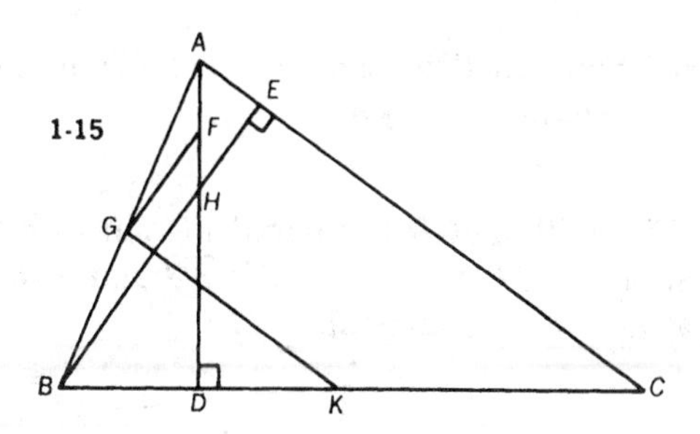

1-15

1-16 In parallelogram $ABCD$, M is the midpoint of $\overline{BC}$. $\overline{DT}$ is drawn from D perpendicular to $\overleftrightarrow{MA}$ as in Fig. 1-16. Prove that $CT = CD$.

Challenge Make the necessary changes in the construction lines, and then prove the theorem for a rectangle.

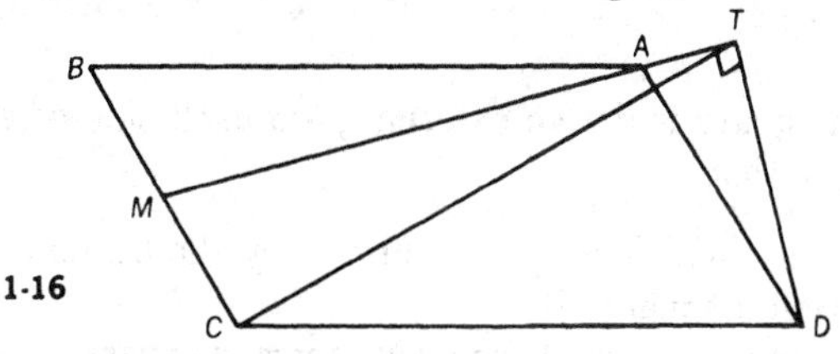

1-16

1-17 Prove that the line segment joining the midpoints of two opposite sides of any quadrilateral bisects the line segment joining the midpoints of the diagonals.

1-18 In any $\triangle ABC$, $\overleftrightarrow{XYZ}$ is any line through the centroid G (Fig. 1-18). Perpendiculars are drawn from each vertex of $\triangle ABC$ to this line. Prove $CY = AX + BZ$.

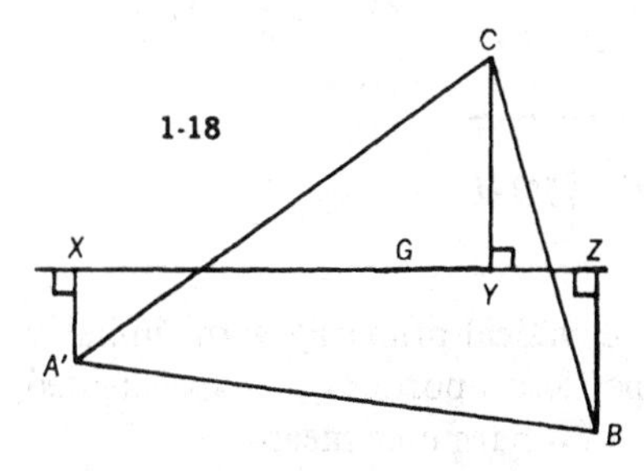

1-18

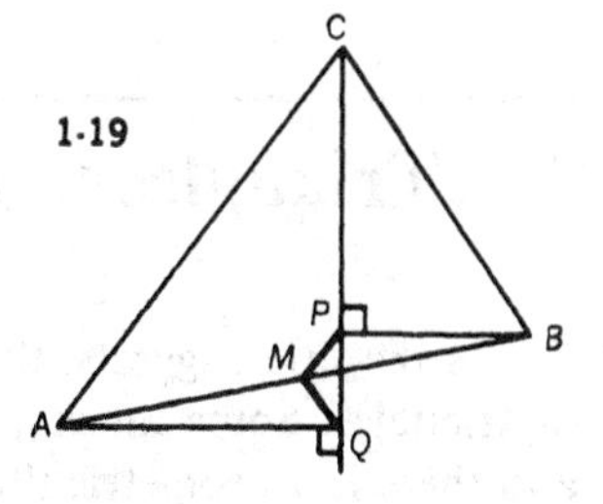

1-19

1-19 In any $\triangle ABC$, $\overleftrightarrow{CPQ}$ is any line through C, interior to $\triangle ABC$ (Fig. 1-19). $\overline{BP}$ is perpendicular to line $\overline{CPQ}$, $\overline{AQ}$ is perpendicular to line $\overline{CPQ}$, and M is the midpoint of $\overline{AB}$. Prove that $MP = MQ$.

Challenge Show that the same result holds if the line through C is exterior to $\triangle ABC$.

1-20 In Fig. 1-20, $ABCD$ is a parallelogram with equilateral triangles ABF and ADE drawn on sides $\overline{AB}$ and $\overline{AD}$, respectively. Prove that $\triangle FCE$ is equilateral.

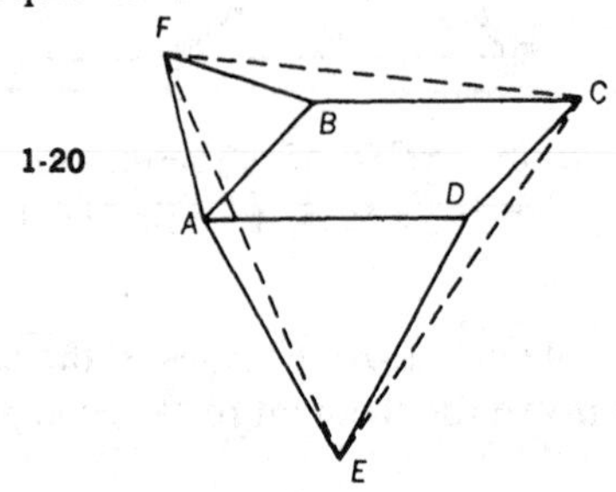

1-20

1-21 If a square is drawn externally on each side of a parallelogram, prove that

(a) the quadrilateral determined by the centers of these squares is itself a square
(b) the diagonals of the newly formed square are concurrent with the diagonals of the original parallelogram.

Challenge Consider other regular polygons drawn externally on the sides of a parallelogram. Study each of these situations!

2. Triangles in Proportion

As the title suggests, these problems deal primarily with similarity of triangles. Some interesting geometric proportions are investigated, and there is a geometric illustration of a harmonic mean.

Do you remember manipulations with proportions such as: if $\frac{a}{b} = \frac{c}{d}$ then $\frac{a-b}{b} = \frac{c-d}{d}$? They are essential to solutions of many problems.

2-1 In $\triangle ABC$, $\overline{DE} \parallel \overline{BC}$, $\overline{FE} \parallel \overline{DC}$, $AF = 4$, and $FD = 6$ (Fig. 2-1). Find DB.

Challenge 1 Find DB if $AF = m_1$ and $FD = m_2$.

Challenge 2 $\overline{FG} \parallel \overline{DE}$, and $\overline{HG} \parallel \overline{FE}$. Find DB if $AH = 2$ and $HF = 4$.

Challenge 3 Find DB if $AH = m_1$ and $HF = m_2$.

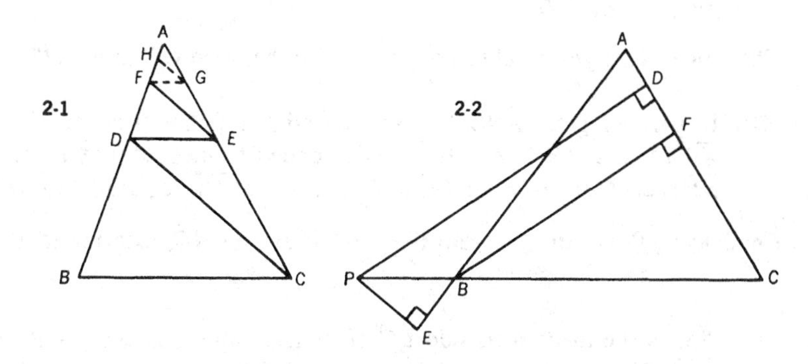

2-2 In isosceles $\triangle ABC$ ($AB = AC$), $\overline{CB}$ is extended through B to P (Fig. 2-2). A line from P, parallel to altitude $\overline{BF}$, meets $\overline{AC}$ at D (where D is between A and F). From P, a perpendicular is drawn to meet the extension of $\overline{AB}$ at E so that B is between E and A. Express BF in terms of PD and PE. Try solving this problem in two different ways.

Challenge Prove that $BF = PD + PE$ when $AB = AC$, P is between B and C, D is between C and F, and a perpendicular from P meets $\overline{AB}$ at E.

2-3 The measure of the longer base of a trapezoid is 97. The measure of the line segment joining the midpoints of the diagonals is 3. Find the measure of the shorter base.

Challenge Find a general solution applicable to any trapezoid.

2-4 In $\triangle ABC$, D is a point on side $\overline{BA}$ such that $BD:DA = 1:2$. E is a point on side $\overline{CB}$ so that $CE:EB = 1:4$. Segments $\overline{DC}$ and $\overline{AE}$ intersect at F. Express $CF:FD$ in terms of two positive relatively prime integers.

Challenge Show that if $BD:DA = m:n$ and $CE:EB = r:s$, then

$$\frac{CF}{FD} = \left(\frac{r}{s}\right)\left(\frac{m+n}{n}\right).$$

2-5 In $\triangle ABC$, $\overline{BE}$ is a median and O is the midpoint of $\overline{BE}$. Draw $\overline{AO}$ and extend it to meet $\overline{BC}$ at D. Draw $\overline{CO}$ and extend it to meet $\overline{BA}$ at F. If $CO = 15$, $OF = 5$, and $AO = 12$, find the measure of $\overline{OD}$.

Challenge Can you establish a relationship between OD and AO?

2-6 In parallelogram $ABCD$, points E and F are chosen on diagonal $\overline{AC}$ so that $AE = FC$. If $\overline{BE}$ is extended to meet $\overline{AD}$ at H, and $\overline{BF}$ is extended to meet $\overline{DC}$ at G, prove that $\overline{HG}$ is parallel to $\overline{AC}$.

Challenge Prove the theorem if E and F are on $\overleftrightarrow{AC}$, exterior to the parallelogram.

2-7 $\overline{AM}$ is the median to side $\overline{BC}$ of $\triangle ABC$, and P is any point on $\overline{AM}$. $\overline{BP}$ extended meets $\overline{AC}$ at E, and $\overline{CP}$ extended meets $\overline{AB}$ at D. Prove that $\overline{DE}$ is parallel to $\overline{BC}$.

Challenge Show that the result holds if P is on $\overleftrightarrow{AM}$, exterior to $\triangle ABC$.

2-8 In $\triangle ABC$, the bisector of $\angle A$ intersects $\overline{BC}$ at D (Fig. 2-8). A perpendicular to $\overline{AD}$ from B intersects $\overline{AD}$ at E. A line segment through E and parallel to $\overline{AC}$ intersects $\overline{BC}$ at G, and $\overline{AB}$ at H. If $AB = 26$, $BC = 28$, $AC = 30$, find the measure of $\overline{DG}$.

Challenge Prove the result for $\overline{CF} \perp \overline{AD}$ where F is on $\overleftrightarrow{AD}$ exterior to $\triangle ABC$.

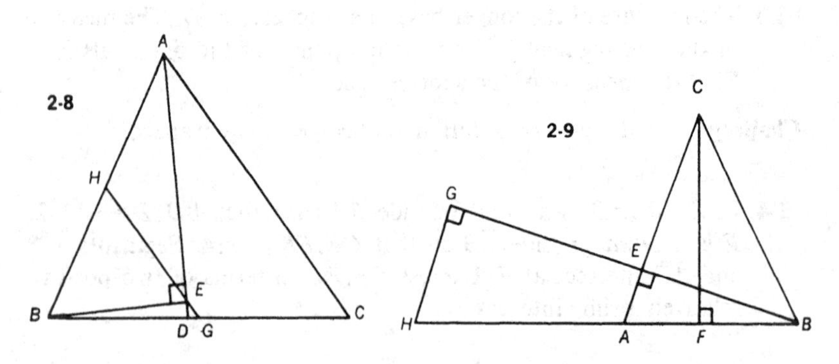

2-9 In $\triangle ABC$, altitude $\overline{BE}$ is extended to G so that $EG =$ the measure of altitude $\overline{CF}$. A line through G and parallel to $\overline{AC}$ meets $\overleftrightarrow{BA}$ at H, as in Fig. 2-9. Prove that $AH = AC$.

Challenge 1 Show that the result holds when $\angle A$ is a right angle.

Challenge 2 Prove the theorem for the case where the measure of altitude $\overline{BE}$ is greater than the measure of altitude $\overline{CF}$, and G is on $\overline{BE}$ (between B and E) so that $EG = CF$.

2-10 In trapezoid $ABCD$ ($\overline{AB} \parallel \overline{DC}$), with diagonals $\overline{AC}$ and $\overline{DB}$ intersecting at P, $\overline{AM}$, a median of $\triangle ADC$, intersects $\overline{BD}$ at E (Fig. 2-10). Through E, a line is drawn parallel to $\overline{DC}$ cutting $\overline{AD}$, $\overline{AC}$, and $\overline{BC}$ at points H, F, and G, respectively. Prove that $HE = EF = FG$.

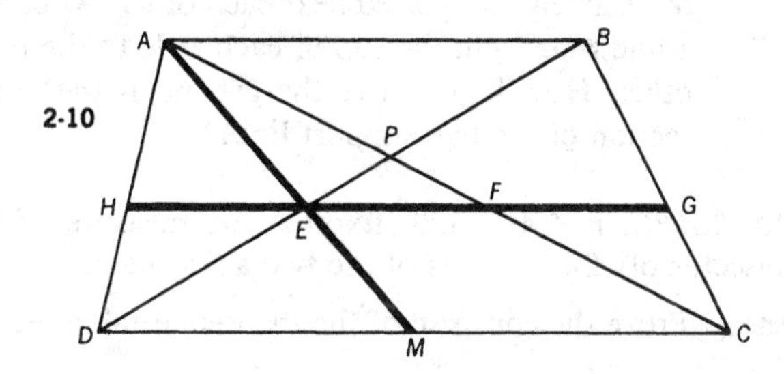

2-11 A line segment $\overline{AB}$ is divided by points K and L in such a way that $(AL)^2 = (AK)(AB)$ (Fig. 2-11). A line segment $\overline{AP}$ is drawn congruent to $\overline{AL}$. Prove that $\overline{PL}$ bisects $\angle KPB$.

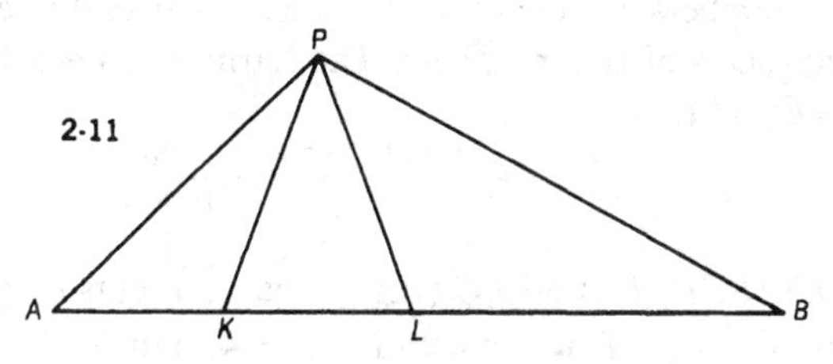

Challenge Investigate the situation when $\angle APB$ is a right angle.

2-12 P is any point on altitude $\overline{CD}$ of $\triangle ABC$. $\overline{AP}$ and $\overline{BP}$ meet sides $\overline{CB}$ and $\overline{CA}$ at points Q and R, respectively. Prove that $\angle QDC \cong \angle RDC$.

2-13 In $\triangle ABC$, Z is any point on base $\overline{AB}$ (Fig. 2-13). $\overline{CZ}$ is drawn. A line is drawn through A parallel to $\overline{CZ}$ meeting $\overleftrightarrow{BC}$ at X. A line is drawn through B parallel to $\overline{CZ}$ meeting $\overleftrightarrow{AC}$ at Y. Prove that $\frac{1}{AX} + \frac{1}{BY} = \frac{1}{CZ}$.

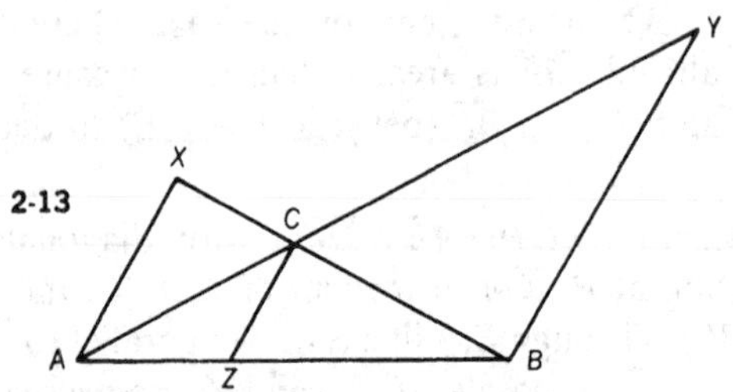

Challenge Two telephone cable poles, 40 feet and 60 feet high, respectively, are placed near each other. As partial support, a line runs from the top of each pole to the bottom of the other. How high above the ground is the point of intersection of the two support lines?

2-14 In $\triangle ABC$, $m\angle A = 120$. Express the measure of the internal bisector of $\angle A$ in terms of the two adjacent sides.

Challenge Prove the converse of the theorem established above.

2-15 Prove that the measure of the segment passing through the point of intersection of the diagonals of a trapezoid and parallel to the bases with its endpoints on the legs, is the harmonic mean between the measures of the parallel sides. The harmonic mean of two numbers is defined as the reciprocal of the average of the reciprocals of two numbers. The harmonic mean between a and b is equal to

$$\left(\frac{a^{-1} + b^{-1}}{2}\right)^{-1} = \frac{2ab}{a + b}.$$

2-16 In $\square ABCD$, E is on $\overline{BC}$ (Fig. 2-16a). $\overline{AE}$ cuts diagonal $\overline{BD}$ at G and $\overleftrightarrow{DC}$ at F. If $AG = 6$ and $GE = 4$, find EF.

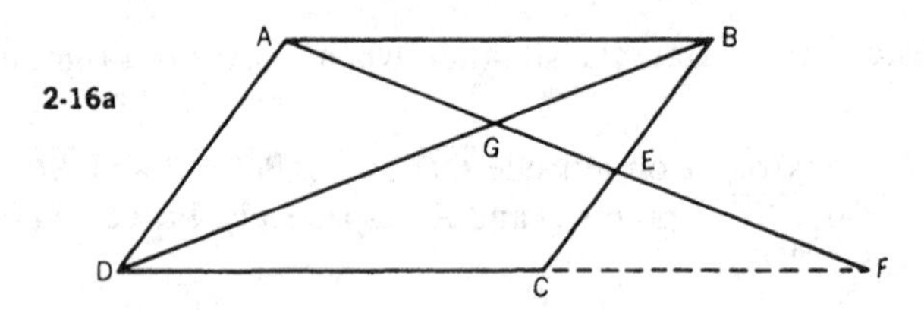

Challenge 1 Show that AG is one-half the harmonic mean between AF and AE.

Challenge 2 Prove the theorem when E is on the extension of $\overline{CB}$ through B (Fig. 2-16b).

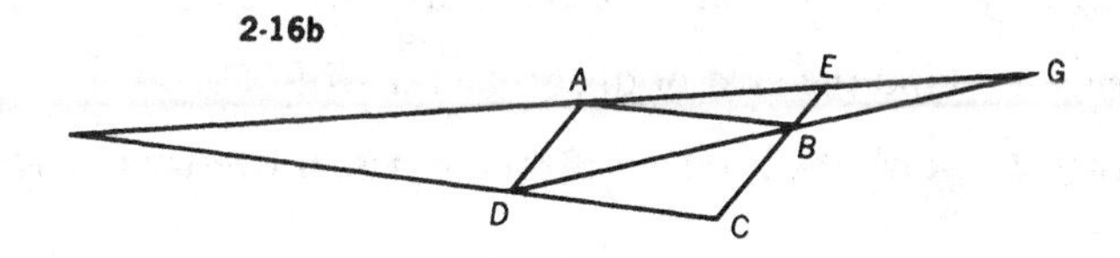

3. The Pythagorean Theorem

You will find two kinds of problems in this section concerning the key result of Euclidean geometry, the theorem of Pythagoras. Some problems involve direct applications of the theorem. Others make use of results that depend on the theorem, such as the relationship between the sides of an isosceles right triangle or a 30–60–90 triangle.

3-1 In any $\triangle ABC$, E is any point on altitude $\overline{AD}$ (Fig. 3-1). Prove that $(AC)^2 - (CE)^2 = (AB)^2 - (EB)^2$.

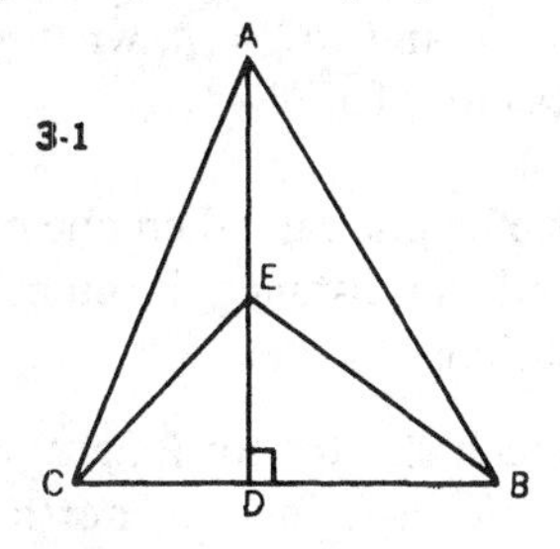

Challenge 1 Show that the result holds if E is on the extension of $\overline{AD}$ through D.

Challenge 2 What change in the theorem results if E is on the extension of $\overline{AD}$ through A?

3-2 In $\triangle ABC$, median $\overline{AD}$ is perpendicular to median $\overline{BE}$. Find AB if $BC = 6$ and $AC = 8$.

Challenge 1 Express AB in general terms for $BC = a$, and $AC = b$.

Challenge 2 Find the ratio of AB to the measure of its median.

3-3 On hypotenuse $\overline{AB}$ of right $\triangle ABC$, draw square $ABLH$ externally. If $AC = 6$ and $BC = 8$, find CH.

Challenge 1 Find the area of quadrilateral $HLBC$.

Challenge 2 Solve the problem if square $ABLH$ overlaps $\triangle ABC$.

3-4 The measures of the sides of a right triangle are 60, 80, and 100. Find the measure of a line segment, drawn from the vertex of the right angle to the hypotenuse, that divides the triangle into two triangles of equal perimeters.

3-5 On sides $\overline{AB}$ and $\overline{DC}$ of rectangle $ABCD$, points F and E are chosen so that $AFCE$ is a rhombus (Fig. 3-5). If $AB = 16$ and $BC = 12$, find EF.

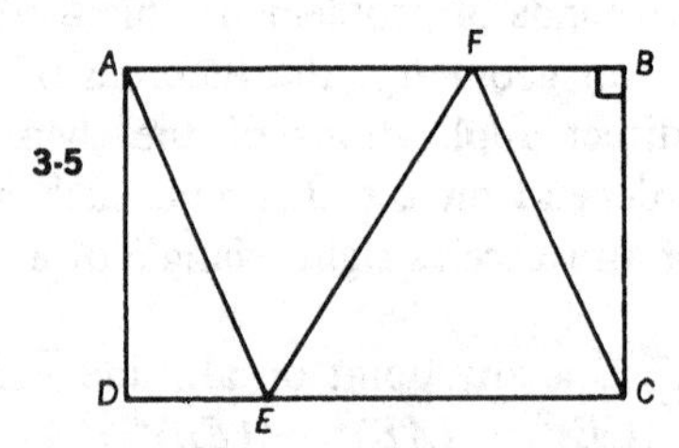

Challenge If $AB = a$ and $BC = b$, what general expression will give the measure of $\overline{EF}$?

3-6 A man walks one mile east, then one mile northeast, then another mile east. Find the distance, in miles, between the man's initial and final positions.

Challenge How much shorter (or longer) is the distance if the course is one mile east, one mile north, then one mile east?

3-7 If the measures of two sides and the included angle of a triangle are 7, $\sqrt{50}$, and 135, respectively, find the measure of the segment joining the midpoints of the two given sides.

Challenge 1 Show that when $m\angle A = 135$,

$$EF = \frac{1}{2}\sqrt{b^2 + c^2 + bc\sqrt{2}},$$

where E and F are midpoints of sides $\overline{AC}$ and $\overline{AB}$, respectively, of $\triangle ABC$.

NOTE: a, b, and c are the lengths of the sides opposite $\angle A$, $\angle B$, and $\angle C$ of $\triangle ABC$.

Challenge 2 Show that when $m\angle A = 120$,

$$EF = \frac{1}{2}\sqrt{b^2 + c^2 + bc\sqrt{1}}.$$

Challenge 3 Show that when $m\angle A = 150$,

$$EF = \frac{1}{2}\sqrt{b^2 + c^2 + bc\sqrt{3}}.$$

Challenge 4 On the basis of these results, predict the values of EF for $m\angle A = 30, 45, 60$, and 90.

3-8 Hypotenuse $\overline{AB}$ of right $\triangle ABC$ is divided into four congruent segments by points G, E, and H, in the order A, G, E, H, B. If $AB = 20$, find the sum of the squares of the measures of the line segments from C to G, E, and H.

Challenge Express the result in general terms when $AB = c$.

3-9 In quadrilateral $ABCD$, $AB = 9$, $BC = 12$, $CD = 13$, $DA = 14$, and diagonal $AC = 15$ (Fig. 3-9). Perpendiculars are drawn from B and D to $\overline{AC}$, meeting $\overline{AC}$ at points P and Q, respectively. Find PQ.

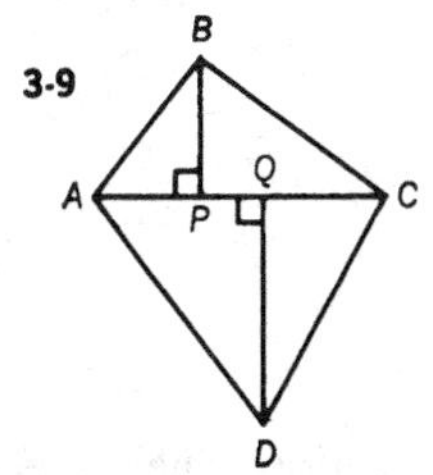

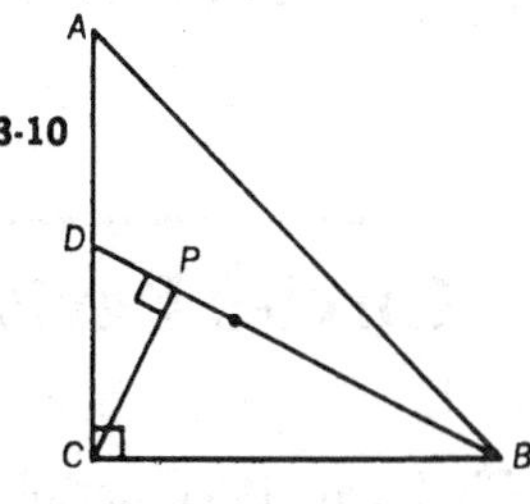

3-10 In $\triangle ABC$, angle C is a right angle (Fig. 3-10). $AC = BC = 1$, and D is the midpoint of $\overline{AC}$. $\overline{BD}$ is drawn, and a line perpendicular to $\overline{BD}$ at P is drawn from C. Find the distance from P to the intersection of the medians of $\triangle ABC$.

Challenge Show that $PG = \dfrac{c\sqrt{10}}{30}$, when G is the centroid, and c is the length of the hypotenuse.

3-11 A right triangle contains a 60° angle. If the measure of the hypotenuse is 4, find the distance from the point of intersection of the 2 legs of the triangle to the point of intersection of the angle bisectors.

3-12 From point P inside $\triangle ABC$, perpendiculars are drawn to the sides meeting $\overline{BC}$, $\overline{CA}$, and $\overline{AB}$, at points D, E, and F, respectively. If $BD = 8$, $DC = 14$, $CE = 13$, $AF = 12$, and $FB = 6$, find AE. Derive a general theorem, and then make use of it to solve this problem.

3-13 For $\triangle ABC$ with medians $\overline{AD}$, $\overline{BE}$, and $\overline{CF}$, let $m = AD + BE + CF$, and let $s = AB + BC + CA$. Prove that $\frac{3}{2}s > m > \frac{3}{4}s$.

3-14 Prove that $\frac{3}{4}(a^2 + b^2 + c^2) = m_a{}^2 + m_b{}^2 + m_c{}^2$. ($m_c$ means the measure of the median drawn to side c.)

Challenge 1 Verify this relation for an equilateral triangle.

Challenge 2 The sum of the squares of the measures of the sides of a triangle is 120. If two of the medians measure 4 and 5, respectively, how long is the third median?

Challenge 3 If $\overline{AE}$ and $\overline{BF}$ are medians drawn to the legs of right $\triangle ABC$, find the numeral value of $\dfrac{(AE)^2 + (BF)^2}{(AB)^2}$.

4. Circles Revisited

Circles are the order of the day in this section. There are problems dealing with arc and angle measurement; others deal with lengths of chords, secants, tangents, and radii; and some problems involve both.

Particular attention should be given to Problems 4-33 thru 4-40, which concern cyclic quadrilaterals (quadrilaterals that may be inscribed in a circle). This often neglected subject has interesting applications. If you are not familiar with it, you might look at the theorems that are listed in Appendix I.

4-1 Two tangents from an external point P are drawn to a circle, meeting the circle at points A and B. A third tangent meets the circle at T, and tangents $\overrightarrow{PA}$ and $\overrightarrow{PB}$ at points Q and R, respectively. Find the perimeter p of $\triangle PQR$.

4-2 $\overline{AB}$ and $\overline{AC}$ are tangent to circle O at B and C, respectively, and $\overline{CE}$ is perpendicular to diameter $\overline{BD}$ (Fig. 4-2). Prove $(BE)(BO) = (AB)(CE)$.

Challenge 1 Find the value of AB when E coincides with O.

Challenge 2 Show that the theorem is true when E is between B and O.

Challenge 3 Show that $\dfrac{AB}{\sqrt{BE}} = \dfrac{BO}{\sqrt{ED}}$.

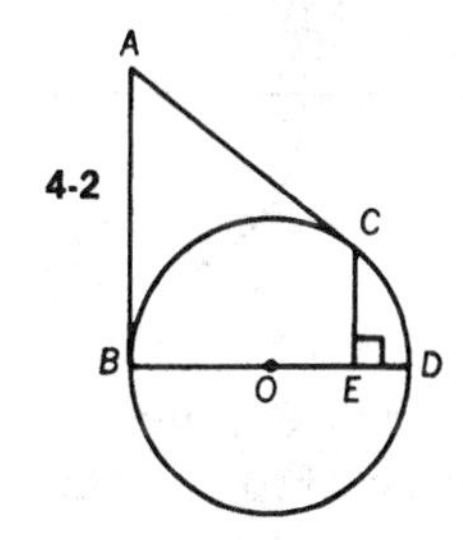

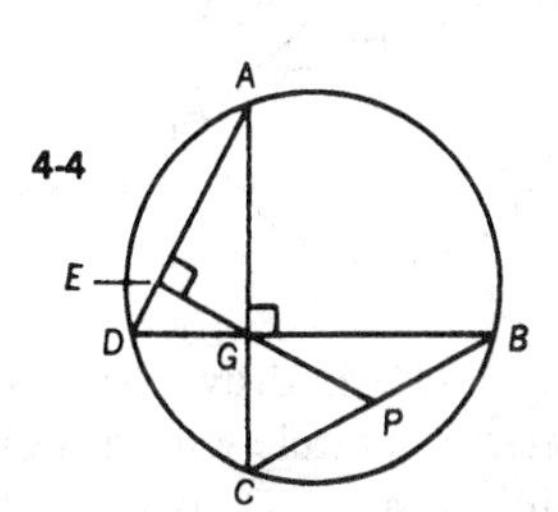

4-3 From an external point P, tangents $\overrightarrow{PA}$ and $\overrightarrow{PB}$ are drawn to a circle. From a point Q on the major (or minor) arc $\overset{\frown}{AB}$, perpendiculars are drawn to $\overline{AB}$, $\overrightarrow{PA}$, and $\overrightarrow{PB}$. Prove that the perpendicular to $\overline{AB}$ is the mean proportional between the other two perpendiculars.

Challenge Show that the theorem is true when the tangents are parallel.

4-4 Chords $\overline{AC}$ and $\overline{DB}$ are perpendicular to each other and intersect at point G (Fig. 4-4). In $\triangle AGD$ the altitude from G meets $\overline{AD}$ at E, and when extended meets $\overline{BC}$ at P. Prove that $BP = PC$.

Challenge One converse of this theorem is as follows. Chords $\overline{AC}$ and $\overline{DB}$ intersect at G. In $\triangle AGD$ the altitude from G meets $\overline{AD}$ at E, and when extended meets $\overline{BC}$ at P so that $BP = PC$. Prove that $\overline{AC} \perp \overline{BD}$.

4-5 Square $ABCD$ is inscribed in a circle. Point E is on the circle. If $AB = 8$, find the value of

$$(AE)^2 + (BE)^2 + (CE)^2 + (DE)^2.$$

Challenge Prove that for $ABCD$, a non-square rectangle, $(AE)^2 + (BE)^2 + (CE)^2 + (DE)^2 = 2d^2$, where d is the measure of the length of a diagonal of the rectangle.

4-6 Radius $\overline{AO}$ is perpendicular to radius $\overline{OB}$, $\overline{MN}$ is parallel to $\overline{AB}$ meeting $\overline{AO}$ at P and $\overline{OB}$ at Q, and the circle at M and N (Fig. 4-6). If $MP = \sqrt{56}$, and $PN = 12$, find the measure of the radius of the circle.

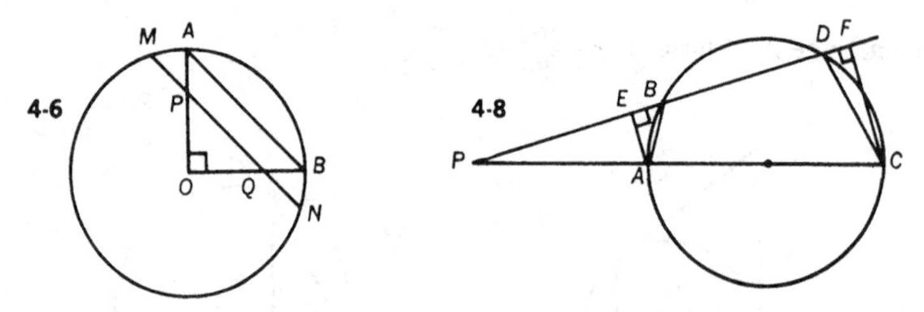

4-7 Chord $\overline{CD}$ is drawn so that its midpoint is 3 inches from the center of a circle with a radius of 6 inches. From A, the midpoint of minor arc $\overset{\frown}{CD}$, any chord $\overline{AB}$ is drawn intersecting $\overline{CD}$ in M. Let v be the range of values of $(AB)(AM)$, as chord $\overline{AB}$ is made to rotate in the circle about the fixed point A. Find v.

4-8 A circle with diameter $\overline{AC}$ is intersected by a secant at points B and D. The secant and the diameter intersect at point P outside the circle, as shown in Fig. 4-8. Perpendiculars $\overline{AE}$ and $\overline{CF}$ are drawn from the extremities of the diameter to the secant. If $EB = 2$, and $BD = 6$, find DF.

Challenge Does $DF = EB$? Prove it!

4-9 A diameter $\overline{CD}$ of a circle is extended through D to external point P. The measure of secant $\overline{CP}$ is 77. From P, another secant is drawn intersecting the circle first at A, then at B. The measure of secant $\overline{PB}$ is 33. The diameter of the circle measures 74. Find the measure of the angle formed by the secants.

Challenge Find the measure of the shorter secant if the measure of the angle between the secants is 45.

4-10 In $\triangle ABC$, in which $AB = 12$, $BC = 18$, and $AC = 25$, a semicircle is drawn so that its diameter lies on $\overline{AC}$, and so that it is tangent to $\overline{AB}$ and $\overline{BC}$. If O is the center of the circle, find the measure of $\overline{AO}$.

Challenge Find the diameter of the semicircle.

4-11 Two parallel tangents to circle O meet the circle at points M and N. A third tangent to circle O, at point P, meets the other two tangents at points K and L. Prove that a circle, whose diameter is $\overline{KL}$, passes through O, the center of the original circle.

Challenge Prove that for different positions of point P, on $\widehat{MN}$, a family of circles is obtained tangent to each other at O.

4-12 $\overline{LM}$ is a chord of a circle, and is bisected at K (Fig. 4-12). $\overline{DKJ}$ is another chord. A semicircle is drawn with diameter $\overline{DJ}$. $\overline{KS}$, perpendicular to $\overline{DJ}$, meets this semicircle at S. Prove $KS = KL$.

Challenge Show that if $\overline{DKJ}$ is a diameter of the first circle, or if $\overline{DKJ}$ coincides with $\overline{LM}$, the theorem is trivial.

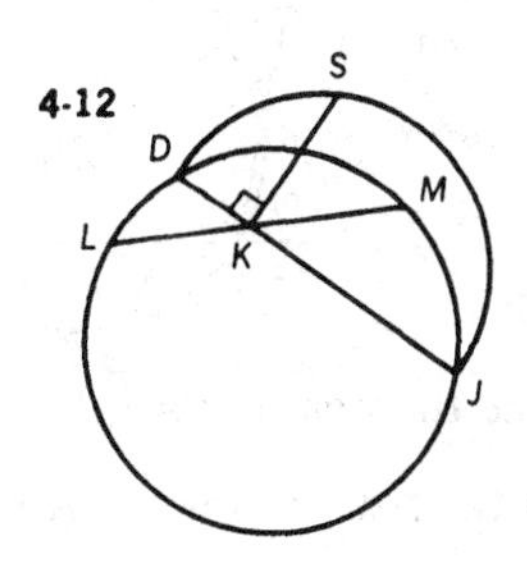

4-13 $\triangle ABC$ is inscribed in a circle with diameter $\overline{AD}$. A tangent to the circle at D cuts $\overline{AB}$ extended at E and $\overline{AC}$ extended at F. If $AB = 4$, $AC = 6$, and $BE = 8$, find CF.

Challenge 1 Find $m\angle DAF$.

Challenge 2 Find BC.

4-14 Altitude $\overline{AD}$ of equilateral $\triangle ABC$ is a diameter of circle O. If the circle intersects $\overline{AB}$ and $\overline{AC}$ at E and F, respectively, find the ratio of $EF : BC$.

Challenge Find the ratio of $EB : BD$.

4-15 Two circles intersect in A and B, and the measure of the common chord $\overline{AB}$ is 10. The line joining the centers cuts the circles in P and Q. If $PQ = 3$ and the measure of the radius of one circle is 13, find the radius of the other circle.

Challenge Find the second radius if $PQ = 2$.

4-16 $ABCD$ is a quadrilateral inscribed in a circle. Diagonal $\overline{BD}$ bisects $\overline{AC}$. If $AB = 10$, $AD = 12$, and $DC = 11$, find BC.

Challenge Solve the problem when diagonal $\overline{BD}$ divides $\overline{AC}$ into two segments, one of which is twice as long as the other.

4-17 A is a point exterior to circle O. $\overline{PT}$ is drawn tangent to the circle so that $PT = PA$. As shown in Fig. 4-17, C is any point on circle O, and $\overline{AC}$ and $\overline{PC}$ intersect the circle at points D and B, respectively. $\overline{AB}$ intersects the circle at E. Prove that $\overline{DE}$ is parallel to $\overline{AP}$.

4-17

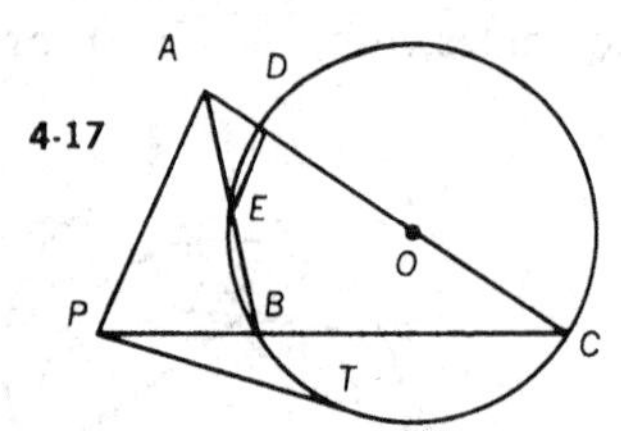

Challenge 1 Prove the theorem for A interior to circle O.

Challenge 2 Explain the situation when A is on circle O.

4-18 $\overline{PA}$ and $\overline{PB}$ are tangents to a circle, and $\overline{PCD}$ is a secant. Chords $\overline{AC}$, $\overline{BC}$, $\overline{BD}$, and $\overline{DA}$ are drawn. If $AC = 9$, $AD = 12$, and $BD = 10$, find BC.

Challenge If in addition to the information given above, $PA = 15$ and $PC = 9$, find AB.

4-19 The altitudes of $\triangle ABC$ meet at O. $\overline{BC}$, the base of the triangle, has a measure of 16. The circumcircle of $\triangle ABC$ has a diameter with a measure of 20. Find AO.

4-20 Two circles are tangent internally at P, and a chord, $\overline{AB}$, of the larger circle is tangent to the smaller circle at C (Fig. 4-20). $\overline{PB}$ and $\overline{PA}$ cut the smaller circle at E and D, respectively. If $AB = 15$, while $PE = 2$ and $PD = 3$, find AC.

Challenge Express AC in terms of AB, PE, and PD.

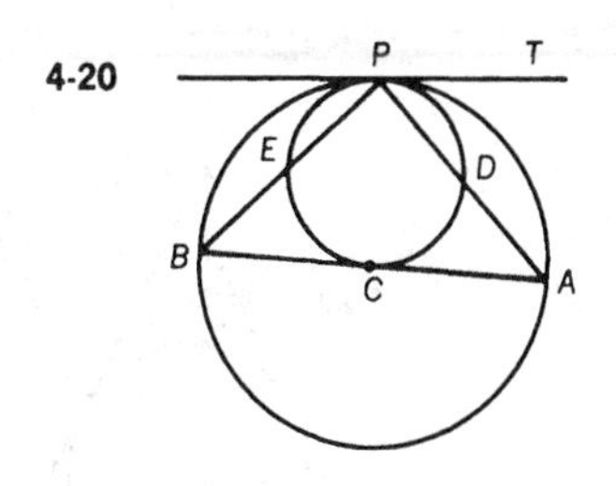

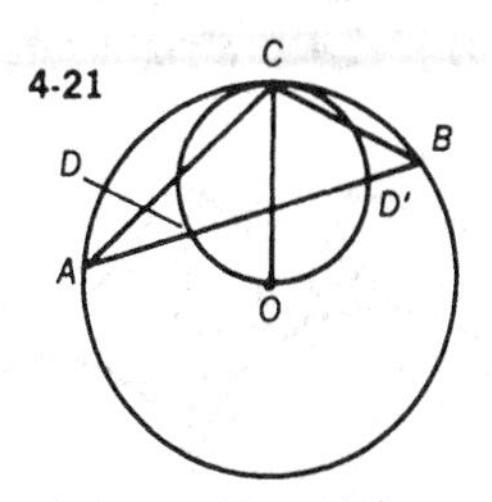

4-21 A circle, center O, is circumscribed about $\triangle ABC$, a triangle in which $\angle C$ is obtuse (Fig. 4-21). With $\overline{OC}$ as diameter, a circle is drawn intersecting $\overline{AB}$ in D and D'. If $AD = 3$, and $DB = 4$, find CD.

Challenge 1 Show that the theorem is or is not true if $m\angle C = 90$.

Challenge 2 Investigate the case for $m\angle C < 90$.

4-22 In circle O, perpendicular chords $\overline{AB}$ and $\overline{CD}$ intersect at E so that $AE = 2$, $EB = 12$, and $CE = 4$. Find the measure of the radius of circle O.

Challenge Find the shortest distance from E to the circle.

4-23 Prove that the sum of the measure of the squares of the segments made by two perpendicular chords is equal to the square of the measure of the diameter of the given circle.

Challenge Prove the theorem for two perpendicular chords meeting outside the circle.

4-24 Two equal circles are tangent externally at T. Chord $\overline{TM}$ in circle O is perpendicular to chord $\overline{TN}$ in circle Q. Prove that $\overline{MN} \parallel \overline{OQ}$ and $MN = OQ$.

Challenge Show that $MN = \sqrt{2(R^2 + r^2)}$ if the circles are unequal, where R and r are the radii of the two circles.

4-25 From point A on the common internal tangent of tangent circles O and O', secants $\overline{AEB}$ and $\overline{ADC}$ are drawn, respectively (Fig. 4-25). If $\overline{DE}$ is the common external tangent, and points C and B are collinear with the centers of the circles, prove

(a) $m\angle 1 = m\angle 2$, and
(b) $\angle A$ is a right angle.

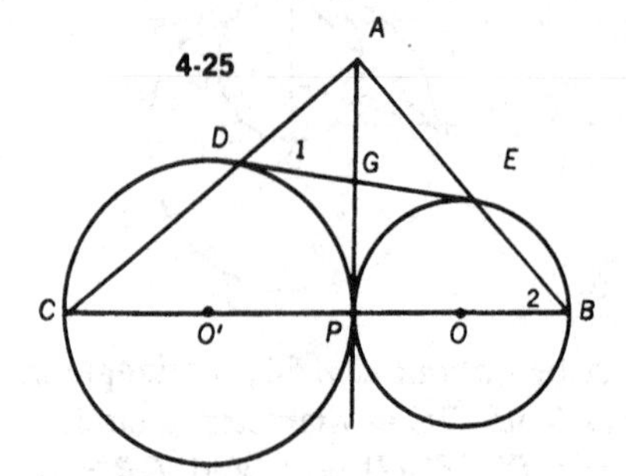

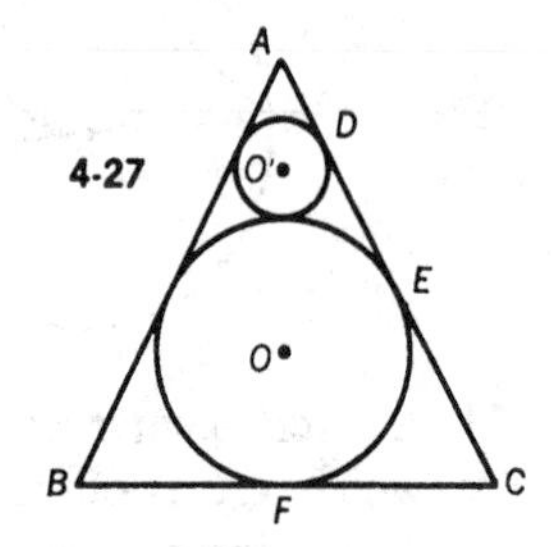

Challenge 1 Prove or disprove that if $\overline{BC}$ does not pass through the centers of the circles, the designated pairs of angles are not equal and $\angle A$ is not a right angle.

Challenge 2 Prove that DE is the mean proportional between the diameters of circles O and O'.

4-26 Two equal intersecting circles O and O' have a common chord $\overline{RS}$. From any point P on $\overline{RS}$ a ray is drawn perpendicular to $\overline{RS}$ cutting circles O and O' at A and B, respectively. Prove that $\overline{AB}$ is parallel to the line of centers, $\overleftrightarrow{OO'}$, and that $AB = OO'$.

4-27 A circle is inscribed in a triangle whose sides are 10, 10, and 12 units in measure (Fig. 4-27). A second, smaller circle is inscribed tangent to the first circle and to the equal sides of the triangle. Find the measure of the radius of the second circle.

Challenge 1 Solve the problem in general terms if $AC = a$, $BC = 2b$.

Challenge 2 Inscribe a third, smaller circle tangent to the second circle and to the equal sides, and find its radius by inspection.

Challenge 3 Extend the legs of the triangle through B and C, and draw a circle tangent to the original circle and to the extensions of the legs. What is its radius?

4-28 A circle with radius 3 is inscribed in a square. Find the radius of the circle that is inscribed between two sides of the square and the original circle.

Challenge Show that the area of the small circle is approximately 3% of the area of the large circle.

4-29 $\overline{AB}$ is a diameter of circle O, as shown in Fig. 4-29. Two circles are drawn with $\overline{AO}$ and $\overline{OB}$ as diameters. In the region between the circumferences, a circle D is inscribed, tangent to the three previous circles. If the measure of the radius of circle D is 8, find AB.

Challenge Prove that the area of the shaded region equals the area of circle E.

4-29

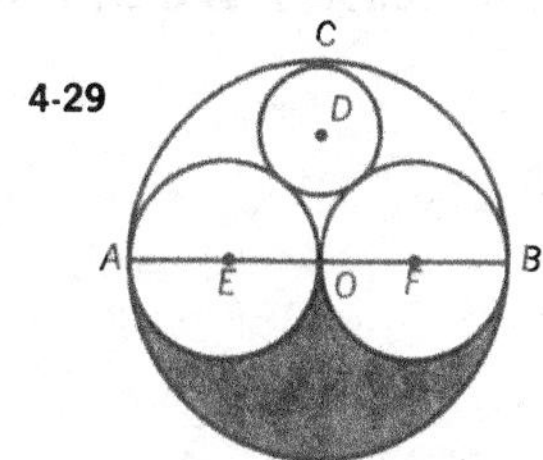

4-31

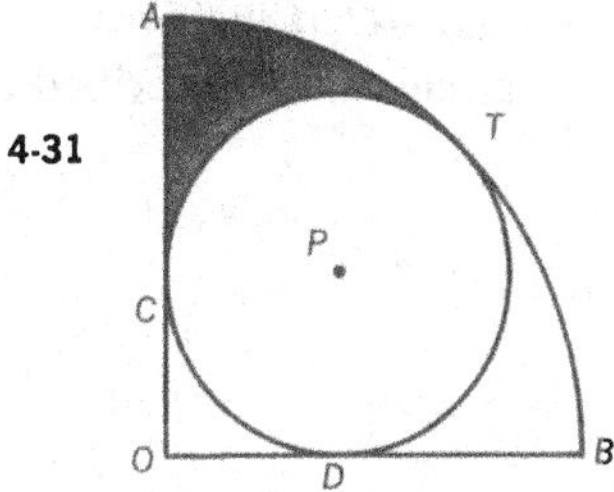

4-30 A carpenter wishes to cut four equal circles from a circular piece of wood whose area is 9π square feet. He wants these circles of wood to be the largest that can possibly be cut from this piece of wood. Find the measure of the radius of each of the four new circles.

Challenge 1 Find the correct radius if the carpenter decides to cut out three equal circles of maximum size.

Challenge 2 Which causes the greater waste of wood, the four circles or the three circles?

4-31 A circle is inscribed in a quadrant of a circle of radius 8 (Fig. 4-31). What is the measure of the radius of the inscribed circle?

Challenge Find the area of the shaded region.

4-32 Three circles intersect. Each pair of circles has a common chord. Prove that these three chords are concurrent.

Challenge Investigate the situation in which one circle is externally tangent to each of two intersecting circles.

4-33 The bisectors of the angles of a quadrilateral are drawn. From each pair of adjacent angles, the two bisectors are extended until they intersect. The line segments connecting the points of intersection form a quadrilateral. Prove that this figure is cyclic (i.e., can be inscribed in a circle).

4-34 In cyclic quadrilateral $ABCD$, perpendiculars to $\overline{AB}$ and $\overline{CD}$ are erected at B and D and extended until they meet sides $\overleftrightarrow{CD}$ and $\overleftrightarrow{AB}$ at B' and D', respectively. Prove $\overline{AC}$ is parallel to $\overline{B'D'}$.

4-35 Perpendiculars $\overline{BD}$ and $\overline{CE}$ are drawn from vertices B and C of $\triangle ABC$ to the interior bisectors of angles C and B, meeting them at D and E, respectively (Fig. 4-35). Prove that $\overleftrightarrow{DE}$ intersects $\overline{AB}$ and $\overline{AC}$ at their respective points of tangency, F and G, with the circle that is inscribed in $\triangle ABC$.

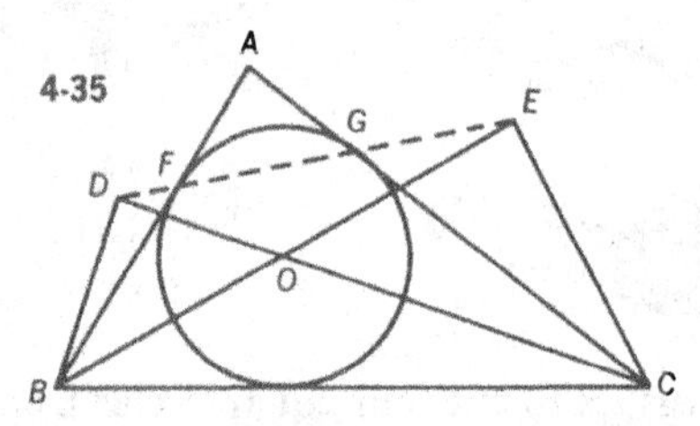

4-36 A line, $\overline{PQ}$, parallel to base $\overline{BC}$ of $\triangle ABC$, cuts $\overline{AB}$ and $\overline{AC}$ at P and Q, respectively (Fig. 4-36). The circle passing through P and tangent to $\overline{AC}$ at Q cuts $\overline{AB}$ again at R. Prove that the points R, Q, C, and B lie on a circle.

Challenge Prove the theorem when P and R coincide.

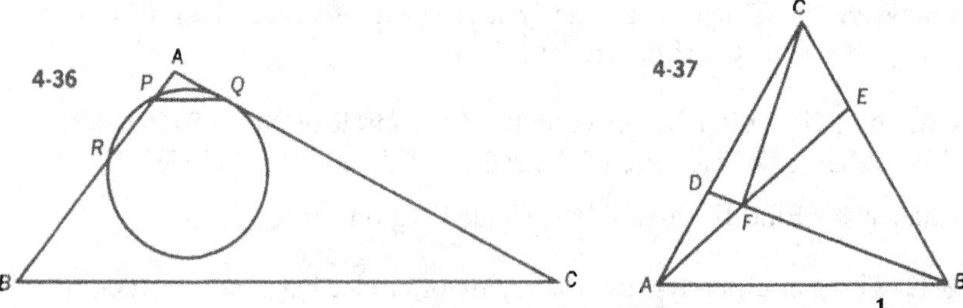

4-37 In equilateral $\triangle ABC$, D is chosen on $\overline{AC}$ so that $AD = \frac{1}{3}(AC)$, and E is chosen on $\overline{BC}$ so that $CE = \frac{1}{3}(BC)$ (Fig. 4-37). $\overline{BD}$ and $\overline{AE}$ intersect at F. Prove that $\angle CFB$ is a right angle.

Challenge Prove or disprove the theorem when $AD = \frac{1}{4}(AC)$ and $CE = \frac{1}{4}(BC)$.

4-38 The measure of the sides of square $ABCD$ is x. F is the midpoint of $\overline{BC}$, and $\overline{AE} \perp \overline{DF}$ (Fig. 4-38). Find BE.

4-38

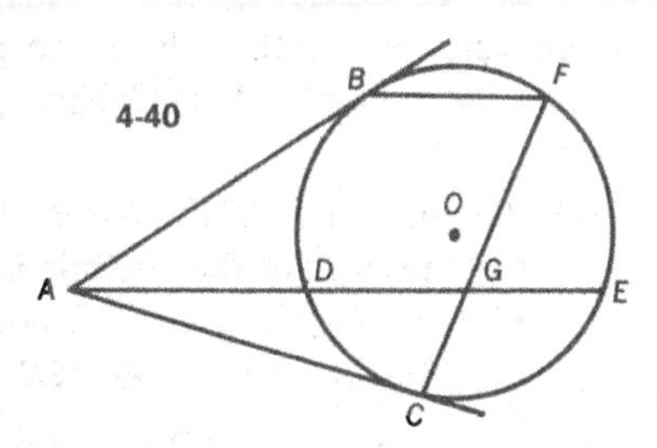

4-40

4-39 If equilateral $\triangle ABC$ is inscribed in a circle, and a point P is chosen on minor arc $\overset{\frown}{AC}$, prove that $PB = PA + PC$.

4-40 From point A, tangents are drawn to circle O, meeting the circle at B and C (Fig. 4-40). Chord $\overline{BF} \parallel$ secant $\overline{ADE}$. Prove that $\overline{FC}$ bisects $\overline{DE}$.

5. Area Relationships

While finding the area of a polygon or circle is a routine matter when a formula can be applied directly, it becomes a challenging task when the given information is "indirect." For example, to find the area of a triangle requires some ingenuity if you know only the measures of its medians. Several problems here explore this kind of situation. The other problems involve a comparison of related areas. To tackle these problems, it may be helpful to keep in mind the following basic relationships. The ratio of the areas of triangles with congruent altitudes is that of their bases. The ratio of the areas of similar triangles is the square of the ratio of the lengths of any corresponding line segments. The same is true for circles, which are all similar, with the additional possibility of comparing the lengths of corresponding arcs. Theorem #56 in Appendix I states another useful relationship.

5-1 As shown in Fig. 5-1, E is on $\overline{AB}$ and C is on $\overline{FG}$. Prove $\square ABCD$ is equal in area to $\square EFGD$.

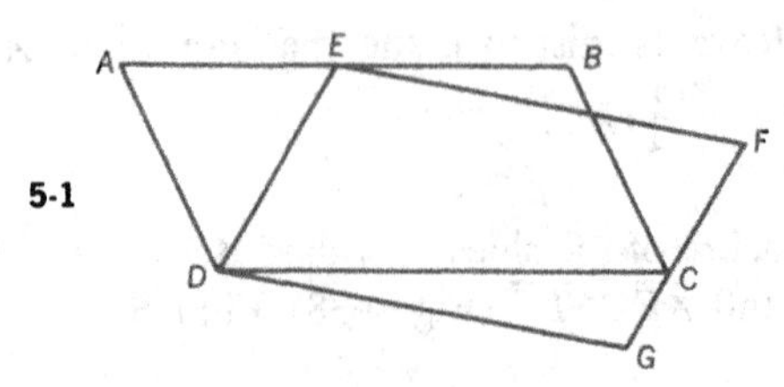

Challenge Prove that the same proposition is true if E lies on the extension of $\overline{AB}$ through B.

5-2 The measures of the bases of trapezoid $ABCD$ are 15 and 9, and the measure of the altitude is 4. Legs $\overline{DA}$ and $\overline{CB}$ are extended to meet at E. If F is the midpoint of $\overline{AD}$, and G is the midpoint of $\overline{BC}$, find the area of $\triangle FGE$.

Challenge Draw $\overline{GL} \parallel \overline{ED}$ and find the ratio of the area of $\triangle GLC$ to the area of $\triangle EDC$.

5-3 The distance from a point A to a line $\overleftrightarrow{BC}$ is 3. Two lines l and l', parallel to $\overleftrightarrow{BC}$, divide $\triangle ABC$ into three parts of equal area. Find the distance between l and l'.

5-4 Find the ratio between the areas of a square inscribed in a circle and an equilateral triangle circumscribed about the same circle.

Challenge 1 Using a similar procedure, find the ratio between the areas of a square circumscribed about a circle and an equilateral triangle inscribed in the same circle.

Challenge 2 Let D represent the difference in area between the circumscribed triangle and the inscribed square. Let K represent the area of the circle. Is the ratio $D:K$ greater than 1, equal to 1, or less than 1?

Challenge 3 Let D represent the difference in area between the circumscribed square and the circle. Let T represent the area of the inscribed equilateral triangle. Find the ratio $D:T$.

5-5 A circle O is tangent to the hypotenuse $\overline{BC}$ of isosceles right $\triangle ABC$. $\overline{AB}$ and $\overline{AC}$ are extended and are tangent to circle O at E and F, respectively, as shown in Fig. 5-5. The area of the triangle is X^2. Find the area of the circle.

5-5

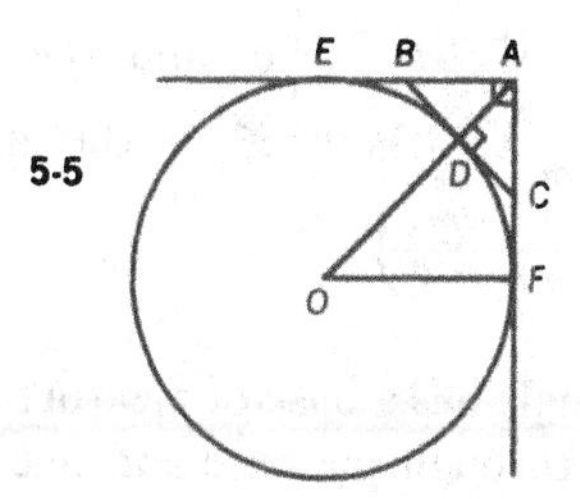

Challenge Find the area of trapezoid $EBCF$.

5-6 $\overline{PQ}$ is the perpendicular bisector of $\overline{AD}$, $\overline{AB} \perp \overline{BC}$, and $\overline{DC} \perp \overline{BC}$ (Fig. 5-6). If $AB = 9$, $BC = 8$, and $DC = 7$, find the area of quadrilateral $APQB$.

5-6

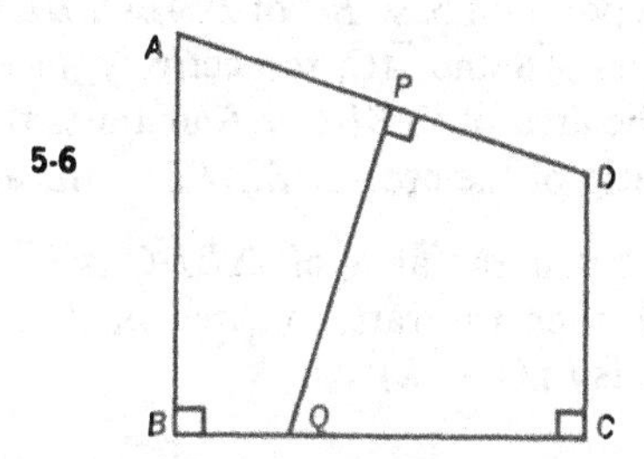

5-7 A triangle has sides that measure 13, 14, and 15. A line perpendicular to the side of measure 14 divides the interior of the triangle into two regions of equal area. Find the measure of the segment of the perpendicular that lies within the triangle.

Challenge Find the area of the trapezoid determined by the perpendicular to the side whose measure is 14, the altitude to that side, and sides of the given triangle.

5-8 In $\triangle ABC$, $AB = 20$, $AC = 22\frac{1}{2}$, and $BC = 27$. Points X and Y are taken on $\overline{AB}$ and $\overline{AC}$, respectively, so that $AX = AY$. If the area of $\triangle AXY = \frac{1}{2}$ area of $\triangle ABC$, find AX.

Challenge Find the ratio of the area of $\triangle BXY$ to that of $\triangle CXY$.

5-9 In $\triangle ABC$, $AB = 7$, $AC = 9$. On $\overline{AB}$, point D is taken so that $BD = 3$. $\overline{DE}$ is drawn cutting $\overline{AC}$ in E so that quadrilateral $BCED$ has $\frac{5}{7}$ the area of $\triangle ABC$. Find CE.

Challenge Show that if $BD = \frac{1}{n}c$, and the area of quadrilateral $BCED = \frac{1}{m}K$, where K is the area of $\triangle ABC$, then $CE = b\left(\frac{n-m}{m(n-1)}\right)$.

5-10 An isosceles triangle has a base of measure 4, and sides measuring 3. A line drawn through the base and one side (but not through any vertex) divides both the perimeter and the area in half. Find the measures of the segments of the base defined by this line.

Challenge Find the measure of the line segment cutting the two sides of the triangle.

5-11 Through D, a point on base $\overline{BC}$ of $\triangle ABC$, $\overline{DE}$ and $\overline{DF}$ are drawn parallel to sides $\overline{AB}$ and $\overline{AC}$, respectively, meeting $\overline{AC}$ at E and $\overline{AB}$ at F. If the area of $\triangle EDC$ is four times the area of $\triangle BFD$, what is the ratio of the area of $\triangle AFE$ to the area of $\triangle ABC$?

Challenge Show that if the area of $\triangle EDC$ is k^2 times the area of $\triangle BFD$, then the ratio of area of $\triangle AFE$ to the area of $\triangle ABC$ is $k:(1+k)^2$.

5-12 Two circles, each of which passes through the center of the other, intersect at points M and N (Fig. 5-12). A line from M intersects the circles at K and L. If $KL = 6$, compute the area of $\triangle KLN$.

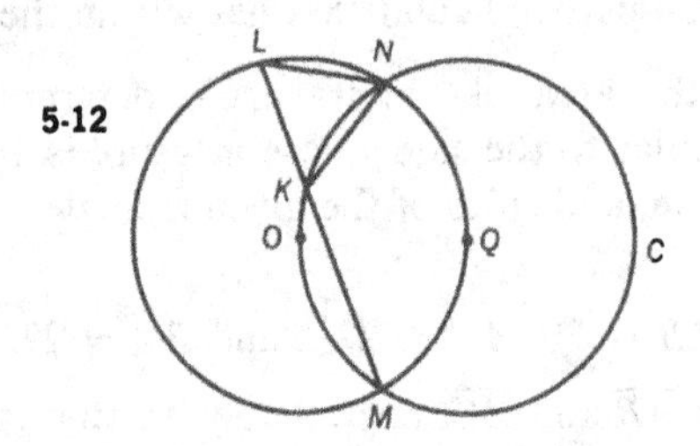

Challenge If r is the measure of the radius of each circle, find the least value and the greatest value of the area of $\triangle KLN$.

5-13 Find the area of a triangle whose medians have measures 39, 42, 45.

5-14 The measures of the sides of a triangle are 13, 14, and 15. A second triangle is formed in which the measures of the three sides are the same as the measures of the medians of the first triangle. What is the area of the second triangle?

Challenge 1 Show that $K(m) = \left(\frac{3}{4}\right) K$ where K represents the area of $\triangle ABC$, and $K(m)$ the area of a triangle with sides m_a, m_b, m_c, the medians of $\triangle ABC$.

Challenge 2 Solve Problem 5-13 using the results of Challenge 1.

5-15 Find the area of a triangle formed by joining the midpoints of the sides of a triangle whose medians have measures 15, 15, and 18.

Challenge Express the required area in terms of $K(m)$, where $K(m)$ is the area of the triangle formed from the medians.

5-16 In $\triangle ABC$, E is the midpoint of $\overline{BC}$, while F is the midpoint of $\overline{AE}$, and $\overleftrightarrow{BF}$ meets AC at D. If the area of $\triangle ABC = 48$, find the area of $\triangle AFD$.

Challenge 1 Solve this problem in general terms.

Challenge 2 Change $AF = \frac{1}{2}AE$ to $AF = \frac{1}{3}AE$, and find a general solution.

5-17 In $\triangle ABC$, D is the midpoint of side $\overline{BC}$, E is the midpoint of $\overline{AD}$, F is the midpoint of $\overline{BE}$, and G is the midpoint of $\overline{FC}$ (Fig. 5-17). What part of the area of $\triangle ABC$ is the area of $\triangle EFG$?

Challenge Solve the problem if $BD = \frac{1}{3}BC, AE = \frac{1}{3}AD, BF = \frac{1}{3}BE$, and $GC = \frac{1}{3}FC$.

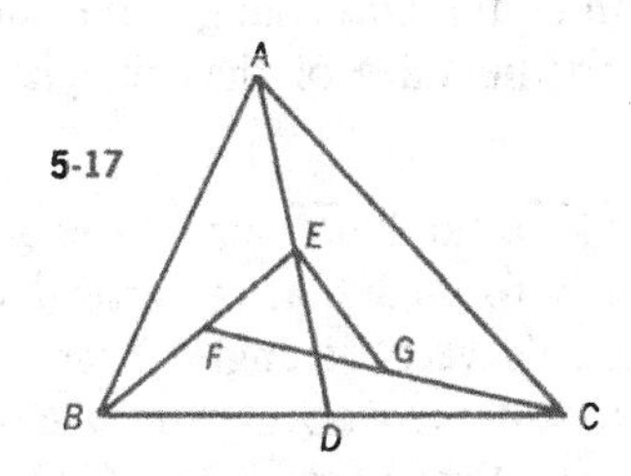

5-17

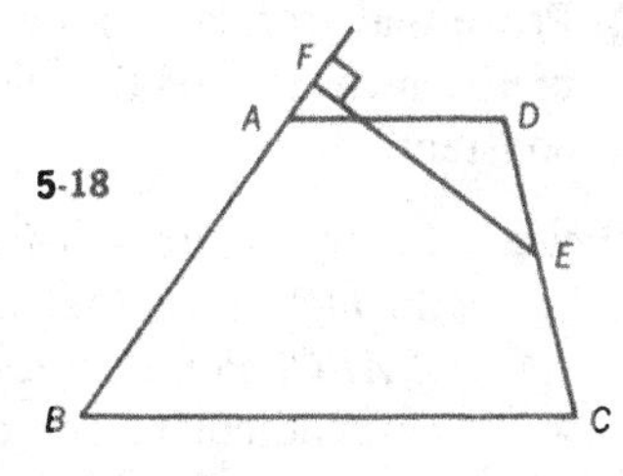

5-18

5-18 In trapezoid $ABCD$ with upper base $\overline{AD}$, lower base $\overline{BC}$, and legs $\overline{AB}$ and $\overline{CD}$, E is the midpoint of $\overline{CD}$ (Fig. 5-18). A perpendicular, $\overline{EF}$, is drawn to $\overline{BA}$ (extended if necessary). If $EF = 24$ and $AB = 30$, find the area of the trapezoid. (Note that the figure is not drawn to scale.)

Challenge Establish a relationship between points F, A, and B such that the area of the trapezoid $ABCD$ is equal to the area of $\triangle FBH$.

5-19 In $\square ABCD$, a line from C cuts diagonal $\overline{BD}$ in E and $\overline{AB}$ in F. If F is the midpoint of $\overline{AB}$, and the area of $\triangle BEC$ is 100, find the area of quadrilateral $AFED$.

Challenge Find the area of $\triangle GEC$ where G is the midpoint of $\overline{BD}$.

5-20 P is any point on side $\overline{AB}$ of $\square ABCD$. $\overline{CP}$ is drawn through P meeting $\overline{DA}$ extended at Q. Prove that the area of $\triangle DPA$ is equal to the area of $\triangle QPB$.

Challenge Prove the theorem for point P on the endpoints of side $\overline{BA}$.

5-21 $\overline{RS}$ is the diameter of a semicircle. Two smaller semicircles, $\widehat{RT}$ and $\widehat{TS}$, are drawn on $\overline{RS}$, and their common internal tangent $\overline{AT}$ intersects the large semicircle at A, as shown in Fig. 5-21. Find the ratio of the area of a semicircle with radius $\overline{AT}$ to the area of the shaded region.

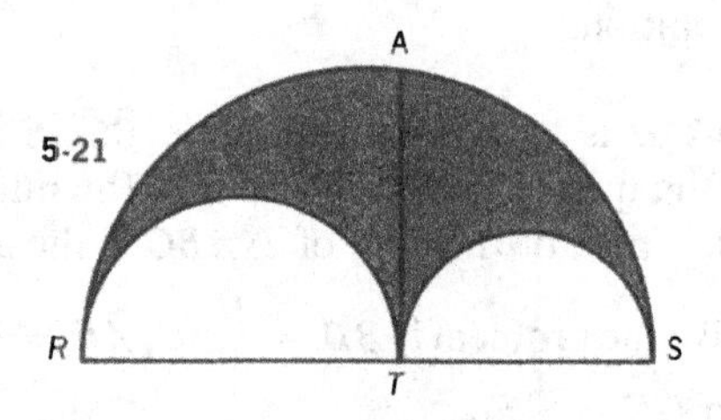

5-22 Prove that from any point inside an equilateral triangle, the sum of the measures of the distances to the sides of the triangle is constant.

Challenge In equilateral $\triangle ABC$, legs $\overline{AB}$ and $\overline{BC}$ are extended through B so that an angle is formed that is vertical to $\angle ABC$. Point P lies within this vertical angle. From P, perpendiculars are drawn to sides $\overline{BC}$, $\overline{AC}$, and $\overline{AB}$ at points Q, R, and S, respectively. Prove that $PR - (PQ + PS)$ equals a constant for $\triangle ABC$.

SECTION II
Further Investigations

6. A Geometric Potpourri

A variety of somewhat difficult problems from elementary Euclidean geometry will be found in this section. Included are Heron's Theorem and its extension to the cyclic quadrilateral, Brahmagupta's Theorem. There are problems often considered classics, such as the butterfly problem and Morley's Theorem. Other famous problems presented are Euler's Theorem and Miquel's Theorem.

Several ways to solve a problem are frequently given in the Solution Part of the book, as many as seven different methods in one case! We urge you to experiment with different methods. After all, 'the right answer' is not the name of the game in Geometry.

6-1 Heron's Formula is used to find the area of any triangle, given only the measures of the sides of the triangle. Derive this famous formula. The area of any triangle $= \sqrt{s(s-a)(s-b)(s-c)}$, where a, b, c are measures of the sides of the triangle and s is the semiperimeter.

Challenge Find the area of a triangle whose sides measure $6, \sqrt{2}, \sqrt{50}$.

6-2 An interesting extension of Heron's Formula to the cyclic quadrilateral is credited to Brahmagupta, an Indian mathematician who lived in the early part of the seventh century. Although Brahmagupta's Formula was once thought to hold for all quadrilaterals, it has been proved to be valid only for cyclic quadrilaterals.

The formula for the area of a cyclic quadrilateral with side measures a, b, c, and d is

$$K = \sqrt{(s-a)(s-b)(s-c)(s-d)},$$

where s is the semiperimeter. Derive this formula.

Challenge 1 Find the area of a cyclic quadrilateral whose sides measure 9, 10, 10, and 21.

Challenge 2 Find the area of a cyclic quadrilateral whose sides measure 15, 24, 7, and 20.

6-3 Sides $\overline{BA}$ and $\overline{CA}$ of $\triangle ABC$ are extended through A to form rhombuses $BATR$ and $CAKN$. (See Fig. 6-3.) $\overline{BN}$ and $\overline{RC}$, intersecting at P, meet $\overline{AB}$ at S and $\overline{AC}$ at M. Draw $\overline{MQ}$ parallel to $\overline{AB}$. **(a)** Prove $AMQS$ is a rhombus and **(b)** prove that the area of $\triangle BPC$ is equal to the area of quadrilateral $ASPM$.

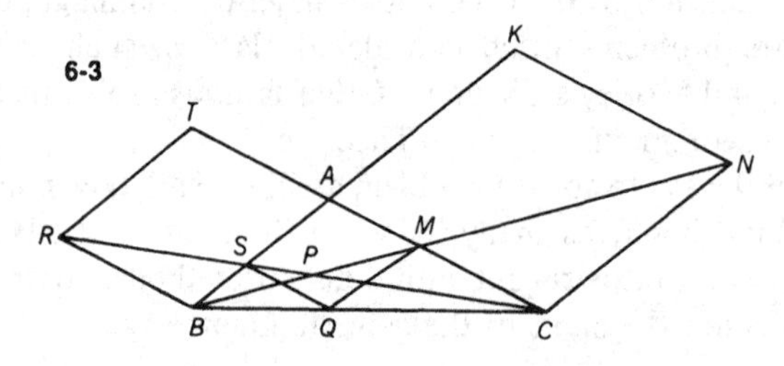

6-4 Two circles with centers A and B intersect at points M and N. Radii $\overline{AP}$ and $\overline{BQ}$ are parallel (on opposite sides of $\overleftrightarrow{AB}$). If the common external tangents meet $\overleftrightarrow{AB}$ at D, and $\overline{PQ}$ meets $\overline{AB}$ at C, prove that $\angle CND$ is a right angle.

6-5 In a triangle whose sides measure 5″, 6″, and 7″, point P is 2″ from the 5″ side and 3″ from the 6″ side. How far is P from the 7″ side?

6-6 Prove that if the measures of the interior bisectors of two angles of a triangle are equal, then the triangle is isosceles.

6-7 In circle O, draw any chord $\overline{AB}$, with midpoint M. Through M two other chords, $\overline{FE}$ and $\overline{CD}$, are drawn. $\overline{CE}$ and $\overline{FD}$ intersect $\overline{AB}$ at Q and P, respectively. Prove that $MP = MQ$. (See Fig. 6-7.) This problem is often referred to as the butterfly problem.

6-8 $\triangle ABC$ is isosceles with $CA = CB$. $m\angle ABD = 60$, $m\angle BAE = 50$, and $m\angle C = 20$. Find the measure of $\angle EDB$ (Fig. 6-8).

6-7

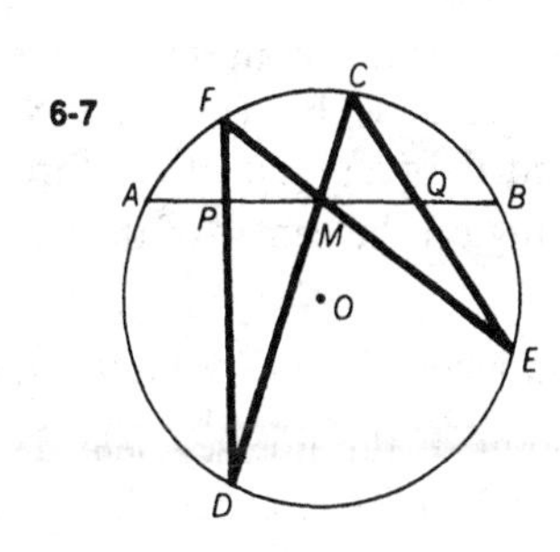

6-8

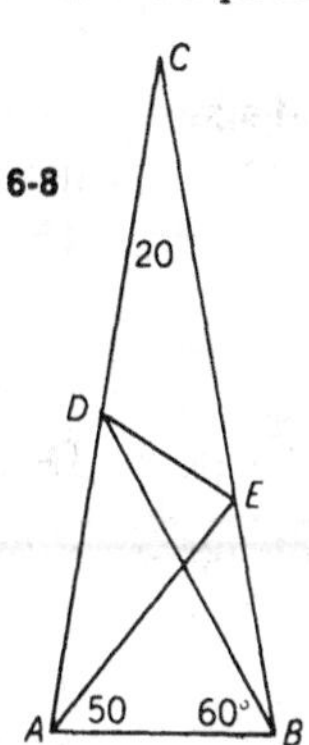

6-9 Find the area of an equilateral triangle containing in its interior a point P, whose distances from the vertices of the triangle are 3, 4, and 5.

6-10 Find the area of a square $ABCD$ containing a point P such that $PA = 3$, $PB = 7$, and $PD = 5$.

Challenge 1 Find the measure of $\overline{PC}$.

Challenge 2 Express PC in terms of PA, PB, and PD.

6-11 If, on each side of a given triangle, an equilateral triangle is constructed externally, prove that the line segments formed by joining a vertex of the given triangle with the remote vertex of the equilateral triangle drawn on the side opposite it are congruent.

Challenge 1 Prove that these lines are concurrent.

Challenge 2 Prove that the circumcenters of the three equilateral triangles determine another equilateral triangle.

6-12 Prove that if the angles of a triangle are trisected, the intersections of the pairs of trisectors adjacent to the same side determine an equilateral triangle. (This theorem was first derived by F. Morley about 1900.)

6-13 Prove that in any triangle the centroid trisects the line segment joining the center of the circumcircle and the orthocenter (i.e. the point of intersection of the altitudes). This theorem was first published by Leonhard Euler in 1765.

Challenge 1 The result of this theorem leads to an interesting problem first published by James Joseph Sylvester (1814–1897). The problem is to find the resultant of the three vectors $\overrightarrow{OA}$, $\overrightarrow{OB}$, and $\overrightarrow{OC}$ acting on the center of the circumcircle O of $\triangle ABC$.

Challenge 2 Describe the situation when $\triangle ABC$ is equilateral.

Challenge 3 Prove that the midpoint of the line segment determined by the circumcenter and the orthocenter is the center of the nine-point circle. The nine-point circle of a triangle is determined by the following nine points; the feet of the altitudes, the midpoints of the sides of the triangle, and the midpoints of the segments from the vertices to the orthocenter.

6-14 Prove that if a point is chosen on each side of a triangle, then the circles determined by each vertex and the points on the adjacent sides pass through a common point (Figs. 6-14a and 6-14b). This theorem was first published by A. Miquel in 1838.

Challenge 1 Prove in Fig. 6-14a, $m\angle BFM = m\angle CEM = m\angle ADM$; or in Fig. 6-14b, $m\angle BFM = m\angle CDM = m\angle GEM$.

Challenge 2 Give the location of M when $AF = FB = BE = EC = CD = DA$.

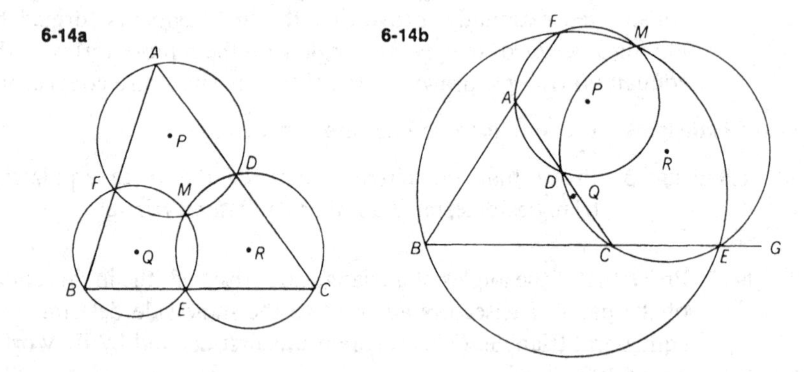

6-15 Prove that the centers of the circles in Problem 6-14 determine a triangle similar to the original triangle.

Challenge Prove that any other triangle whose sides pass through the intersections of the above three circles, P, Q, and R (two at a time), is similar to $\triangle ABC$.

7. Ptolemy and the Cyclic Quadrilateral

One of the great works of the second Alexandrian period was a collection of earlier studies, mainly in astronomy, by Claudius Ptolemaeus (better known as Ptolemy). Included in this work, the *Almagest*, is a theorem stating that in a cyclic (inscribed) quadrilateral the sum of the products of the opposite sides equals the product of the diagonals. This powerful theorem of Ptolemy enables us to solve problems which would otherwise be difficult to handle. The theorem and some of its consequences are explored here.

7-1 Prove that in a cyclic quadrilateral the product of the diagonals is equal to the sum of the products of the pairs of opposite sides (Ptolemy's Theorem).

Challenge 1 Prove that if the product of the diagonals of a quadrilateral equals the sum of the products of the pairs of opposite sides, then the quadrilateral is cyclic. This is the converse of Ptolemy's Theorem.

Challenge 2 To what familiar result does Ptolemy's Theorem lead when the cyclic quadrilateral is a rectangle?

Challenge 3 Find the diagonal, d, of the trapezoid with bases a and b, and equal legs c.

7-2 E is a point on side $\overline{AD}$ of rectangle $ABCD$, so that $DE = 6$, while $DA = 8$, and $DC = 6$. If $\overline{CE}$ extended meets the circumcircle of the rectangle at F, find the measure of chord $\overline{DF}$. (See Fig. 7-2.)

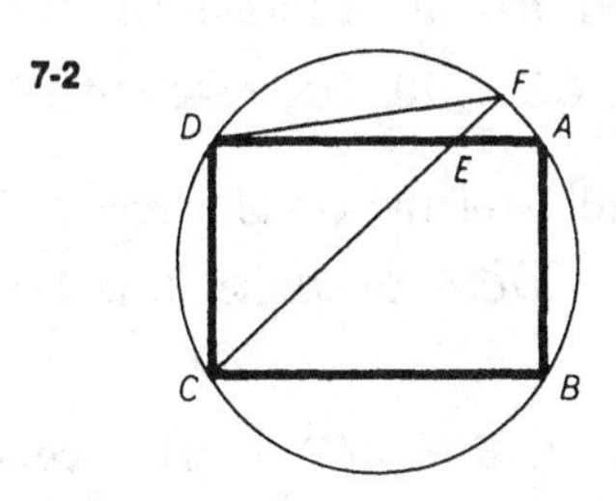

Challenge Find the measure of $\overline{FB}$.

7-3 On side $\overline{AB}$ of square $ABCD$, right $\triangle ABF$, with hypotenuse $\overline{AB}$, is drawn externally to the square. If $AF = 6$ and $BF = 8$ find EF, where E is the point of intersection of the diagonals of the square.

Challenge Find EF, when F is inside square $ABCD$.

7-4 Point P on side $\overline{AB}$ of right $\triangle ABC$ is placed so that $BP = PA = 2$. Point Q is on hypotenuse $\overline{AC}$ so that $\overline{PQ}$ is perpendicular to $\overline{AC}$. If $CB = 3$, find the measure of $\overline{BQ}$, using Ptolemy's Theorem. (See Fig. 7-4.)

Challenge 1 Find the area of quadrilateral $CBPQ$.

Challenge 2 As P is translated from B to A along $\overline{BA}$, find the range of values of BQ, where $\overline{PQ}$ remains perpendicular to $\overline{CA}$.

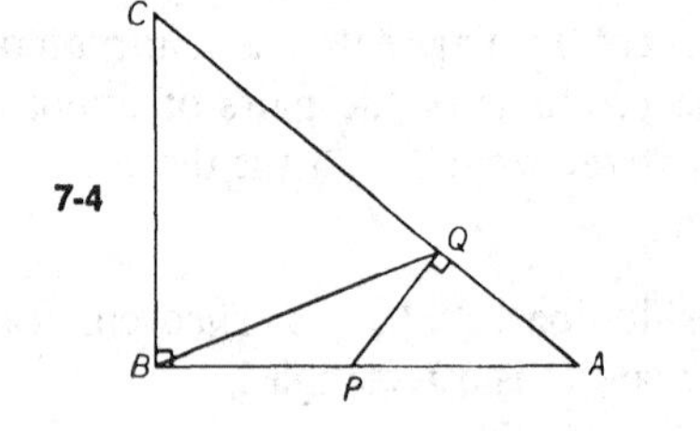

7-4

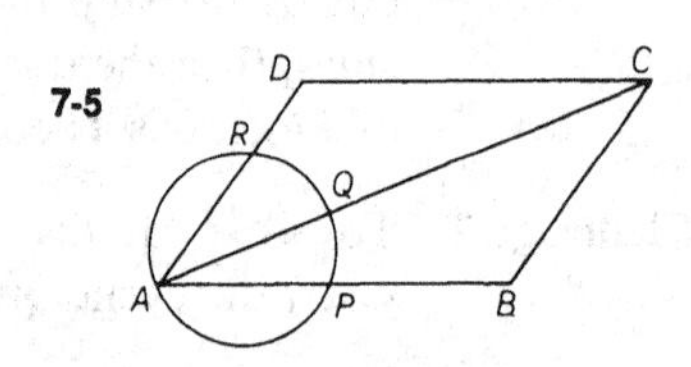

7-5

7-5 If any circle passing through vertex A of parallelogram $ABCD$ intersects sides $\overline{AB}$, and $\overline{AD}$ at points P and R, respectively, and diagonal $\overline{AC}$ at point Q, prove that $(AQ)(AC) = (AP)(AB) + (AR)(AD)$. (See Fig. 7-5.)

Challenge Prove the theorem valid when the circle passes through C.

7-6 Diagonals $\overline{AC}$ and $\overline{BD}$ of quadrilateral $ABCD$ meet at E. If $AE = 2$, $BE = 5$, $CE = 10$, $DE = 4$, and $BC = \frac{15}{2}$, find AB.

Challenge Find the radius of the circumcircle if the measure of the distance from $\overline{DC}$ to the center O is $2\frac{1}{2}$.

7-7 If isosceles $\triangle ABC$ ($AB = AC$) is inscribed in a circle, and a point P is on $\widehat{BC}$, prove that $\dfrac{PA}{PB + PC} = \dfrac{AC}{BC}$, a constant for the given triangle.

7-8 If equilateral $\triangle ABC$ is inscribed in a circle, and a point P is on $\overset{\frown}{BC}$, prove that $PA = PB + PC$.

7-9 If square $ABCD$ is inscribed in a circle, and a point P is on $\overset{\frown}{BC}$, prove that $\frac{PA + PC}{PB + PD} = \frac{PD}{PA}$.

7-10 If regular pentagon $ABCDE$ is inscribed in a circle, and point P is on $\overset{\frown}{BC}$, prove that $PA + PD = PB + PC + PE$.

7-11 If regular hexagon $ABCDEF$ is inscribed in a circle, and point P is on $\overset{\frown}{BC}$, prove that $PE + PF = PA + PB + PC + PD$.

Challenge Derive analogues for other regular polygons.

7-12 Equilateral $\triangle ADC$ is drawn externally on side $\overline{AC}$ of $\triangle ABC$. Point P is taken on $\overline{BD}$. Find $m\angle APC$ such that $BD = PA + PB + PC$.

Challenge Investigate the case where $\triangle ADC$ is drawn internally on side $\overline{AC}$ of $\triangle ABC$.

7-13 A line drawn from vertex A of equilateral $\triangle ABC$, meets $\overline{BC}$ at D and the circumcircle at P. Prove that $\frac{1}{PD} = \frac{1}{PB} + \frac{1}{PC}$.

Challenge 1 If $BP = 5$ and $PC = 20$, find AD.

Challenge 2 If $m\overset{\frown}{BP}:m\overset{\frown}{PC} = 1:3$, find the radius of the circle in challenge I.

7-14 Express in terms of the sides of a cyclic quadrilateral the ratio of the diagonals.

Challenge Verify the result for an isosceles trapezoid.

7-15 A point P is chosen inside parallelogram $ABCD$ such that $\angle APB$ is supplementary to $\angle CPD$. Prove that $(AB)(AD) = (BP)(DP) + (AP)(CP)$.

7-16 A triangle inscribed in a circle of radius 5, has two sides measuring 5 and 6, respectively. Find the measure of the third side of the triangle.

Challenge Generalize the result of this problem for any triangle.

8. Menelaus and Ceva: Collinearity and Concurrency

Proofs of theorems dealing with collinearity and concurrency are ordinarily clumsy, lengthy, and, as a result, unpopular. With the aid of two famous theorems, they may be shortened.

The first theorem is credited to Menelaus of Alexandria (about 100 A.D.). In 1678, Giovanni Ceva, an Italian mathematician, published Menelaus' Theorem and a second one of his own, related to it. The problems in this section concern either Menelaus' Theorem, Ceva's Theorem, or both. Among the applications investigated are theorems of Gerard Desargues, Blaise Pascal, and Pappus of Alexandria. A rule of thumb for these problems is: try to use Menelaus' Theorem for collinearity and Ceva's Theorem for concurrency.

8-1 Points P, Q, and R are taken on sides $\overline{AC}$, $\overline{AB}$, and $\overline{BC}$ (extended if necessary) of $\triangle ABC$. Prove that if these points are collinear,

$$\frac{AQ}{QB} \cdot \frac{BR}{RC} \cdot \frac{CP}{PA} = -1.$$

This theorem, together with its converse, which is given in the Challenge that follows, constitute the classic theorem known as Menelaus' Theorem. (See Fig. 8-1a and Fig. 8-1b.)

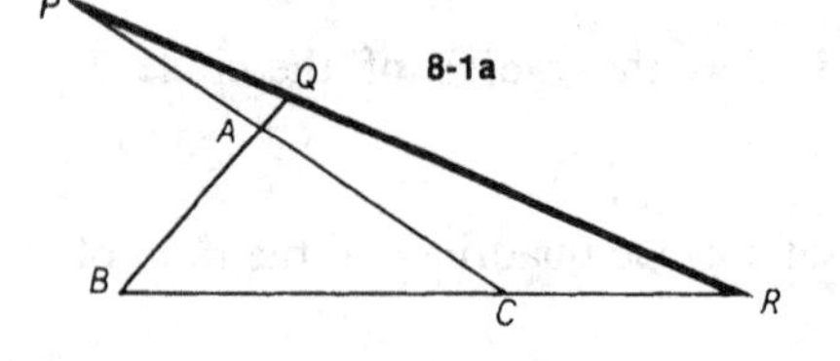

8-1a

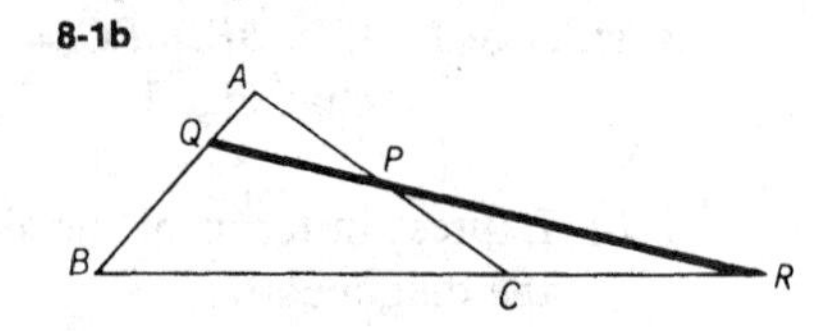

8-1b

Challenge In $\triangle ABC$ points P, Q, and R are situated respectively on sides $\overline{AC}$, $\overline{AB}$, and $\overline{BC}$ (extended when necessary). Prove that if

$$\frac{AQ}{QB} \cdot \frac{BR}{RC} \cdot \frac{CP}{PA} = -1,$$

then P, Q, and R are collinear. This is part of Menelaus' Theorem.

8-2 Prove that three lines drawn from the vertices A, B, and C of $\triangle ABC$ meeting the opposite sides in points L, M, and N, respectively, are concurrent if and only if $\dfrac{AN}{NB} \cdot \dfrac{BL}{LC} \cdot \dfrac{CM}{MA} = 1.$

This is known as Ceva's Theorem. (See Fig. 8-2a, and Fig. 8-2b.)

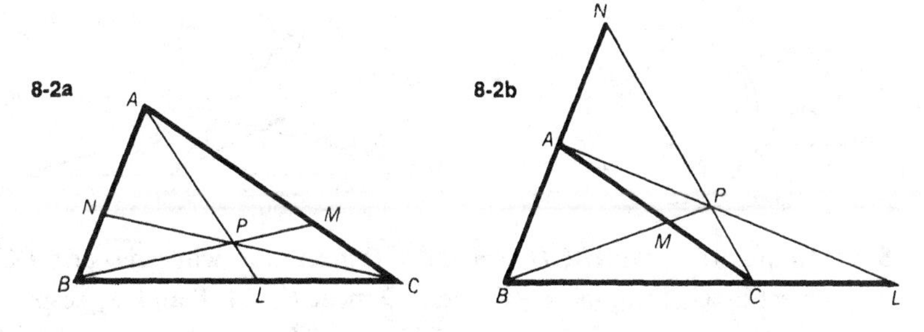

8-3 Prove that the medians of any triangle are concurrent.

8-4 Prove that the altitudes of any triangle are concurrent.

Challenge Investigate the difficulty in applying this proof to a right triangle by Ceva's Theorem.

8-5 Prove that the interior angle bisectors of a triangle are concurrent.

8-6 Prove that the interior angle bisectors of two angles of a non-isosceles triangle and the exterior angle bisector of the third angle meet the opposite sides in three collinear points.

8-7 Prove that the exterior angle bisectors of any non-isosceles triangle meet the opposite sides in three collinear points.

8-8 In right $\triangle ABC$, P and Q are on $\overline{BC}$ and $\overline{AC}$, respectively, such that $CP = CQ = 2$. Through the point of intersection, R, of $\overline{AP}$ and $\overline{BQ}$, a line is drawn also passing through C and meeting $\overline{AB}$ at S. $\overline{PQ}$ extended meets $\overleftrightarrow{AB}$ at T. If the hypotenuse $AB = 10$ and $AC = 8$, find TS.

Challenge 1 By how much is TS decreased if P is taken at the midpoint of $\overline{BC}$?

Challenge 2 What is the minimum value of TS?

8-9 A circle through vertices B and C of $\triangle ABC$ meets $\overline{AB}$ at P and $\overline{AC}$ at R. If $\overleftrightarrow{PR}$ meets $\overleftrightarrow{BC}$ at Q, prove that $\frac{QC}{QB} = \frac{(RC)(AC)}{(PB)(AB)}$. (See Fig. 8-9.)

Challenge Investigate the case where the points P and R are on the extremities of $\overline{BA}$ and $\overline{CA}$, respectively.

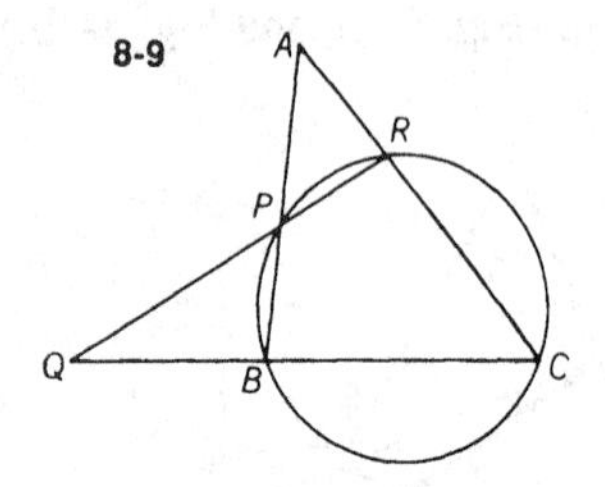

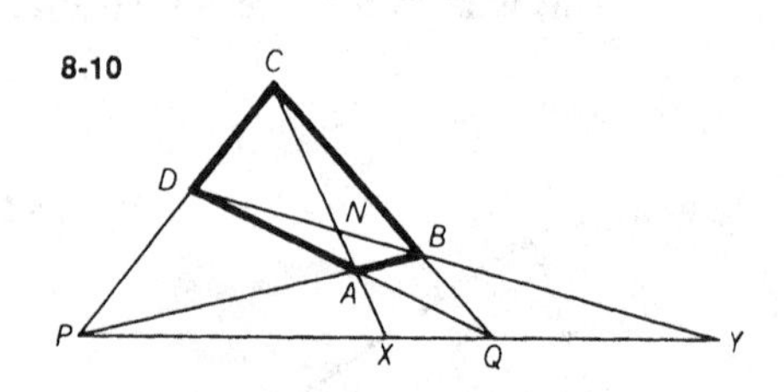

8-10 In quadrilateral $ABCD$, $\overleftrightarrow{AB}$ and $\overleftrightarrow{CD}$ meet at P, while $\overleftrightarrow{AD}$ and $\overleftrightarrow{BC}$ meet at Q. Diagonals $\overleftrightarrow{AC}$ and $\overleftrightarrow{BD}$ meet $\overleftrightarrow{PQ}$ at X and Y, respectively (Fig. 8-10). Prove that $\frac{PX}{XQ} = -\frac{PY}{YQ}$.

8-11 Prove that a line drawn through the centroid, G, of $\triangle ABC$, cuts sides $\overline{AB}$ and $\overline{AC}$ at points M and N, respectively, so that $(AM)(NC) + (AN)(MB) = (AM)(AN)$. (See Fig. 8-11.)

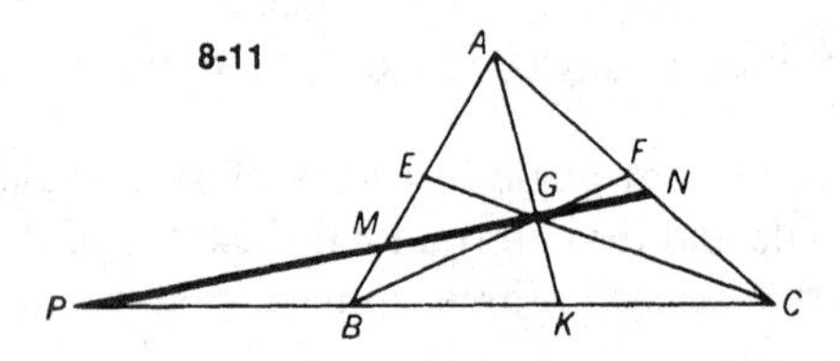

8-12 In $\triangle ABC$, points L, M, and N lie on $\overline{BC}$, $\overline{AC}$, and $\overline{AB}$, respectively, and $\overline{AL}$, $\overline{BM}$, and $\overline{CN}$ are concurrent.

(a) Find the numerical value of $\frac{PL}{AL} + \frac{PM}{BM} + \frac{PN}{CN}$.

(b) Find the numerical value of $\frac{AP}{AL} + \frac{BP}{BM} + \frac{CP}{CN}$.

8-13 Congruent line segments $\overline{AE}$ and $\overline{AF}$ are taken on sides $\overline{AB}$ and $\overline{AC}$, respectively, of $\triangle ABC$. The median $\overline{AM}$ intersects $\overline{EF}$ at point Q. Prove that $\frac{QE}{QF} = \frac{AC}{AB}$.

8-14 In $\triangle ABC$, $\overleftrightarrow{AL}$, $\overleftrightarrow{BM}$, and $\overleftrightarrow{CN}$ are concurrent at P. Express the ratio $\frac{AP}{PL}$ in terms of segments made by the concurrent lines on the sides of $\triangle ABC$. (See Fig. 8-2a, and Fig. 8-2b.)

Challenge Complete the expressions for $\frac{BP}{PM}$ and $\frac{CP}{PN}$.

8-15 Side $\overline{AB}$ of square $ABCD$ is extended to P so that $BP = 2(AB)$. With M, the midpoint of $\overline{DC}$, $\overline{BM}$ is drawn meeting $\overline{AC}$ at Q. $\overline{PQ}$ meets $\overline{BC}$ at R. Using Menelaus' theorem, find the ratio $\frac{CR}{RB}$. (See Fig. 8-15.)

Challenge 1 Find $\frac{CR}{RB}$, when $BP = AB$.

Challenge 2 Find $\frac{CR}{RB}$, when $BP = k \cdot AB$.

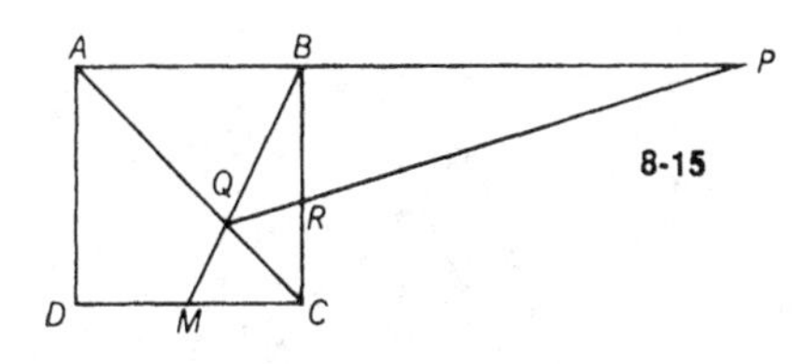

8-15

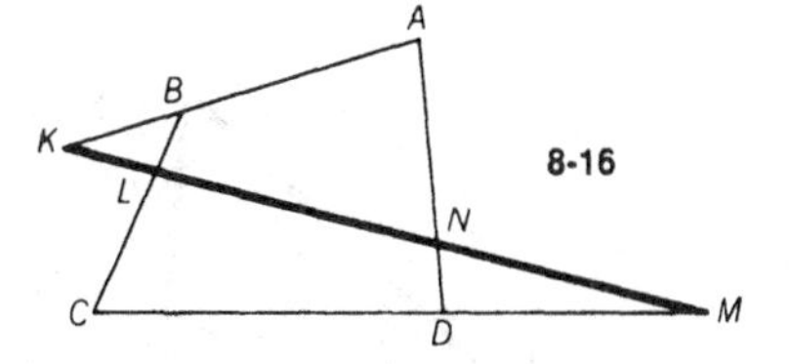

8-16

8-16 Sides $\overleftrightarrow{AB}$, $\overleftrightarrow{BC}$, $\overleftrightarrow{CD}$, and $\overleftrightarrow{DA}$ of quadrilateral $ABCD$ are cut by a straight line at points K, L, M, and N, respectively. (See Fig. 8-16.) Prove that $\frac{BL}{LC} \cdot \frac{AK}{KB} \cdot \frac{DN}{NA} \cdot \frac{CM}{MD} = 1$.

Challenge 1 Prove the theorem for parallelogram $ABCD$.

Challenge 2 Extend this theorem to other polygons.

8-17 Tangents to the circumcircle of $\triangle ABC$ at points A, B, and C meet sides $\overleftrightarrow{BC}$, $\overleftrightarrow{AC}$, and $\overleftrightarrow{AB}$ at points P, Q, and R, respectively. Prove that points P, Q, and R are collinear. (See Fig. 8-17.)

8-17

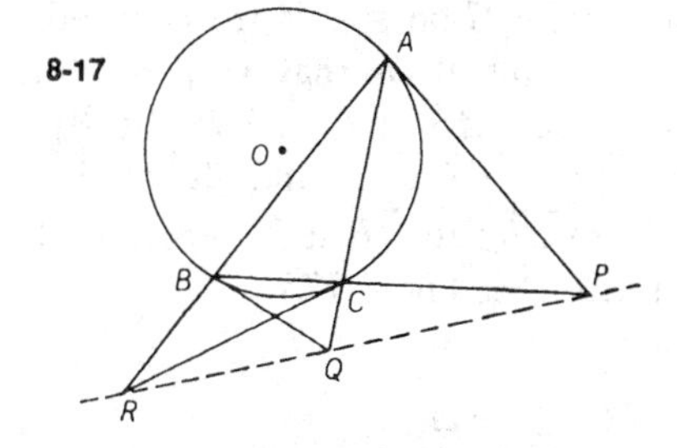

8-18

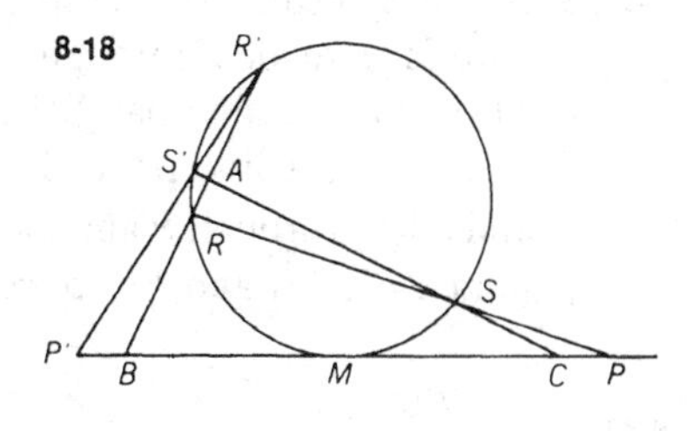

8-18 A circle is tangent to side $\overline{BC}$, of $\triangle ABC$ at M, its midpoint, and cuts $\overleftrightarrow{AB}$ and $\overleftrightarrow{AC}$ at points R, R', and S, S', respectively. If $\overline{RS}$ and $\overline{R'S'}$ are each extended to meet $\overleftrightarrow{BC}$ at points P and P' respectively, prove that $(BP)(BP') = (CP)(CP')$. (See Fig. 8-18.)

Challenge 1 Show that the result implies that $CP = BP'$.

Challenge 2 Investigate the situation when $\triangle ABC$ is equilateral.

8-19 In $\triangle ABC$ (Fig. 8-19) P, Q, and R are the midpoints of the sides $\overline{AB}$, $\overline{BC}$, and $\overline{AC}$. Lines $\overleftrightarrow{AN}$, $\overleftrightarrow{BL}$, and $\overleftrightarrow{CM}$ are concurrent, meeting the opposite sides in N, L, and M, respectively. If $\overleftrightarrow{PL}$ meets $\overleftrightarrow{BC}$ at J, $\overleftrightarrow{MQ}$ meets $\overleftrightarrow{AC}$ at I, and $\overleftrightarrow{RN}$ meets $\overleftrightarrow{AB}$ at H, prove that H, I, and J are collinear.

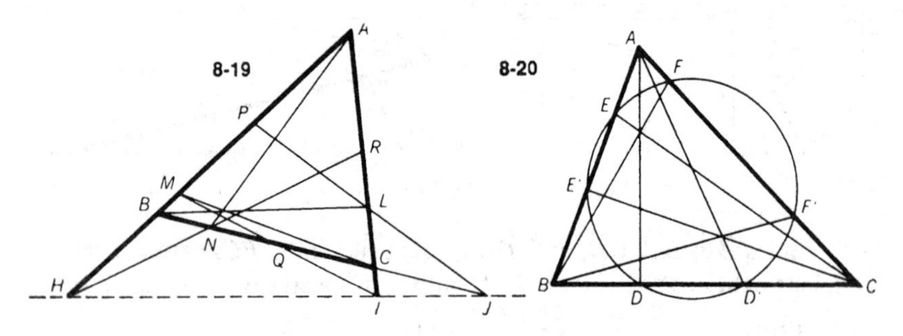

8-20 $\triangle ABC$ cuts a circle at points E, E', D, D', F, F', as in Fig. 8-20. Prove that if $\overline{AD}$, $\overline{BF}$, and $\overline{CE}$ are concurrent, then $\overline{AD'}$, $\overline{BF'}$, and $\overline{CE'}$ are also concurrent.

8-21 Prove that the three pairs of common external tangents to three circles, taken two at a time, meet in three collinear points.

8-22 $\overline{AM}$ is a median of $\triangle ABC$, and point G on $\overline{AM}$ is the centroid. $\overline{AM}$ is extended through M to point P so that $GM = MP$. Through P, a line parallel to $\overline{AC}$ cuts $\overline{AB}$ at Q and $\overline{BC}$ at P_1; through P, a line parallel to $\overline{AB}$ cuts $\overline{CB}$ at N and $\overline{AC}$ at P_2; and a line through P and parallel to $\overline{CB}$ cuts $\overleftrightarrow{AB}$ at P_3. Prove that points P_1, P_2, and P_3 are collinear. (See Fig. 8-22.)

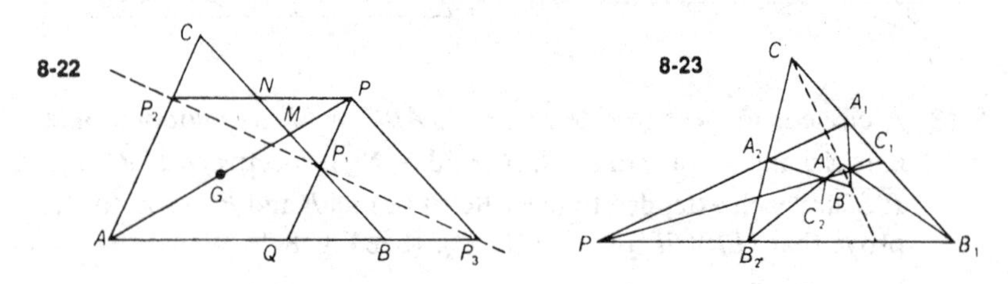

8-23 If $\triangle A_1B_1C_1$ and $\triangle A_2B_2C_2$ are situated so that the lines joining the corresponding vertices, $\overleftrightarrow{A_1A_2}$, $\overleftrightarrow{B_1B_2}$, and $\overleftrightarrow{C_1C_2}$, are concurrent (Fig. 8-23), then the pairs of corresponding sides intersect in three collinear points. (Desargues' Theorem)

Challenge Prove the converse.

8-24 A circle inscribed in $\triangle ABC$ is tangent to sides $\overline{BC}$, $\overline{CA}$, and $\overline{AB}$ at points L, M, and N, respectively. If $\overline{MN}$ extended meets $\overleftrightarrow{BC}$ at P,

(a) prove that $\frac{BL}{LC} = -\frac{BP}{PC}$

(b) prove that if $\overleftrightarrow{NL}$ meets $\overleftrightarrow{AC}$ at Q and $\overleftrightarrow{ML}$ meets $\overleftrightarrow{AB}$ at R, then P, Q, and R are collinear.

8-25 In $\triangle ABC$, where $\overline{CD}$ is the altitude to $\overline{AB}$ and P is any point on $\overline{DC}$, $\overline{AP}$ meets $\overline{CB}$ at Q, and $\overline{BP}$ meets $\overline{CA}$ at R. Prove that $m\angle RDC = m\angle QDC$, using Ceva's Theorem.

8-26 In $\triangle ABC$ points F, E, and D are the feet of the altitudes drawn from the vertices A, B, and C, respectively. The sides of the pedal $\triangle FED$, $\overline{EF}$, $\overline{DF}$, and $\overline{DE}$, when extended, meet the sides of $\triangle ABC$, $\overleftrightarrow{AB}$, $\overleftrightarrow{AC}$, and $\overleftrightarrow{BC}$ (extended) at points M, N, and L, respectively. Prove that M, N, and L are collinear. (See Fig. 8-26.)

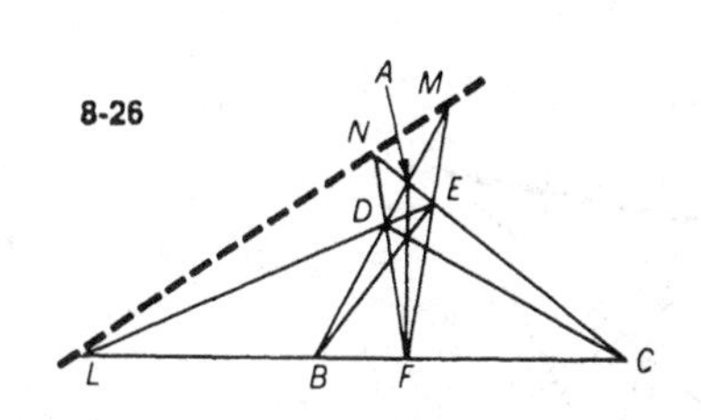

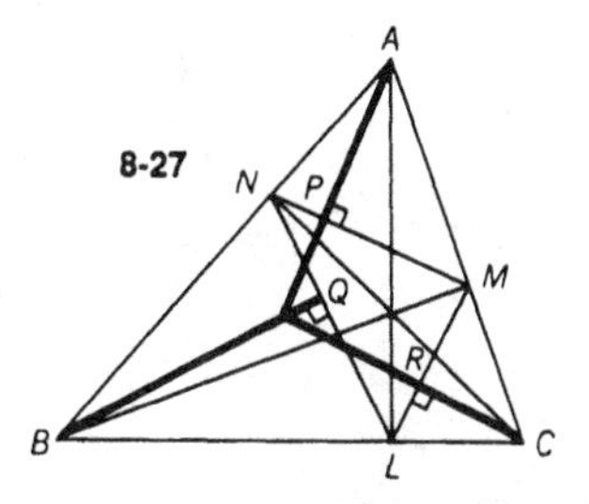

8-27 In $\triangle ABC$ (Fig. 3-27), L, M, and N are the feet of the altitudes from vertices A, B, and C. Prove that the perpendiculars from A, B, and C to $\overline{MN}$, $\overline{LN}$, and $\overline{LM}$, respectively, are concurrent.

Challenge Prove that $\overline{PL}$, $\overline{QM}$, and $\overline{RN}$ are concurrent.

8-28 Prove that the perpendicular bisectors of the interior angle bisectors of any triangle meet the sides opposite the angles being bisected in three collinear points.

8-29 Figure 8-29a shows a hexagon $ABCDEF$ whose pairs of opposite sides are: [$\overline{AB}$, $\overline{DE}$], [$\overline{CB}$, $\overline{EF}$], and [$\overline{CD}$, $\overline{AF}$]. If we place points A, B, C, D, E, and F in any order on a circle, the above pairs of opposite sides intersect at points L, M, and N respectively. Prove that L, M, and N are collinear. Fig. 8-29b shows one arrangement of the six points, A, B, C, D, E, and F on a circle.

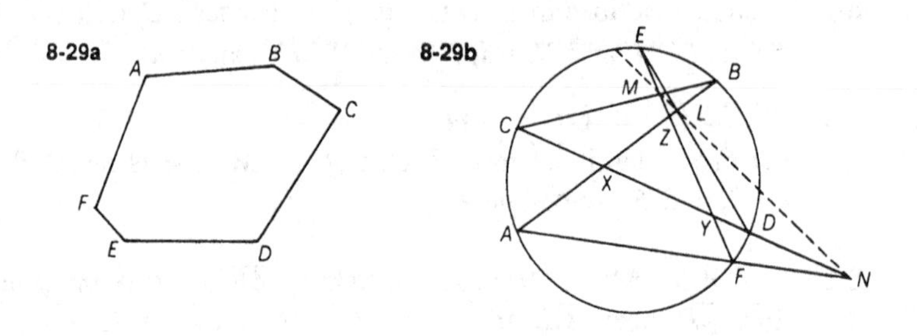

8-29a 8-29b

Challenge 1 Prove the theorem for another arrangement of the points A, B, C, D, E, and F on a circle.

Challenge 2 Can you explain this theorem when one pair of opposite sides are parallel?

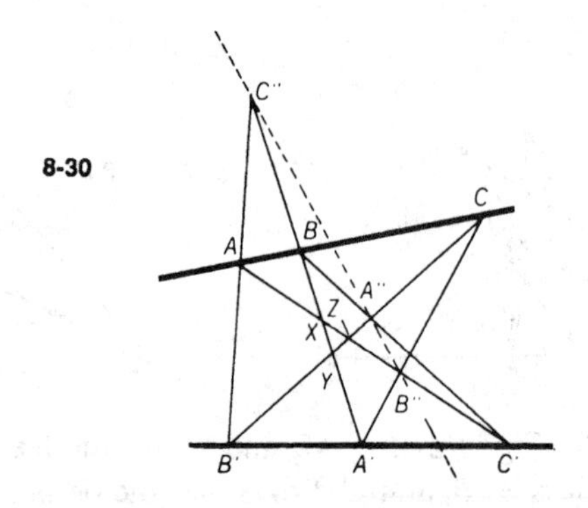

8-30

8-30 Points A, B, and C are on one line and points A', B', and C' are on another line (in any order). (Fig. 8-30) If $\overleftrightarrow{AB'}$ and $\overleftrightarrow{A'B}$ meet at C'', while $\overleftrightarrow{AC'}$ and $\overleftrightarrow{A'C}$ meet at B'', and $\overleftrightarrow{BC'}$ and $\overleftrightarrow{B'C}$ meet at A'', prove that points A'', B'', and C'' are collinear.
(This theorem was first published by Pappus of Alexandria about 300 A.D.)

9. The Simson Line

If perpendiculars are drawn from a point on the circumcircle of a triangle to its sides, their feet lie on a line. Although this famous line was discovered by William Wallace in 1797, careless misquotes have, in time, attributed it to Robert Simson (1687–1768). The following problems present several properties and applications of the Simson Line.

9-1 Prove that the feet of the perpendiculars drawn from any point on the circumcircle of a given triangle to the sides of the triangle are collinear. (Simson's Theorem)

Challenge 1 State and prove the converse of Simson's Theorem.

Challenge 2 Which points on the circumcircle of a given triangle lie on their own Simson Lines with respect to the given triangle?

9-2 Altitude $\overleftrightarrow{AD}$ of $\triangle ABC$ meets the circumcircle at P. (Fig. 9-2) Prove that the Simson Line of P with respect to $\triangle ABC$ is parallel to the line tangent to the circle at A.

Challenge Investigate the special case where $BA = CA$.

9-3 From point P on the circumcircle of $\triangle ABC$, perpendiculars $\overline{PX}$, $\overline{PY}$, and $\overline{PZ}$ are drawn to sides $\overleftrightarrow{AC}$, $\overleftrightarrow{AB}$, and $\overleftrightarrow{BC}$, respectively. Prove that $(PA)(PZ) = (PB)(PX)$.

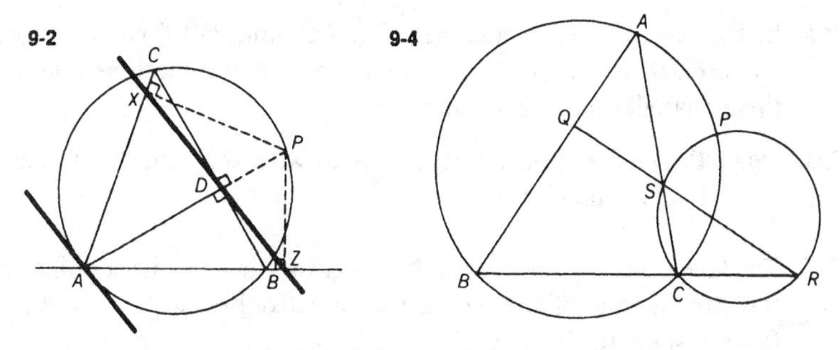

9-4 In Fig. 9-4, sides $\overleftrightarrow{AB}$, $\overleftrightarrow{BC}$, and $\overleftrightarrow{CA}$ of $\triangle ABC$ are cut by a transversal at points Q, R, and S, respectively. The circumcircles of $\triangle ABC$ and $\triangle SCR$ intersect at P. Prove that quadrilateral $APSQ$ is cyclic.

9-5 In Fig. 9-5, right $\triangle ABC$, with right angle at A, is inscribed in circle O. The Simson Line of point P, with respect to $\triangle ABC$, meets $\overline{PA}$ at M. Prove that $\overline{MO}$ is perpendicular to $\overline{PA}$.

Challenge Show that $\overline{PA}$ is a side of the inscribed hexagon if $m\angle AOM = 30$.

9-6 From a point P on the circumference of circle O, three chords are drawn meeting the circle at points A, B, and C. Prove that the three points of intersection of the three circles with $\overline{PA}$, $\overline{PB}$, and $\overline{PC}$ as diameters, are collinear.

Challenge Prove the converse.

9-7 P is any point on the circumcircle of cyclic quadrilateral $ABCD$. If $\overline{PK}$, $\overline{PL}$, $\overline{PM}$, and $\overline{PN}$ are the perpendiculars from P to sides $\overleftrightarrow{AB}$, $\overleftrightarrow{BC}$, $\overleftrightarrow{CD}$, and $\overleftrightarrow{DA}$, respectively, prove that $(PK)(PM) = (PL)(PN)$.

9-5

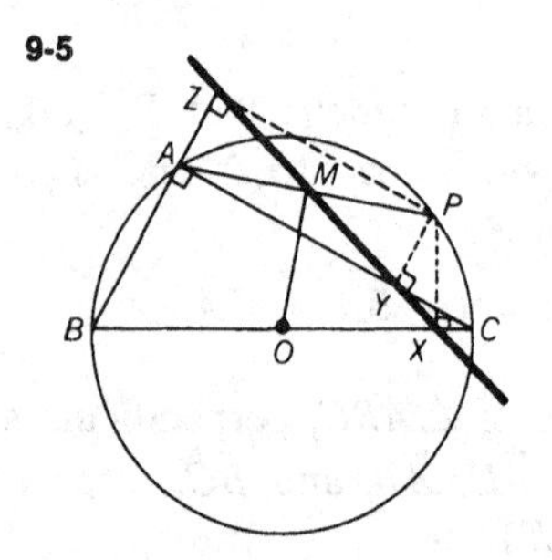

9-8

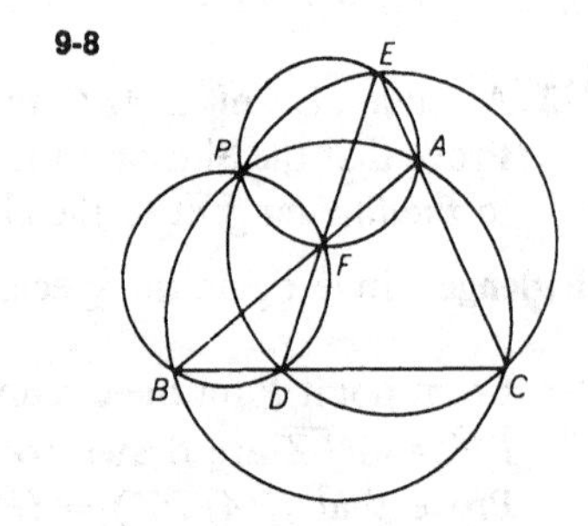

9-8 In Fig. 9-8, line segments $\overline{AB}$, $\overline{BC}$, $\overline{EC}$, and $\overline{ED}$ form triangles ABC, FBD, EFA, and EDC. Prove that the four circumcircles of these triangles meet at a common point.

Challenge Prove that point P is concyclic with the centers of these four circumcircles.

9-9 The line joining the orthocenter of a given triangle with a point on the circumcircle of the triangle is bisected by the Simson Line (with respect to that point).

9-10 The measure of the angle determined by the Simson Lines of two given points on the circumcircle of a given triangle is equal to one-half the measure of the arc determined by the two points.

Challenge Prove that if three points are chosen at random on a circle, the triangle formed by these three points is similar to the triangle formed by the Simson Lines of these points with respect to any inscribed triangle.

9-11 If two triangles are inscribed in the same circle, a single point on the circumcircle determines a Simson Line for each triangle. Prove that the angle formed by these two Simson Lines is constant, regardless of the position of the point.

9-12 In the circumcircle of $\triangle ABC$, chord $\overline{PQ}$ is drawn parallel to side $\overline{BC}$. Prove that the Simson Lines of $\triangle ABC$, with respect to points P and Q, are concurrent with the altitude $\overline{AD}$ of $\triangle ABC$.

10. The Theorem of Stewart

The geometry student usually feels at ease with medians, angle bisectors, and altitudes of triangles. What about 'internal line segments' (segments with endpoints on a vertex and its opposite side) that are neither medians, angle bisectors, nor altitudes? As the problems in this section show, much can be learned about such segments thanks to Stewart's Theorem. Named after Matthew Stewart who published it in 1745, this theorem describes the relationship between an 'internal line segment', the side to which it is drawn, the two parts of this side, and the other sides of the triangle.

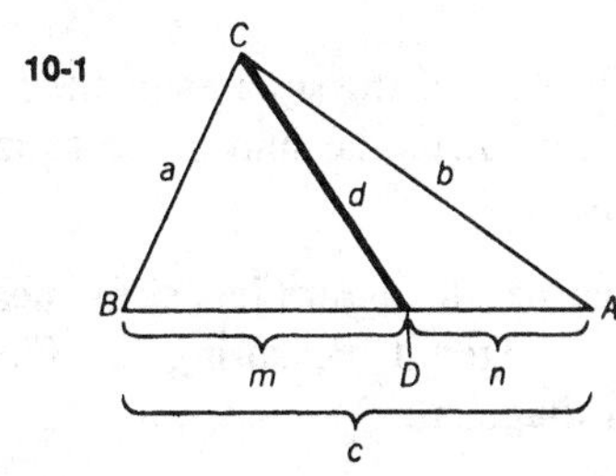

10-1 A classic theorem known as Stewart's Theorem, is very useful as a means of finding the measure of any line segment from the vertex of a triangle to the opposite side. Using the letter designations in Fig. 10-1, the theorem states the following relationship: $a^2n + b^2m = c(d^2 + mn)$. Prove the validity of the theorem.

Challenge If a line from C meets $\overleftrightarrow{AB}$ at F, where F is not between A and B, prove that

$$(BC)^2(AF) - (AC)^2(BF) = AB[(CF)^2 - (AF)(BF)].$$

10-2 In an isosceles triangle with two sides of measure 17, a line measuring 16 is drawn from the vertex to the base. If one segment of the base, as cut by this line, exceeds the other by 8, find the measures of the two segments.

10-3 In $\triangle ABC$, point E is on $\overline{AB}$, so that $AE = \frac{1}{2}EB$. Find CE if $AC = 4$, $CB = 5$, and $AB = 6$.

Challenge Find the measure of the segment from E to the midpoint of $\overline{CB}$.

10-4 Prove that the sum of the squares of the distances from the vertex of the right angle, in a right triangle, to the trisection points along the hypotenuse, is equal to $\frac{5}{9}$ the square of the measure of the hypotenuse.

Challenge 1 Verify that the median to the hypotenuse of a right triangle is equal in measure to one-half the hypotenuse. Use Stewart's Theorem.

Challenge 2 Try to predict, from the results of Problem 10-4 and Challenge 1, the value of the sum of the squares for a quadrisection of the hypotenuse.

10-5 Prove that the sum of the squares of the measures of the sides of a parallelogram equals the sum of the squares of the measures of the diagonals.

Challenge A given parallelogram has sides measuring 7 and 9, and a shorter diagonal measuring 8. Find the measure of the longer diagonal.

10-6 Using Stewart's Theorem, prove that in any triangle the square of the measure of the internal bisectors of any angle is equal to the product of the measures of the sides forming the bisected angle decreased by the product of the measures of the segments of the side to which this bisector is drawn.

Challenge 1 Can you also prove the theorem in Problem 10-6 without using Stewart's Theorem?

Challenge 2 Prove that in $\triangle ABC$, $t_a = \frac{bc}{b+c}\sqrt{2}$, when $\angle BAC$ is a right angle.

10-7 The two shorter sides of a triangle measure 9 and 18. If the internal angle bisector drawn to the longest side measures 8, find the measure of the longest side of the triangle.

Challenge Find the measure of a side of a triangle if the other two sides and the bisector of the included angle have measures 12, 15, and 10, respectively.

10-8 In a right triangle, the bisector of the right angle divides the hypotenuse into segments that measure 3 and 4. Find the measure of the angle bisector of the larger acute angle of the right triangle.

10-9 In a 30–60–90 right triangle, if the measure of the hypotenuse is 4, find the distance from the vertex of the right angle to the point of intersection of the angle bisectors.

SOLUTIONS

1. Congruence and Parallelism

1-1 *In any* $\triangle ABC$, E *and* D *are interior points of* $\overline{AC}$ *and* $\overline{BC}$, *respectively (Fig. S1-1a).* $\overline{AF}$ *bisects* $\angle CAD$, *and* $\overline{BF}$ *bisects* $\angle CBE$. *Prove* $m\angle AEB + m\angle ADB = 2m\angle AFB$.

$$m\angle AFB = 180 - [(x + w) + (y + z)] \ (\#13) \qquad \text{(I)}$$

$$m\angle AEB = 180 - [(2x + w) + z] \ (\#13)$$
$$m\angle ADB = 180 - [(2y + z) + w] \ (\#13)$$
$$\overline{m\angle AEB + m\angle ADB = 360 - [2x + 2y + 2z + 2w]}$$

$$2m\angle AFB = 2[180 - (x + y + z + w)] \qquad \text{(twice I)}$$
$$2m\angle AFB = 360 - [2x + 2y + 2z + 2w]$$

Therefore, $m\angle AEB + m\angle ADB = 2m\angle AFB$.

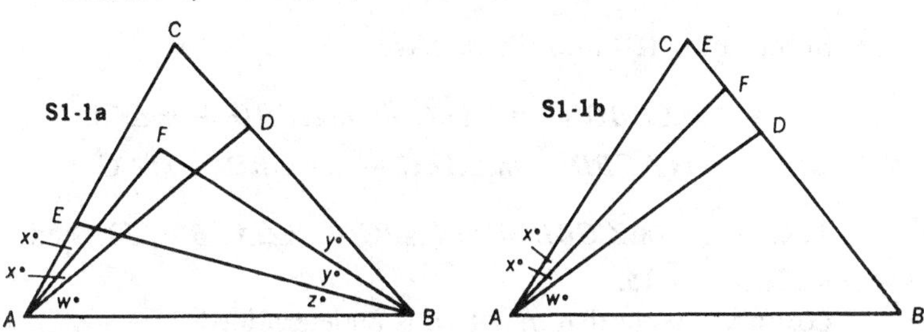

Challenge 1 *Prove that this result holds if* E *coincides with* C *(Fig. S1-1b).*

PROOF:

We must show that $m\angle AEB + m\angle ADB = 2m\angle AFB$.

Let $m\angle CAF = m\angle FAD = x$.

$$\text{Since } \angle ADB \text{ is an exterior angle of } \triangle AFD,$$
$$m\angle ADB = m\angle AFD + x \ (\#12).$$
$$\text{Similarly, in } \triangle AEF,$$
$$m\angle AFD = m\angle AEB + x \ (\#12).$$
$$m\angle ADB + m\angle AEB + x = 2m\angle AFD + x,$$
$$\text{thus, } m\angle AEB + m\angle ADB = 2m\angle AFB.$$

1-2 *In* $\triangle$ABC, *a point* D *is on* $\overline{\text{AC}}$ *so that* AB = AD (*Fig.* S1-2). m$\angle$ABC − m$\angle$ACB = 30. *Find* m$\angle$CBD.

$$m\angle CBD = m\angle ABC - m\angle ABD$$
$$\text{Since } AB = AD, m\angle ABD = m\angle ADB \ (\#5).$$

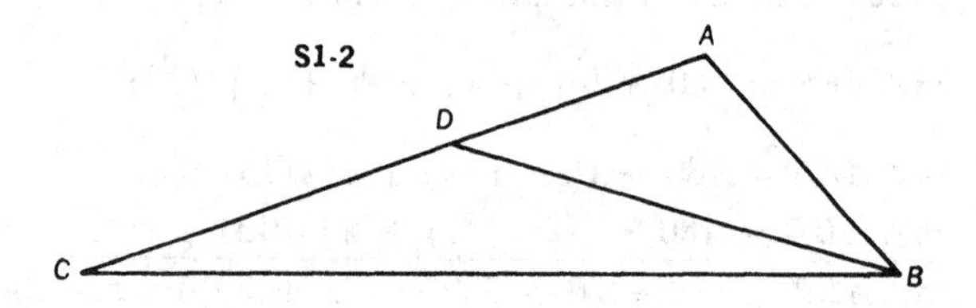

Therefore, by substitution,

$$m\angle CBD = m\angle ABC - m\angle ADB. \qquad \text{(I)}$$
$$\text{But } m\angle ADB = m\angle CBD + m\angle C \ (\#12). \qquad \text{(II)}$$

Substituting (II) into (I), we have

$$m\angle CBD = m\angle ABC - [m\angle CBD + m\angle C].$$
$$m\angle CBD = m\angle ABC - m\angle CBD - m\angle C$$

Therefore, $2m\angle CBD = m\angle ABC - m\angle ACB = 30$, and $m\angle CBD = 15$.

COMMENT: Note that $m\angle ACB$ is undetermined.

1-3 *The interior bisector of* $\angle$B, *and the exterior bisector of* $\angle$C *of* $\triangle$ABC *meet at* D (*Fig.* S1-3). *Through* D, *a line parallel to* $\overline{\text{CB}}$ *meets* $\overline{\text{AC}}$ *at* L *and* $\overline{\text{AB}}$ *at* M. *If the measures of legs* $\overline{\text{LC}}$ *and* $\overline{\text{MB}}$ *of trapezoid* CLMB *are* 5 *and* 7, *respectively, find the measure of base* $\overline{\text{LM}}$. *Prove your result.*

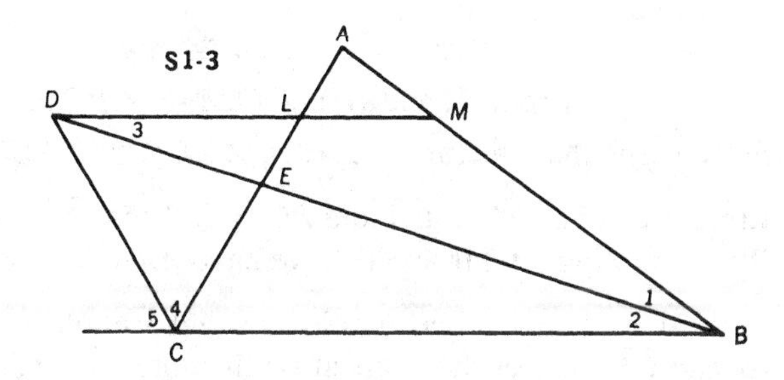

$m\angle 1 = m\angle 2$ and $m\angle 2 = m\angle 3$ (#8).

Therefore, $m\angle 1 = m\angle 3$ (transitivity).

In isosceles $\triangle DMB$, $DM = MB$ (#5).

Similarly, $m\angle 4 = m\angle 5$ and $m\angle 5 = m\angle LDC$ (#8).

Therefore, $m\angle 4 = m\angle LDC$ (transitivity).

Thus, in isosceles $\triangle DLC$, $DL = CL$ (#5).

Since $DM = DL + LM$, by substitution,

$MB = LC + LM$, or $LM = MB - LC$.

Since $LC = 5$ and $MB = 7$, $LM = 2$.

Challenge *Find* LM *if* $\triangle$ABC *is equilateral.*

ANSWER: Zero

1-4 *In right* $\triangle$ABC, $\overline{\text{CF}}$ *is the median drawn to hypotenuse* $\overline{\text{AB}}$, $\overline{\text{CE}}$ *is the bisector of* $\angle$ACB, *and* $\overline{\text{CD}}$ *is the altitude to* $\overline{\text{AB}}$ (*Fig. S*1-4*a*). *Prove that* $\angle$DCE $\cong$ $\angle$ECF.

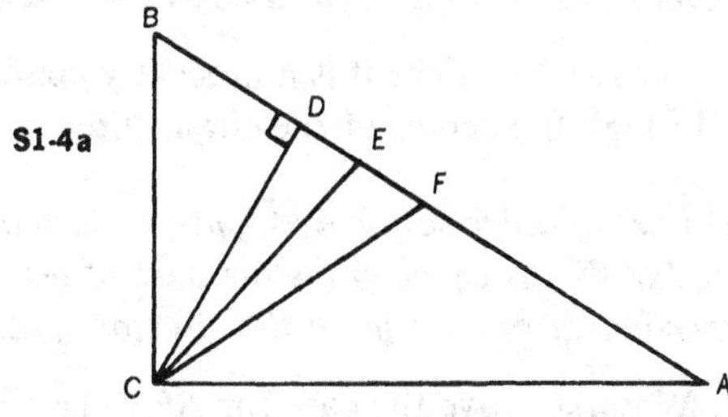

METHOD I: In right $\triangle ABC$, $CF = \frac{1}{2}AB = FA$ (#27).

Since $\triangle CFA$ is isosceles, $\angle FCA \cong \angle A$ (#5). (I)

In right $\triangle BDC$, $\angle B$ is complementary to $\angle BCD$.

In right $\triangle ABC$, $\angle B$ is complementary to $\angle A$.

Therefore, $\angle BCD \cong \angle A$. (II)

From (I) and (II), $\angle FCA \cong \angle BCD$. (III)

Since $\overline{CE}$ is the bisector of $\angle ACB$, $\angle ACE \cong \angle BCE$. (IV)

In right $\triangle ABC$, $\overline{PC} \perp \overline{CA}$ and $PC = BC$ and $\overline{AE}$ bisects $\angle A$. By subtracting (III) from (IV), we have $\angle DCE \cong \angle ECF$.

METHOD II: Let a circle be circumscribed about right $\triangle ABC$. Extend $\overline{CE}$ to meet the circle at G; then draw $\overline{FG}$ (Fig. S1-4b).

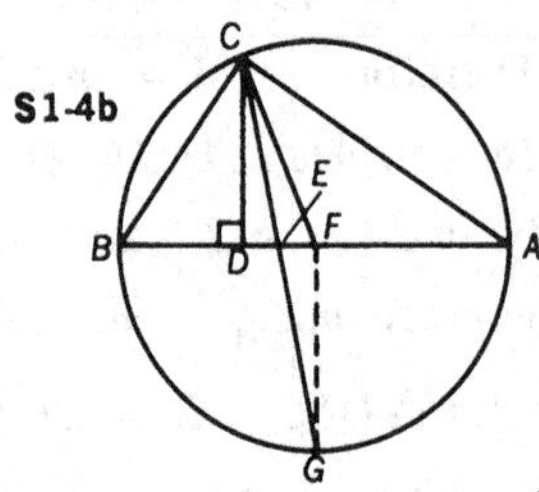

Since $\overline{CE}$ bisects $\angle ACB$, it also bisects $\widehat{AGB}$. Thus, G is the midpoint of $\widehat{AGB}$, and $\overline{FG} \perp \overline{AB}$. Since both $\overline{FG}$ and $\overline{CD}$ are perpendicular to $\overline{AB}$,

$\overline{FG} \parallel \overline{CD}$ (#9), and $\angle DCE \cong \angle FGE$ (#8). (I)

Since radius $\overline{CF} \cong$ radius $\overline{FG}$, $\triangle CFG$ is isosceles and

$\angle ECF \cong \angle FGE$. (II)

Thus, by transitivity, from (I) and (II), $\angle DCE \cong \angle ECF$.

Challenge *Does this result hold for a non-right triangle?*

ANSWER: No, since it is a necessary condition that $\overline{BA}$ pass through the center of the circumcircle.

1-5 *The measure of a line segment* $\overline{\mathrm{PC}}$, *perpendicular to hypotenuse* $\overline{\mathrm{AC}}$ *of right* $\triangle\mathrm{ABC}$, *is equal to the measure of leg* $\overline{\mathrm{BC}}$. *Show* $\overline{\mathrm{BP}}$ *may be perpendicular or parallel to the bisector of* $\angle\mathrm{A}$.

CASE I: We first prove the case for $\overline{BP} \parallel \overline{AE}$ (Fig. S1-5a).
In right $\triangle ABC$, $\overline{PC} \perp \overline{AC}$, $PC = BC$, and $\overline{AE}$ bisects $\angle A$. $\angle CEA$ is complementary to $\angle CAE$, while $\angle BDA$ is complementary to $\angle DAB$ (#14).
Since $\angle CAE \cong \angle DAB$, $\angle BDA \cong \angle CEA$. However, $\angle BDA \cong \angle EDC$ (#1). Therefore, $\angle CED \cong \angle CDE$, and $\triangle CED$ is

isosceles (#5). Since isosceles triangles CED and CPB share the same vertex angle, they are mutually equiangular. Thus, since $\angle CED \cong \angle CPB$, $\overline{EA} \parallel \overline{PB}$ (#7).

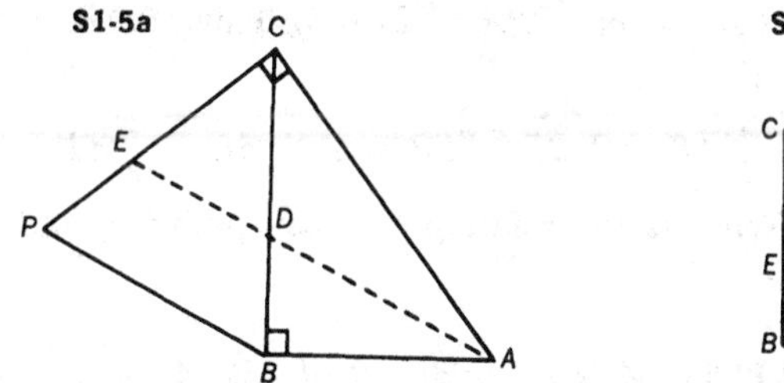

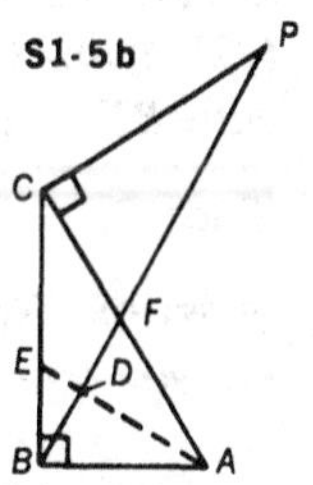

CASE II: We now prove the case for $\overline{AE} \perp \overline{BP}$ (Fig. S1-5b). $\angle CPF$ is complementary to $\angle CFP$ (#14). Since $\angle CFP \cong \angle BFA$ (#1), $\angle CPF$ is complementary to $\angle BFA$. However, in $\triangle CPB$, $\overline{CP} \cong \overline{CB}$ and $\angle CPB \cong \angle CBP$ (#5); hence, $\angle CBP$ is complementary to $\angle BFA$. But $\angle CBP$ is complementary to $\angle PBA$. Therefore, $\angle BFA \cong \angle FBA$ (both are complementary to $\angle CBP$). Now we have $\triangle FAB$ isosceles with $\overline{AD}$ an angle bisector; thus, $\overline{AD} \perp \overline{BFP}$ since the bisector of the vertex angle of an isosceles triangle is perpendicular to the base.

1-6 *Prove the following: if, in* $\triangle$ABC, *median* $\overline{\text{AM}}$ *is such that* m$\angle$BAC *is divided in the ratio* 1:2, *and* $\overline{\text{AM}}$ *is extended through* M *to* D *so that* $\angle$DBA *is a right angle, then* AC $= \frac{1}{2}$AD (*Fig. S*1-6).

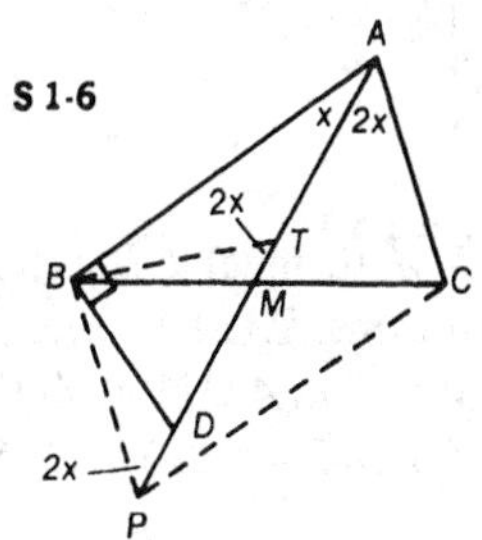

Let $m\angle BAM = x$; then $m\angle MAC = 2x$. Choose point P on $\overline{AD}$ so that $AM = MP$.

Since $BM = MC$, $ACPB$ is a parallelogram (#21f). Thus, $BP = AC$.

Let T be the midpoint of $\overline{AD}$ making $\overline{BT}$ the median of right $\triangle ABD$.

It follows that $BT = \frac{1}{2}AD$, or $BT = AT$ (#27), and, consequently, $m\angle TBA = x$. $\angle BTP$ is an exterior angle of isosceles $\triangle BTA$. Therefore, $m\angle BTP = 2x$ (#12). However, since $\overline{BP} \parallel \overline{AC}$ (#21a), $m\angle CAP = m\angle BPA = 2x$ (#8). Thus, $\triangle TBP$ is isosceles with $BT = BP$.

Since $BT = \frac{1}{2}AD$ and $BT = BP = AC$, $AC = \frac{1}{2}AD$.

QUESTION: What is the relation between points P and D when $m\angle A = 90$?

1-7 *In square* ABCD, M *is the midpoint of* $\overline{\text{AB}}$. *A line perpendicular to* $\overline{\text{MC}}$ *at* M *meets* $\overline{\text{AD}}$ *at* K. *Prove that* $\angle\text{BCM} \cong \angle\text{KCM}$.

METHOD I: Draw $\overline{ML} \parallel \overline{AD}$ (Fig. S1-7a). Since $AM = MB$ and $\overline{AD} \parallel \overline{ML} \parallel \overline{BC}$, $KP = PC$ (#24). Consider right $\triangle KMC$; $\overline{MP}$ is a median. Therefore, $MP = PC$ (#27). Since $\triangle MPC$ is isosceles, $m\angle 1 = m\angle 2$. However, since $\overline{ML} \parallel \overline{BC}$, $m\angle 1 = m\angle 3$ (#8), thus, $m\angle 2 = m\angle 3$; that is, $\angle BCM \cong \angle KCM$.

S1-7a

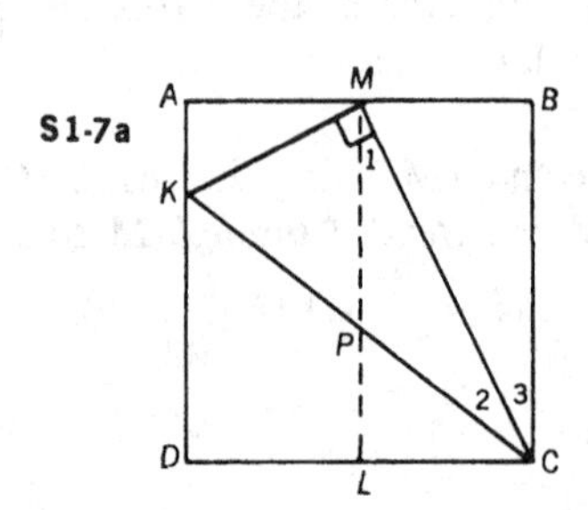

S1-7b

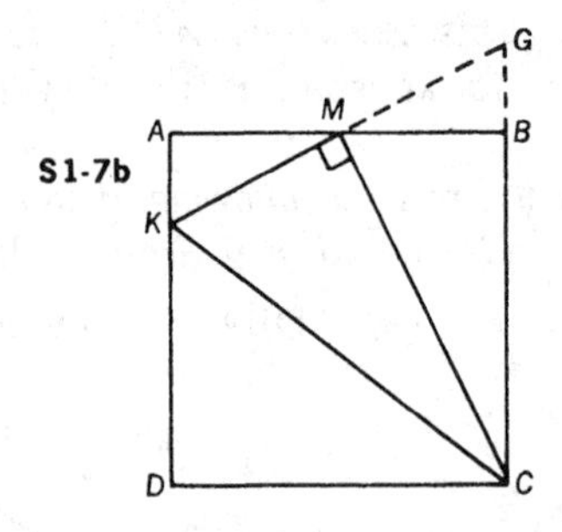

METHOD II: Extend $\overline{KM}$ to meet $\overline{CB}$ extended at G (Fig. S1-7b). Since $AM = MB$ and $m\angle KAM = m\angle MBG$ (right angles) and $m\angle AMK = m\angle GMB$, $\triangle AMK \cong \triangle BMG$ (A.S.A.). Then, $KM = MG$. Now, $\triangle KMC \cong \triangle GMC$ (S.A.S.), and $\angle BCM \cong \angle KCM$.

METHOD III: Other methods may easily be found. Here is one without auxiliary constructions in which similarity is employed (Fig. S1-7c).

$AM = MB = \frac{1}{2}s$, where $BC = s$.

$\angle AMK$ is complementary to $\angle BMC$, and $\angle BCM$ is complementary to $\angle BMC$ (#14).

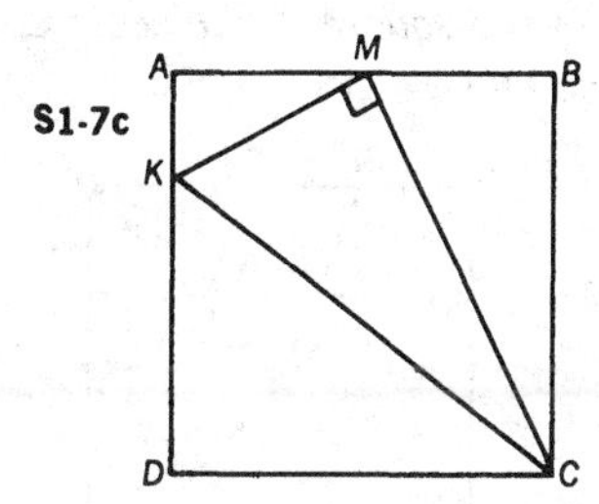

Therefore, right $\triangle MAK \sim$ right $\triangle CBM$, and $AK = \frac{1}{4}s$.

In right $\triangle MAK$, $MK = \frac{s\sqrt{5}}{4}$ (#55), while in right $\triangle CBM$, $MC = \frac{s\sqrt{5}}{2}$ (#55).

Therefore, since $\frac{MK}{MC} = \frac{\frac{\sqrt{5}}{4}}{\frac{\sqrt{5}}{2}} = \frac{1}{2} = \frac{MB}{BC}$, right $\triangle MKC \sim$ right $\triangle BMC$ (#50), and $\angle BCM \cong \angle KCM$.

1-8 *Given any* $\triangle$ABC, $\overline{\text{AE}}$ *bisects* $\angle$BAC, $\overline{\text{BD}}$ *bisects* $\angle$ABC, $\overline{\text{CP}} \perp \overline{\text{BD}}$, *and* $\overline{\text{CQ}} \perp \overline{\text{AE}}$, *prove that* $\overline{\text{PQ}}$ *is parallel to* $\overline{\text{AB}}$.

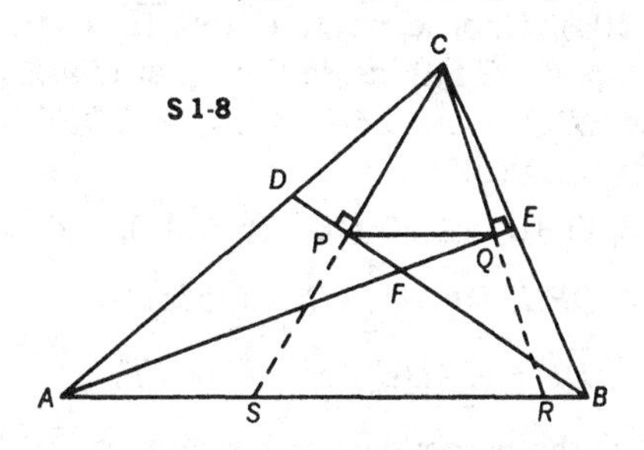

Extend $\overline{CP}$ and $\overline{CQ}$ to meet $\overline{AB}$ at S and R, respectively (Fig. S1-8). It may be shown that $\triangle CPB \cong \triangle SPB$, and $\triangle CQA \cong \triangle RQA$ (A.S.A.).

It then follows that $CP = SP$ and $CQ = RQ$ or P and Q are midpoints of $\overline{CS}$ and $\overline{CR}$, respectively. Therefore, in $\triangle CSR$, $\overline{PQ} \parallel \overline{SR}$ (#26). Thus, $\overline{PQ} \parallel \overline{AB}$.

Challenge *Identify the points* P *and* Q *when* $\triangle$ABC *is equilateral.*

ANSWER: P and Q are the midpoints of $\overline{CA}$ and $\overline{CB}$, respectively.

1-9 *Given that* ABCD *is a square,* $\overline{CF}$ *bisects* $\angle ACD$, *and* $\overline{BPQ} \perp \overline{CF}$ (*Fig. S*1-9), *prove* $DQ = 2PE$.

S 1-9

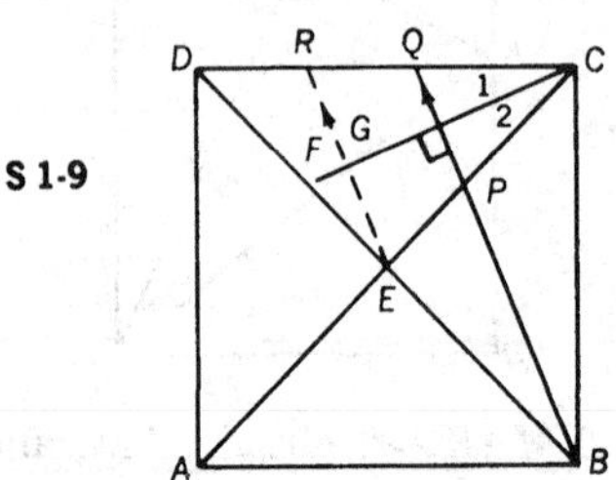

Draw $\overline{RE} \parallel \overline{BPQ}$. Since E is the midpoint of $\overline{DB}$ (#21n) in $\triangle DQB$, $DR = QR$ (#25). Since $\overline{RE} \perp \overline{CF}$ (#10), $\triangle RGC \cong \triangle EGC$, and $\angle CRG = \angle CEG$.

Therefore, $RQPE$ is an isosceles trapezoid (#23), and $PE = QR$. $2RQ = DQ$ and, therefore, $DQ = 2PE$.

1-10 *Given square* ABCD *with* $m\angle EDC = m\angle ECD = 15$, *prove* $\triangle ABE$ *is equilateral.*

METHOD I: In square $ABCD$, draw $\overline{AF}$ perpendicular to $\overline{DE}$ (Fig. S1-10a). Choose point G on $\overline{AF}$ so that $m\angle FDG = 60$. Why does point G fall inside the square? $m\angle AGD = 150$ (#12). Since $m\angle EDC = m\angle ECD = 15$, $m\angle DEC = 150$ (#13); thus, $\angle AGD \cong \angle DEC$.

Therefore, $\triangle AGD \cong \triangle DEC$ (S.A.A.), and $DE = DG$.

In right $\triangle DFG$, $DF = \frac{1}{2}(DG)$ (#55c).

Therefore, $DF = \frac{1}{2}(DE)$, or $DF = EF$.

Since $\overline{AF}$ is the perpendicular bisector of $\overline{DE}$, $AD = AE$ (#18).

A similar argument shows $BC = BE$. Therefore, $AE = BE = AB$ (all are equal to the measure of a side of square $ABCD$).

S1-10a

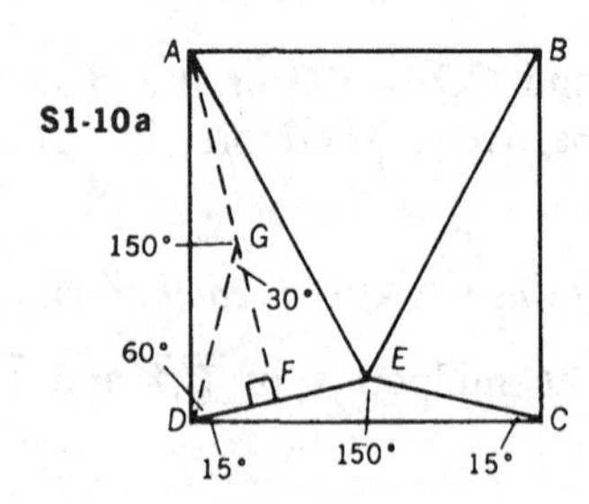

S1-10b

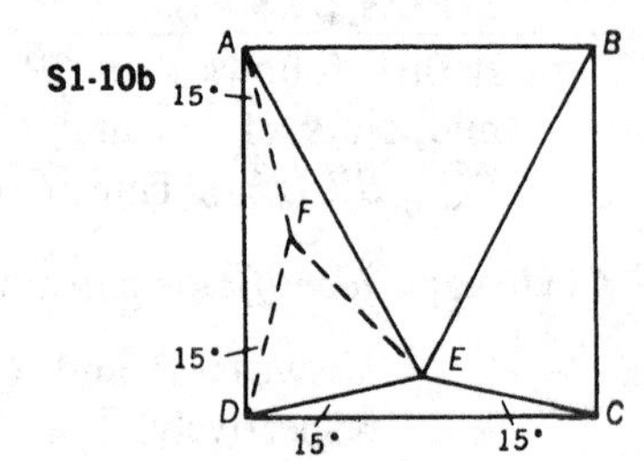

METHOD II: In square $ABCD$, with $m\angle EDC = m\angle ECD = 15$, draw $\triangle AFD$ on $\overline{AD}$ such that $m\angle FAD = m\angle FDA = 15$. Then draw $\overline{FE}$ (Fig. S1-10b).

$\triangle FAD \cong \triangle EDC$ (A.S.A.), and $DE = DF$.

Since $\angle ADC$ is a right angle, $m\angle FDE = 60$ and $\triangle FDE$ is equilateral so that $DF = DE = FE$. Since $m\angle DFE = 60$ and $m\angle AFD = 150$ (#13), $m\angle AFE = 150$. Thus, $m\angle FAE = 15$ and $m\angle DAE = 30$.

Therefore, $m\angle EAB = 60$. In a similar fashion it may be proved that $m\angle ABE = 60$; thus, $\triangle ABE$ is equilateral.

METHOD III: In square $ABCD$, with $m\angle EDC = m\angle ECD = 15$, draw equilateral $\triangle DFC$ on $\overline{DC}$ externally; then draw $\overline{EF}$ (Fig. S1-10c).

$\overline{EF}$ is the perpendicular bisector of $\overline{DC}$ (#18).

Since $AD = FD$, and $m\angle ADE = m\angle FDE = 75$, $\triangle ADE \cong \triangle FDE$ (S.A.S.).

Since $m\angle DFE = 30$, $m\angle DAE = 30$.

Therefore, $m\angle BAE = 60$.

In a similar fashion, it may be proved that $m\angle ABE = 60$; thus, $\triangle ABE$ is equilateral.

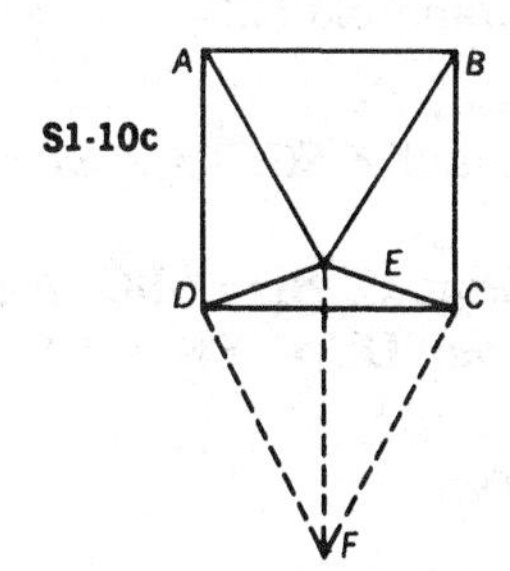

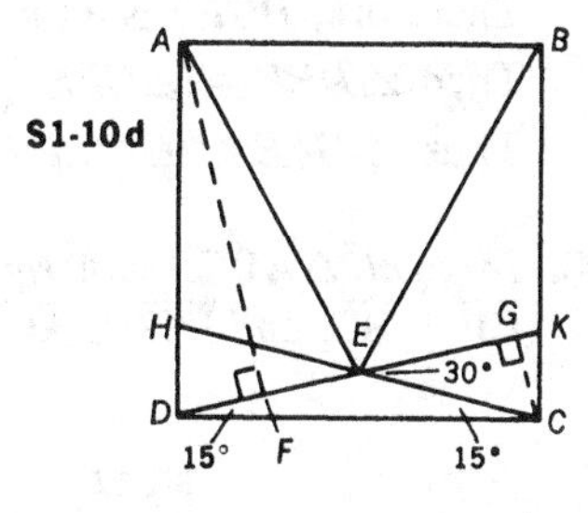

METHOD IV: Extend $\overline{DE}$ and $\overline{CE}$ to meet $\overline{BC}$ and $\overline{AD}$ at K and H, respectively (Fig. S1-10d). In square $ABCD$, $m\angle KDC = m\angle HCD = 15$, therefore, $ED = EC$ (#5).

Draw $\overline{AF}$ and $\overline{CG}$ perpendicular to $\overline{DK}$.

In right $\triangle DGC$, $m\angle GCD = 75$ (#14), while $m\angle ADF = 75$ also. Thus, $\triangle ADF \cong \triangle DCG$, and $DF = CG$. $m\angle GEC = 30$ (#12). In $\triangle GEC$, $CG = \frac{1}{2}(EC)$ (#55c). Therefore, $CG = \frac{1}{2}(ED)$, or $DF = \frac{1}{2}(ED)$.

Since $\overline{AF}$ is the perpendicular bisector of $\overline{DE}$, $AD = AE$ (#18). In a similar fashion, it may be proved that $BE = BC$; therefore, $\triangle ABE$ is equilateral.

1-11 *In any* $\triangle$ABC, D, E, *and* F *are midpoints of the sides* $\overline{\text{AC}}$, $\overline{\text{AB}}$, *and* $\overline{\text{BC}}$, *respectively.* $\overline{\text{BG}}$ *is an altitude of* $\triangle$ABC (*Fig. S*1-11). *Prove that* $\angle$EGF $\cong$ $\angle$EDF.

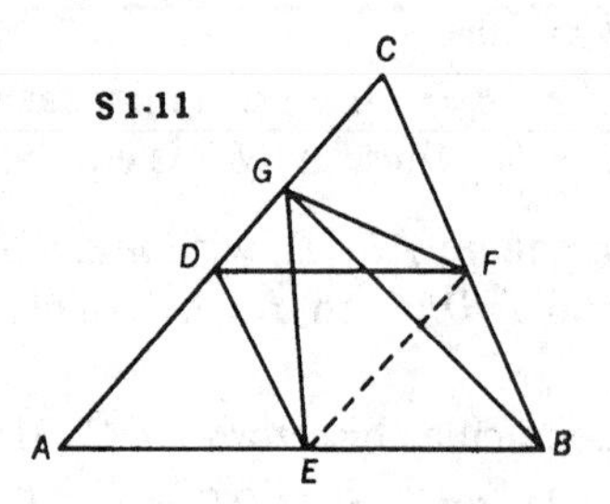

$\overline{GF}$ is the median to hypotenuse $\overline{CB}$ of right $\triangle CGB$, therefore, $GF = \frac{1}{2}(CB)$ (#27).
$DE = \frac{1}{2}CB$ (#26), therefore, $DE = GF$.
Join midpoints E and F. Thus, $\overline{EF} \parallel \overline{AC}$ (#26).
Therefore, $DGFE$ is an isosceles trapezoid (#23).
Then $\angle DEF \cong \angle GFE$.
Thus, $\triangle GFE \cong \triangle DEF$ (S.A.S.), and $\angle EGF \cong \angle EDF$.

1-12 *In right* $\triangle$ABC, *with right angle at* C, BD = BC, AE = AC, $\overline{\text{EF}} \perp \overline{\text{BC}}$, *and* $\overline{\text{DG}} \perp \overline{\text{AC}}$. *Prove that* DE = EF + DG.

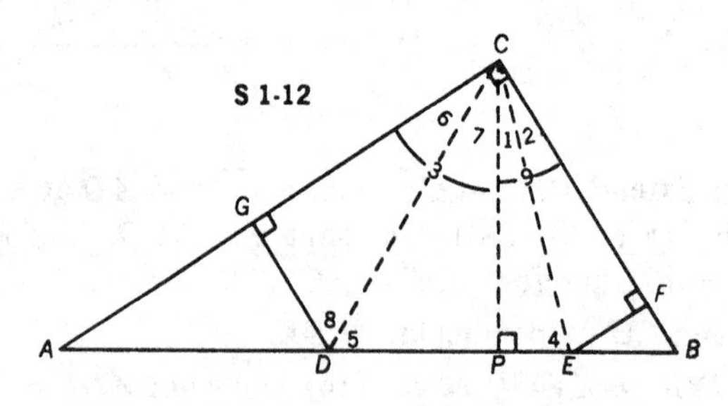

Draw $\overline{CP} \perp \overline{AB}$, also draw $\overline{CE}$ and $\overline{CD}$ (Fig. S1-12).

$$m\angle 3 + m\angle 1 + m\angle 2 = 90$$
$$m\angle 3 + m\angle 1 = m\angle 4 \text{ (\#5)}$$

By substitution, $m\angle 4 + m\angle 2 = 90$;
but in right $\triangle CPE$, $m\angle 4 + m\angle 1 = 90$ (#14).
Thus, $\angle 1 \cong \angle 2$ (both are complementary to $\angle 4$), and right $\triangle CPE \cong$ right $\triangle CFE$, and $PE = EF$.
Similarly, $m\angle 9 + m\angle 7 + m\angle 6 = 90$,
$m\angle 9 + m\angle 7 = m\angle 5$ (#5).
By substitution, $m\angle 5 + m\angle 6 = 90$.
However, in right $\triangle CPD$, $m\angle 5 + m\angle 7 = 90$ (#14).
Thus, $\angle 6 \cong \angle 7$ (both are complementary to $\angle 5$), and right $\triangle CPD \cong$ right $\triangle CGD$, and $DP = DG$.
Since $DE = DP + PE$, we get $DE = DG + EF$.

1-13 *Prove that the sum of the measures of the perpendiculars from any point on a side of a rectangle to the diagonals is constant.*

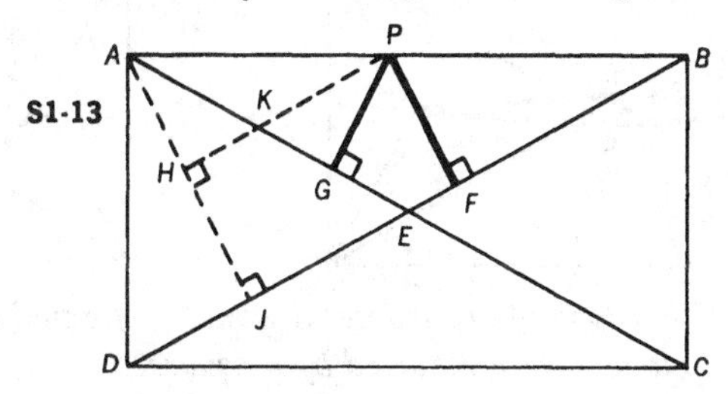

Let P be any point on side $\overline{AB}$ of rectangle $ABCD$ (Fig. S1-13). $\overline{PG}$ and $\overline{PF}$ are perpendiculars to the diagonals.
Draw $\overline{AJ}$ perpendicular to $\overline{DB}$, and then $\overline{PH}$ perpendicular to $\overline{AJ}$.
Since $PHJF$ is a rectangle (a quadrilateral with three right angles), we get $PF = HJ$.
Since $\overline{PH}$ and $\overline{BD}$ are both perpendicular to $\overline{AJ}$, $\overline{PH}$ is parallel to $\overline{BD}$ (#9).
Thus, $\angle APH \cong \angle ABD$ (#7).
Since $AE = EB$ (#21f, 21h), $\angle CAB \cong \angle ABD$ (#5). Thus, by transitivity, $\angle EAP \cong \angle APH$; also in $\triangle APK$, $AK = PK$ (#5). Since $\angle AKH \cong \angle PKG$ (#1), right $\triangle AHK \cong$ right $\triangle PGK$ (S.A.A.). Hence, $AH = PG$ and, by addition, $PF + PG = HJ + AH = AJ$, a constant.

1-14 *The trisectors of the angles of a rectangle are drawn. For each pair of adjacent angles, those trisectors that are closest to the* *enclosed side are extended until a point of intersection is established. The line segments connecting those points of intersection form a quadrilateral. Prove that the quadrilateral is a rhombus.*

As a result of the trisections,
isosceles $\triangle AHD \cong$ isosceles $\triangle BFC$, and
isosceles $\triangle AGB \cong$ isosceles $\triangle DEC$ (Fig. S1-14a).
Since $AH = HD = FB = FC$, and $AG = GB = DE = CE$,
and $\angle HAG \cong \angle GBF \cong \angle FCE \cong \angle HDE \cong \frac{1}{3}$ right angle,
$\triangle HAG \cong \triangle FBG \cong \triangle FCE \cong \triangle HDE$ (S.A.S.).
Therefore, $HG = FG = FE = HE$, and $EFGH$ is a rhombus (#21-I).

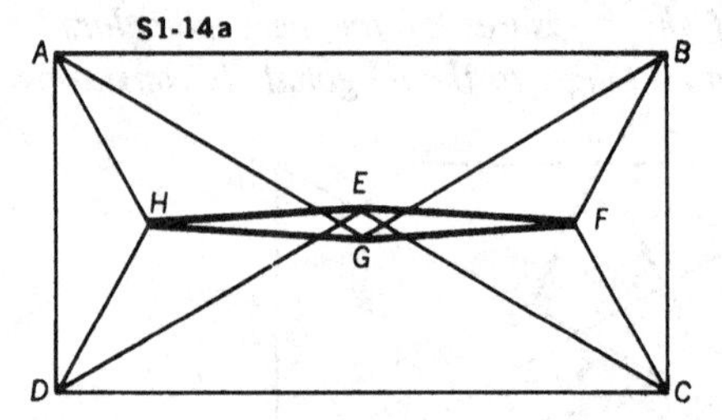

S1-14a

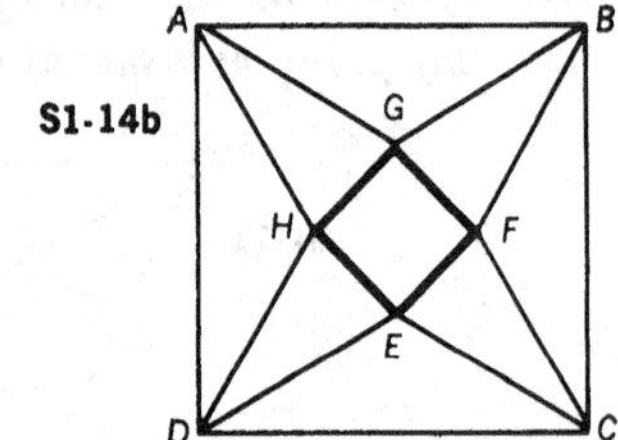

S1-14b

Challenge 1 *What type of quadrilateral would be formed if the original rectangle were replaced by a square?*

Consider $ABCD$ to be a square (Fig. S1-14b). All of the above still holds true; thus we still maintain a rhombus. However, we now can easily show $\triangle AHG$ to be isosceles, $m\angle AGH = m\angle AHG = 75$.

Similarly, $m\angle BGF = 75$. $m\angle AGB = 120$, since $m\angle GAB = m\angle GBA = 30$.

Therefore, $m\angle HGF = 90$. We now have a rhombus with one right angle; hence, a square.

1-15 *In Fig. S1-15, $\overline{BE}$ and $\overline{AD}$ are altitudes of $\triangle ABC$. F, G, and K are midpoints of $\overline{AH}$, $\overline{AB}$, and $\overline{BC}$, respectively. Prove that $\angle FGK$ is a right angle.*

In $\triangle AHB$, $\overline{GF} \parallel \overline{BH}$ (#26).
And in $\triangle ABC$, $\overline{GK} \parallel \overline{AC}$ (#26).
Since $\overline{BE} \perp \overline{AC}$, $\overline{BE} \perp \overline{GK}$,
then $\overline{GF} \perp \overline{GK}$ (#10); that is, $\angle FGK$ is a right angle.

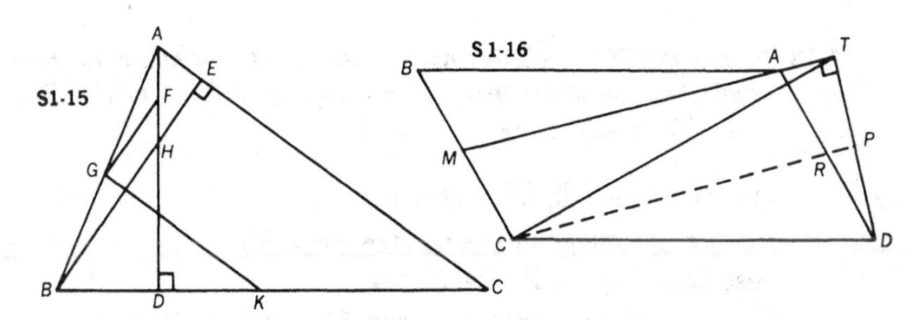

1-16 *In parallelogram* ABCD, M *is the midpoint of* $\overline{BC}$. $\overline{DT}$ *is drawn from* D *perpendicular to* $\overleftrightarrow{MA}$ (*Fig. S1-16*). *Prove* CT = CD.

Let R be the midpoint of $\overline{AD}$; draw $\overline{CR}$ and extend it to meet $\overline{TD}$ at P. Since $AR = \frac{1}{2}AD$, and $MC = \frac{1}{2}BC$, $AR = MC$. Since $\overline{AR} \parallel \overline{MC}$, $ARCM$ is a parallelogram (#22). Thus, $\overline{CP} \parallel \overline{MT}$. In $\triangle ATD$, since $\overline{RP} \parallel \overline{AT}$ and passes through the midpoint of $\overline{AD}$, it must also pass through the midpoint of $\overline{TD}$ (#25). Since $\overline{MT} \parallel \overline{CP}$, and $\overline{MT} \perp \overline{TD}$, $\overline{CP} \perp \overline{TD}$ (#10). Thus, $\overline{CP}$ is the perpendicular bisector of $\overline{TD}$, and $CT = CD$ (#18).

1-17 *Prove that the line segment joining the midpoints of two opposite sides of any quadrilateral bisects the line segment joining the midpoints of the diagonals.*

$ABCD$ is any quadrilateral. K, L, P, and Q are midpoints of $\overline{AD}$, $\overline{BC}$, $\overline{BD}$, and $\overline{AC}$, respectively. We are to prove that $\overline{KL}$ bisects $\overline{PQ}$. Draw $\overline{KP}$ and $\overline{QL}$ (Fig. S1-17).

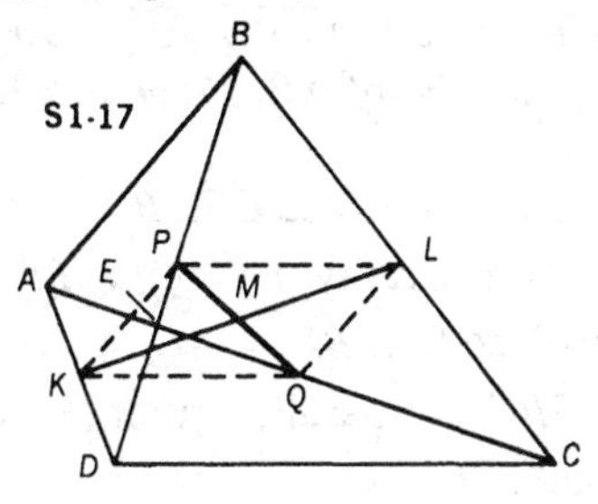

In $\triangle ADB$, $KP = \frac{1}{2}AB$, and $\overline{KP} \parallel \overline{AB}$ (#26).

Similarly, in $\triangle ACB$, $QL = \frac{1}{2}AB$, and $\overline{QL} \parallel \overline{AB}$ (#26).

By transitivity, $KP = QL$, and $\overline{KP} \parallel \overline{QL}$. It then follows that $KPLQ$ is a parallelogram (#22), and so $PM = QM$ (#21f).

1-18 *In any* $\triangle$ABC, $\overleftrightarrow{XYZ}$ *is any line through the centroid* G. *Perpendiculars are drawn from each vertex of* $\triangle$ABC *to this line. Prove* CY = AX + BZ.

Draw medians $\overline{CD}$, $\overline{AF}$, and $\overline{BH}$.
From E, the midpoint of $\overline{CG}$, draw $\overline{EP} \perp \overline{XZ}$.
Also draw $\overline{DQ} \perp \overline{XZ}$ (Fig. S1-18).
Since $\angle CGY \cong \angle QGD$ (#1), and $EC = EG = DG$ (#29), $\triangle QGD \cong \triangle PGE$, and $QD = EP$.
$\overline{AX} \parallel \overline{BZ}$ (#9), therefore, $\overline{QD}$ is the median of trapezoid $AXZB$, and $QD = \frac{1}{2}(AX + BZ)$ (#28).
$EP = \frac{1}{2}CY$ (#25, #26), therefore, $\frac{1}{2}CY = \frac{1}{2}(AX + BZ)$ (transitivity), and $CY = AX + BZ$.

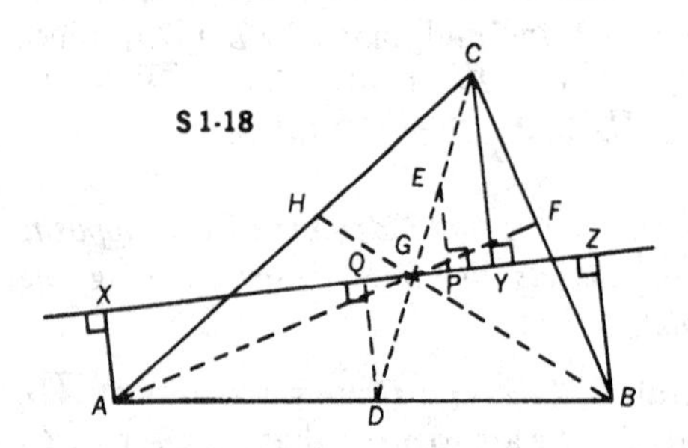

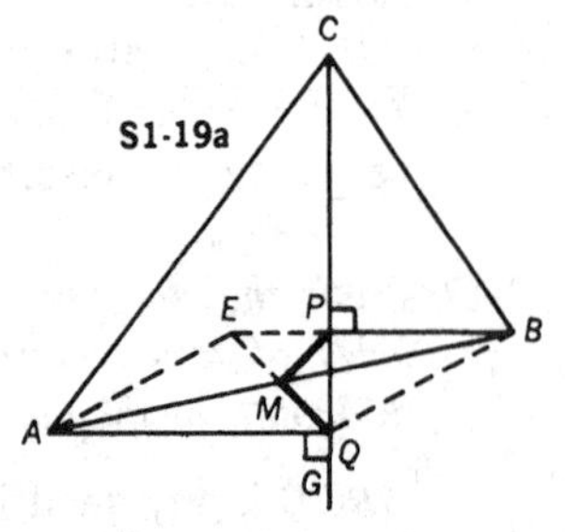

1-19 *In any* $\triangle$ABC, $\overleftrightarrow{CPQ}$ *is any line through* C *interior to* $\triangle$ABC. $\overline{BP}$ *is perpendicular to line* $\overline{CPQ}$, $\overline{AQ}$ *is perpendicular to line* $\overline{CPQ}$, *and* M *is the midpoint of* $\overline{AB}$. *Prove that* MP = MQ.

Since $\overline{BP} \perp \overline{CG}$ and $\overline{AQ} \perp \overline{CG}$, $\overline{BP} \parallel \overline{AQ}$, (#9).
Without loss of generality, let $AQ > BP$ (Fig. S1-19a).
Extend $\overline{BP}$ to E so that $BE = AQ$.
Therefore, $AEBQ$ is a parallelogram (#22).
Draw diagonal $\overline{EQ}$.
$\overline{EQ}$ must pass through M, the midpoint of AB, since the diagonals of the parallelogram bisect each other. Consequently, M is also the midpoint of EQ.
In right $\triangle EPQ$, $\overline{MP}$ is the median to hypotenuse $\overline{EQ}$.
Therefore, $MP = \frac{1}{2}EQ = MQ$ (#27).

Challenge *Show that the same result holds if the line through* C *is exterior to* △ABC.

Extend $\overline{PB}$ through B to E so that $BE = AQ$ (Fig. S1-19b). Since $\overline{AQ} \parallel \overline{PE}$ (#9), quadrilateral $AEBQ$ is a parallelogram (#22).

Thus, if M is the midpoint of $\overline{AB}$, it must also be the midpoint of $\overline{QE}$ (#21f).

Therefore, in right $\triangle QPE$, $MP = \frac{1}{2} EQ = MQ$ (#27).

S1-19b

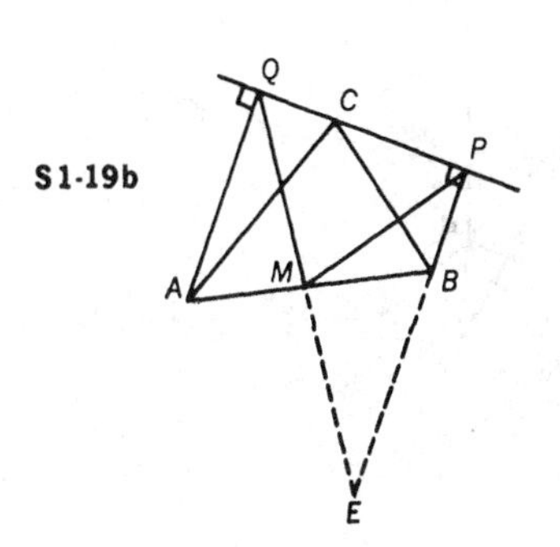

S1-20

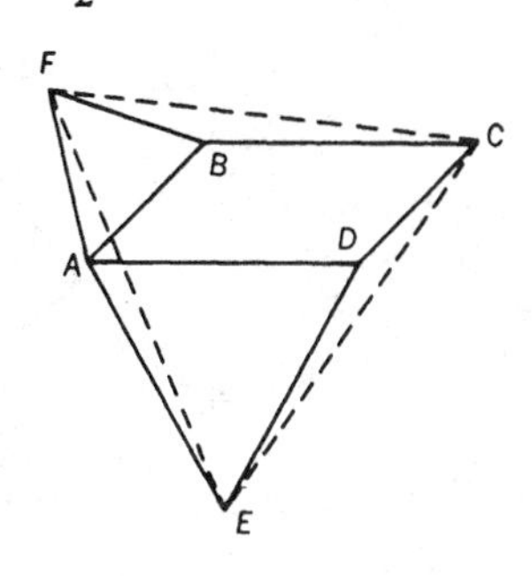

1-20 *In Fig. S*1-20, ABCD *is a parallelogram with equilateral triangles* ABF *and* ADE *drawn on sides* $\overline{AB}$ *and* $\overline{AD}$, *respectively. Prove that* △FCE *is equilateral.*

In order to prove $\triangle FCE$ equilateral, we must show $\triangle AFE \cong \triangle BFC \cong \triangle DCE$ so that we may get $FE = FC = CE$.

Since $AB = DC$ (#21b), and $AB = AF = BF$ (sides of an equilateral triangle are equal), $DC = AF = BF$. Similarly, $AE = DE = BC$.

We have $\angle ADC \cong \angle ABC$.

$m\angle EDC = 360 - m\angle ADE - m\angle ADC = 360 - m\angle ABF - m\angle ABC = m\angle FBC$.

Now $m\angle BAD = 180 - m\angle ADC$ (#21d),

and $m\angle FAE = m\angle FAB + m\angle BAD + m\angle DAE = 120 + m\angle BAD = 120 + 180 - m\angle ADC = 300 - m\angle ADC = m\angle EDC$.

Thus, $\triangle AFE \cong \triangle BFC \cong \triangle DCE$ (S.A.S.), and the conclusion follows.

1-21 *If a square is drawn externally on each side of a parallelogram, prove that*

(a) *the quadrilateral, determined by the centers of these squares, is itself a square*

(b) *the diagonals of the newly formed square are concurrent with the diagonals of the original parallelogram.*

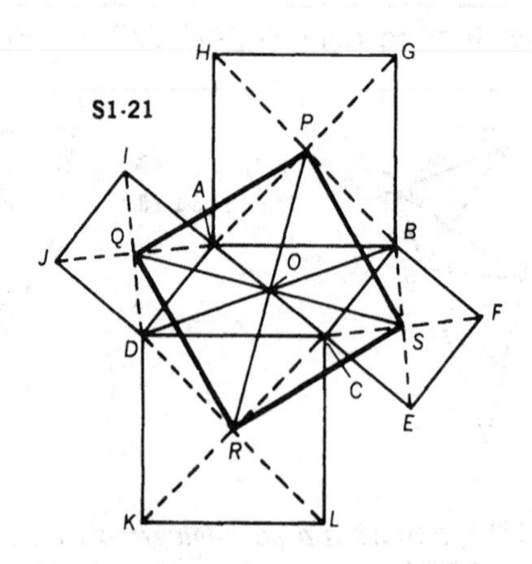

(a) $ABCD$ is a parallelogram.

Points P, Q, R, and S are the centers of the four squares $ABGH$, $DAIJ$, $DCLK$, and $CBFE$, respectively (Fig. S1-21).

$PA = DR$ and $AQ = QD$ (each is one-half a diagonal).

$\angle ADC$ is supplementary to $\angle DAB$ (#21d), and

$\angle IAH$ is supplementary to $\angle DAB$ (since $\angle IAD \cong \angle HAB \cong$ right angle). Therefore, $\angle ADC \cong \angle IAH$.

Since $m\angle RDC = m\angle QDA = m\angle HAP = m\angle QAI = 45$, $\angle RDQ \cong \angle QAP$. Thus, $\triangle RDQ \cong \triangle PAQ$ (S.A.S.), and $QR = QP$.

In a similar fashion, it may be proved that $QP = PS$ and $PS = RS$.

Therefore, $PQRS$ is a rhombus.

Since $\triangle RDQ \cong \triangle PAQ$, $\angle DQR \cong \angle AQP$;

therefore, $\angle PQR \cong \angle DQA$ (by addition).

Since $\angle DQA \cong$ right angle, $\angle PQR \cong$ right angle, and $PQRS$ is a square.

(b) To prove that the diagonals of square $PQRS$ are concurrent with the diagonals of parallelogram $ABCD$, we must prove that a diagonal of the square and a diagonal of the parallelogram bisect each other. In other words, we prove that the diagonals of the square and the diagonals of the parallelogram all share the same midpoint, (i.e., point O).

$\angle BAC \cong \angle ACD$ (#8), and

$m\angle PAB = m\angle RCD = 45$; therefore, $\angle PAC \cong \angle RCA$.

Since $\angle AOP \cong \angle COR$ (#1), and $AP = CR$, $\triangle AOP \cong \triangle COR$ (S.A.A.).

Thus, $AO = CO$, and $PO = RO$.

Since $\overline{DB}$ passes through the midpoint of $\overline{AC}$ (#21f), and, similarly, $\overline{QS}$ passes through the midpoint of $\overline{PR}$, and since $\overline{AC}$ and $\overline{PR}$ share the same midpoint (i.e., O), we have shown that $\overline{AC}$, $\overline{PR}$, $\overline{DB}$, and $\overline{QS}$ are concurrent (i.e., all pass through point O).

2. Triangles in Proportion

2-1 *In* $\triangle ABC$, $\overline{DE} \parallel \overline{BC}$, $\overline{FE} \parallel \overline{DC}$, $AF = 4$, *and* $FD = 6$ (*Fig. S2-1*). *Find* DB.

S2-1

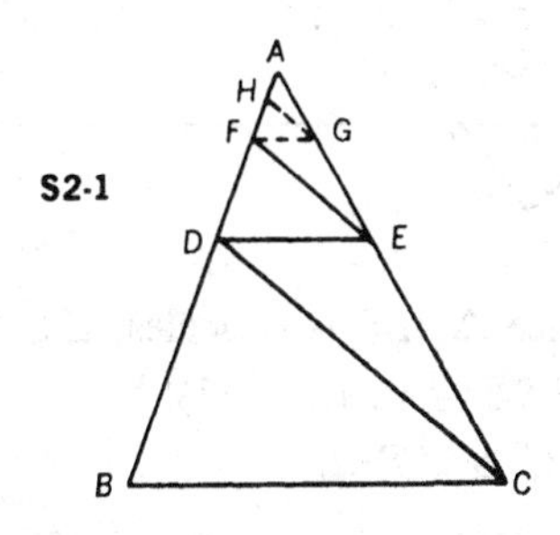

In $\triangle ADC$, $\frac{AF}{FD} = \frac{AE}{EC}$ (#46). So, $\frac{2}{3} = \frac{AE}{EC}$. (I)

However in $\triangle ABC$, $\frac{AD}{DB} = \frac{AE}{EC}$ (#46),

$$\text{and } \frac{10}{DB} = \frac{AE}{EC}. \qquad \text{(II)}$$

From (I) and (II), $\frac{2}{3} = \frac{10}{DB}$. Thus, $DB = 15$.

Challenge 1 *Find* DB *if* AF $= m_1$ *and* FD $= m_2$.

ANSWER: $DB = \frac{m_2}{m_1}(m_1 + m_2)$

Challenge 2 *In Fig. S2-1,* $\overline{FG} \parallel \overline{DE}$, *and* $\overline{HG} \parallel \overline{FE}$. *Find* DB *if* AH = 2 *and* HF = 4.

ANSWER: $DB = 36$

Challenge 3 *Find* DB *if* AH $= m_1$ *and* HF $= m_2$.

ANSWER: $DB = \frac{m_2}{m_1^2}(m_1 + m_2)^2$

2-2 *In isosceles* $\triangle$ABC (AB = AC), $\overline{CB}$ *is extended through* B *to* P (*Fig. S2-2*). *A line from* P, *parallel to altitude* $\overline{BF}$, *meets* $\overline{AC}$ *at* D (*where* D *is between* A *and* F). *From* P, *a perpendicular is drawn to meet the extension of* $\overline{AB}$ *at* E *so that* B *is between* E *and* A. *Express* BF *in terms of* PD *and* PE. *Try solving this problem in two different ways.*

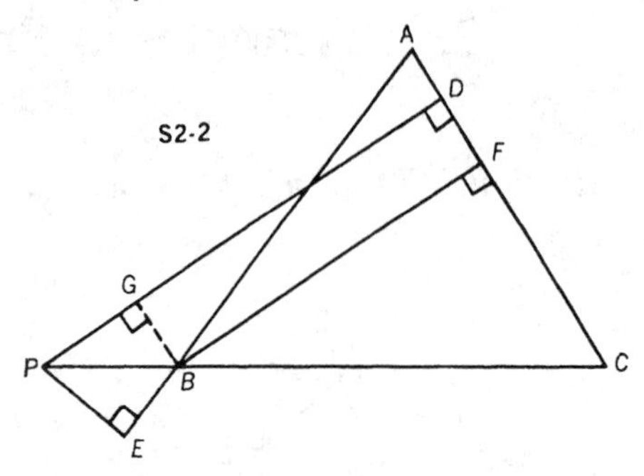

METHOD I: Since $\triangle ABC$ is isosceles, $\angle C \cong \angle ABC$.
However, $\angle PBE \cong \angle ABC$ (#1).
Therefore, $\angle C \cong \angle PBE$.

Thus, right $\triangle BFC \sim$ right $\triangle PEB$ (#48), and $\frac{PE}{BF} = \frac{PB}{BC}$.

In $\triangle PDC$, since $\overline{BF}$ is parallel to $\overline{PD}$, $\frac{PD}{BF} = \frac{PB + BC}{BC}$ (#49).

Using a theorem on proportions, we get

$$\frac{PD - BF}{BF} = \frac{PB + BC - BC}{BC} = \frac{PB}{BC}.$$

Therefore, $\frac{PD - BF}{BF} = \frac{PE}{BF}$.

Thus, $PD - BF = PE$, and $BF = PD - PE$.

METHOD II: Since $\overline{PD}$ is parallel to $\overline{BF}$, and $\overline{BF}$ is perpendicular to $\overline{AC}$, $\overline{PD}$ is perpendicular to $\overline{AC}$ (#10).

Draw a line from B perpendicular to $\overline{PD}$ at G.

$\angle ABC \cong \angle ACB$ (#5), and $\angle ABC \cong \angle PBE$ (#1);

therefore, $\angle ACB \cong \angle PBE$ (transitivity).

$\angle E$ and $\angle F$ are right angles; thus, $\triangle PBE$ and $\triangle BCF$ are mutually equiangular and, therefore, $\angle EPB \cong \angle FBC$.

Also, since $\overline{BF} \parallel \overline{PD}$, $\angle FBC \cong \angle DPC$ (#7).

By transitivity, $\angle GPB(\angle DPC) \cong \angle EPB$.

Thus, $\triangle GPB \cong \triangle EPB$ (A.A.S.), and $PG = PE$.

Since quadrilateral $GBFD$ is a rectangle (a quadrilateral with three right angles is a rectangle), $BF = GD$.

However, since $GD = PD - PG$, by substitution we get, $BF = PD - PE$.

2-3 *The measure of the longer base of a trapezoid is 97. The measure of the line segment joining the midpoints of the diagonals is 3 (Fig. S2-3). Find the measure of the shorter base. (Note that the figure is not drawn to scale.)*

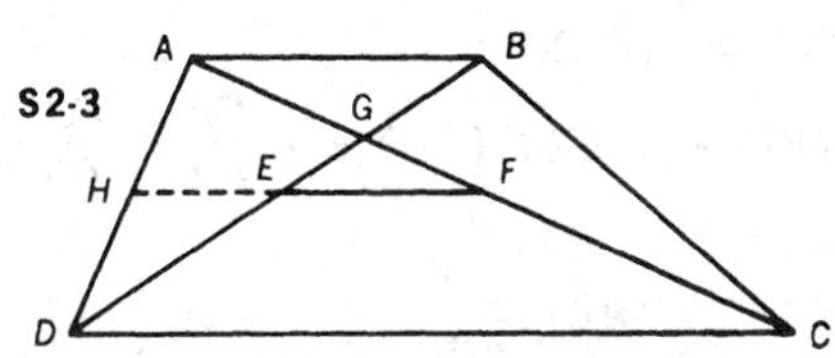

METHOD I: Since E and F are the midpoints of $\overline{DB}$ and $\overline{AC}$, respectively, $\overline{EF}$ must be parallel to $\overline{DC}$ and $\overline{AB}$ (#24).

Since $\overline{EF}$ is parallel to $\overline{DC}$, $\triangle EGF \sim \triangle DGC$ (#49), and $\frac{GC}{GF} = \frac{DC}{EF}$.

However, since $DC = 97$ and $EF = 3$, $\frac{GC}{GF} = \frac{97}{3}$.

Then, $\frac{GC - GF}{GF} = \frac{97 - 3}{3}$, or $\frac{FC}{GF} = \frac{94}{3}$.

Since $FC = FA$, $\frac{FA}{GF} = \frac{94}{3}$, or $\frac{GA}{GF} = \frac{91}{3}$.

Since $\triangle AGB \sim \triangle FGE$ (#48), $\frac{GA}{GF} = \frac{AB}{EF}$.

Thus, $\frac{91}{3} = \frac{AB}{3}$, and $AB = 91$.

METHOD II: Extend $\overline{FE}$ to meet $\overline{AD}$ at H. In $\triangle ADC$, $HF = \frac{1}{2}(DC)$ (#25, #26).

Since $DC = 97$, $HF = \frac{97}{2}$.

Since $EF = 3$, $HE = \frac{91}{2}$.

In $\triangle ADB$, $HE = \frac{1}{2}(AB)$ (#25, #26).

Hence, $AB = 91$.

Challenge *Find a general solution applicable to any trapezoid.*

ANSWER: $b - 2d$, where b is the length of the longer base and d is the length of the line joining the midpoints of the diagonals.

2-4 *In* $\triangle$ABC, D *is a point on side* $\overline{\text{BA}}$ *such that* BD:DA = 1:2 (*Fig.* S2-4). E *is a point on side* $\overline{\text{CB}}$ *so that* CE:EB = 1:4. *Segments* $\overline{\text{DC}}$ *and* $\overline{\text{AE}}$ *intersect at* F. *Express* CF:FD *in terms of two positive relatively prime integers.*

Draw $\overline{DG} \parallel \overline{BC}$.

$\triangle ADG \sim \triangle ABE$ (#49), and $\frac{AD}{AB} = \frac{DG}{BE} = \frac{2}{3}$.

Then $DG = \frac{2}{3}(BE)$.

But $\triangle DGF \sim \triangle CEF$ (#48), and $\frac{CF}{FD} = \frac{EC}{DG}$.

Since $EC = \frac{1}{4}(BE)$, $\frac{CF}{FD} = \frac{\frac{1}{4}(BE)}{\frac{2}{3}(BE)} = \frac{3}{8}$.

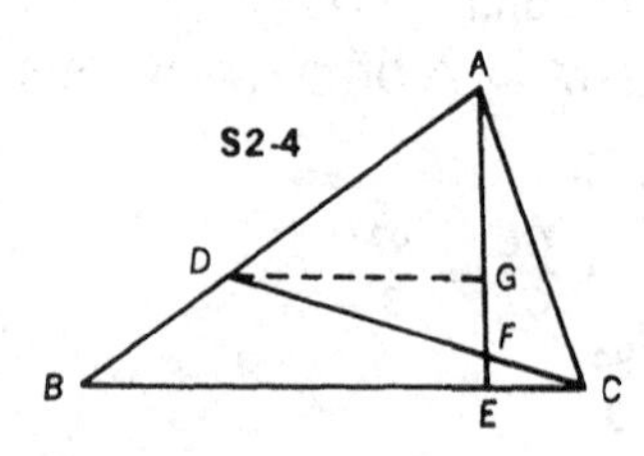

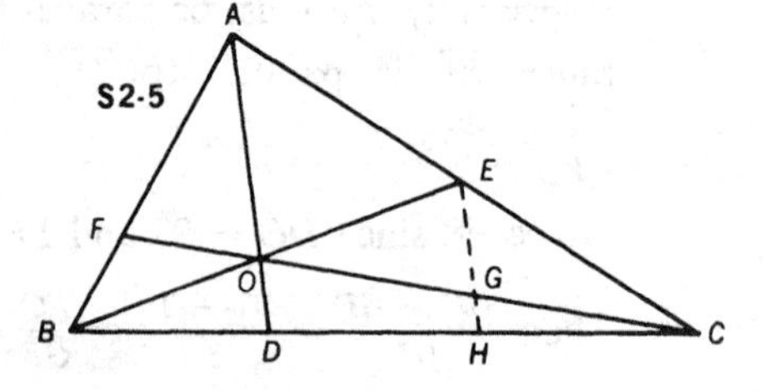

2-5 *In △ABC, $\overline{BE}$ is a median and O is the midpoint of $\overline{BE}$ (Fig. S2-5). Draw $\overline{AO}$ and extend it to meet $\overline{BC}$ at D. Draw $\overline{CO}$ and extend it to meet $\overline{BA}$ at F. If CO = 15, OF = 5, and AO = 12, find the measure of $\overline{OD}$.*

Draw $\overline{EH}$ parallel to $\overline{AD}$. Since E is the midpoint of $\overline{AC}$, $EG = \frac{1}{2}(AO) = 6$ (#25, #26). Since H is the midpoint of $\overline{CD}$, $GH = \frac{1}{2}(OD)$ (#25, #26). In $\triangle BEH$, $\overline{OD}$ is parallel to $\overline{EH}$ and O is the midpoint of $\overline{BE}$; therefore, $OD = \frac{1}{2}EH$ (#25, #26).

Then $OD = \frac{1}{2}[EG + GH]$, so $OD = \frac{1}{2}[6 + \frac{1}{2}OD] = 4$.

Note that the measures of $\overline{CO}$ and $\overline{OF}$ were not necessary for the solution of this problem.

Challenge *Can you establish a relationship between* OD *and* AO?

ANSWER: $OD = \frac{1}{3}AO$, regardless of the measures of $\overline{CO}$, $\overline{OF}$, and $\overline{AO}$.

2-6 *In parallelogram* ABCD, *points* E *and* F *are chosen on diagonal* $\overline{AC}$ *so that* AE = FC *(Fig. S2-6). If* $\overline{BE}$ *is extended to meet* $\overline{AD}$ *at* H, *and* $\overline{BF}$ *is extended to meet* $\overline{DC}$ *at* G, *prove that* $\overline{HG}$ *is parallel to* $\overline{AC}$.

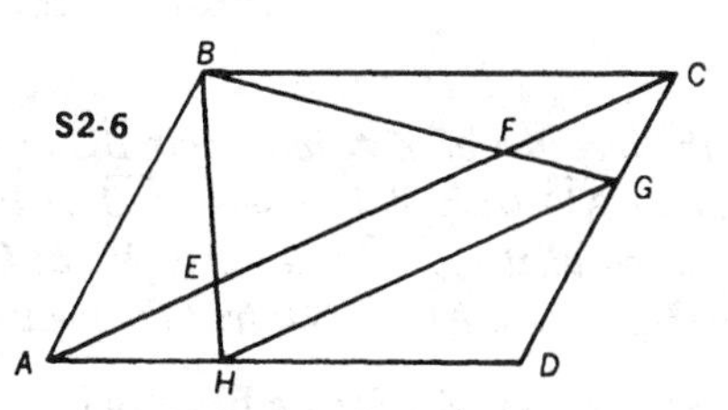

In $\square ABCD$, $AE = FC$.

Since $\angle BEC \cong \angle HEA$ (#1), and $\angle HAC \cong \angle ACB$ (#8), $\triangle HEA \sim \triangle BEC$ (#48), and $\frac{AE}{EF + FC} = \frac{HE}{BE}$.

Similarly, $\triangle BFA \sim \triangle GFC$ (#48), and $\frac{FC}{AE + EF} = \frac{FG}{BF}$.

However, since $FC = AE$, $\frac{HE}{BE} = \frac{FG}{BF}$ (transitivity).

Therefore, in $\triangle HBG$, $\overline{HG} \parallel \overline{EF}$ (#46), or $\overline{HG} \parallel \overline{AC}$.

2-7 $\overline{\text{AM}}$ *is the median to side* $\overline{\text{BC}}$ *of* $\triangle$ABC, *and* P *is any point on* $\overline{\text{AM}}$ *(Fig. S2-7).* $\overline{\text{BP}}$ *extended meets* $\overline{\text{AC}}$ *at* E, *and* $\overline{\text{CP}}$ *extended meets* $\overline{\text{AB}}$ *at* D. *Prove that* $\overline{\text{DE}}$ *is parallel to* $\overline{\text{BC}}$.

Extend $\overline{APM}$ to G so that $PM = MG$. Then draw $\overline{BG}$ and $\overline{CG}$. Since $BM = MC$, $\overline{PG}$ and $\overline{BC}$ bisect each other, making $BPCG$ a parallelogram (#21f). Thus, $\overline{PC} \parallel \overline{BG}$ and $\overline{BP} \parallel \overline{GC}$ (#21a), or $\overline{BE} \parallel \overline{GC}$ and $\overline{DC} \parallel \overline{BG}$. It follows that $\overline{DP} \parallel \overline{BG}$.

Therefore, in $\triangle ABG$, $\frac{AD}{DB} = \frac{AP}{PG}$ (#46). (I)

Similarly, in $\triangle AGC$ where $\overline{PE} \parallel \overline{GC}$, $\frac{AE}{EC} = \frac{AP}{PG}$ (#46). (II)

From (I) and (II), $\frac{AD}{DB} = \frac{AE}{EC}$.

Therefore, $\overline{DE}$ is parallel to $\overline{BC}$, since in $\triangle ABC$, $\overline{DE}$ cuts sides $\overline{AB}$ and $\overline{AC}$ proportionally (#46).

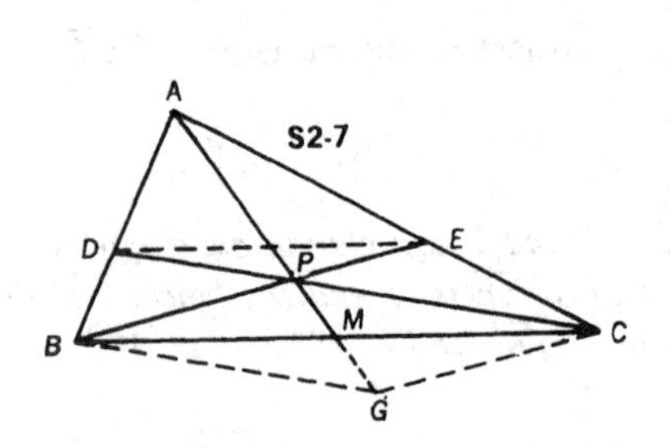

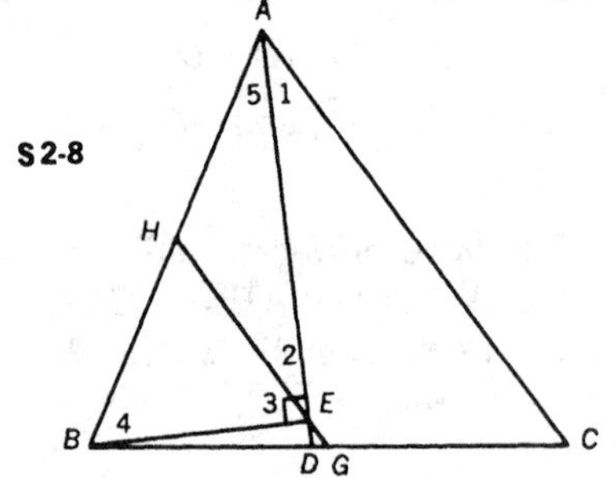

2-8 *In* $\triangle$ABC, *the bisector of* $\angle$A *intersects* $\overline{\text{BC}}$ *at* D (Fig. *S2-8*). *A perpendicular to* $\overline{\text{AD}}$ *from* B *intersects* $\overline{\text{AD}}$ *at* E. *A line segment through* E *and parallel to* $\overline{\text{AC}}$ *intersects* $\overline{\text{BC}}$ *at* G, *and* $\overline{\text{AB}}$ *at* H. *If* AB = 26, BC = 28, AC = 30, *find the measure of* $\overline{\text{DG}}$.

$\angle 1 \cong \angle 2$ (#8), $\angle 1 \cong \angle 5$ (angle bisector), therefore, $\angle 2 \cong \angle 5$.

In $\triangle AHE$, $AH = HE$ (#5).

In right $\triangle AEB$, $\angle 4$ is complementary to $\angle 5$ (#14), and $\angle 3$ is complementary to $\angle 2$.

Since $\angle 2 \cong \angle 5$, $\angle 3$ is complementary to $\angle 5$.

Therefore, since both $\angle 3$ and $\angle 4$ are complementary to $\angle 5$, they are congruent. Thus, in $\triangle BHE$, $BH = HE$ (#5) and, therefore, $BH = AH$. In $\triangle ABC$, since $\overline{HG} \parallel \overline{AC}$ and H is the midpoint of $\overline{AB}$, G is the midpoint of $\overline{BC}$ (#25), and $BG = 14$.

In $\triangle ABC$, $\overline{AD}$ is an angle bisector, therefore,

$$\frac{AB}{AC} = \frac{BD}{DC} \text{ (\#47).}$$

Let $BD = x$; then $DC = 28 - x$. By substituting,

$$\frac{26}{30} = \frac{x}{28 - x}, \text{ and } x = 13 = BD.$$

Since $BG = 14$, and $BD = 13$, then $DG = 1$.

2-9 *In* $\triangle$ABC, *altitude* $\overline{\text{BE}}$ *is extended to* G *so that* EG = *the measure of altitude* $\overline{\text{CF}}$. *A line through* G *and parallel to* $\overline{\text{AC}}$ *meets* $\overleftrightarrow{\text{BA}}$ *at* H (*Fig. S2-9*). *Prove that* AH = AC.

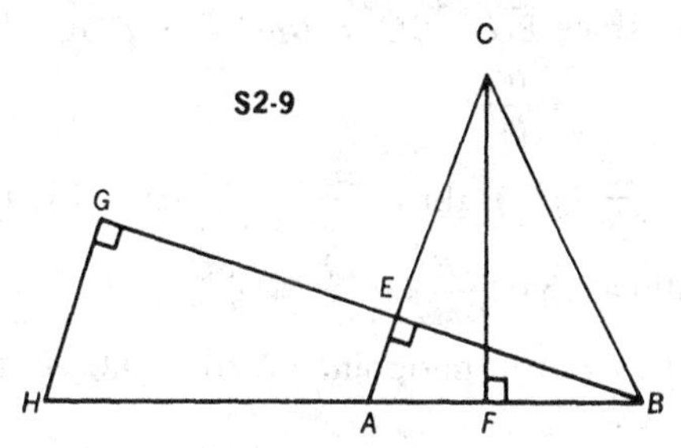

Since $\overline{BE} \perp \overline{AC}$ and $\overline{HG} \parallel \overline{AC}$, $\overline{HG} \perp \overline{BG}$.
$\angle H \cong \angle BAC$ (#7)
Since $\angle AFC$ is also a right angle, $\triangle AFC \sim \triangle HGB$ (#48),

$$\text{and } \frac{AC}{FC} = \frac{BH}{GB}. \qquad \text{(I)}$$

In $\triangle BHG$, $\overline{AE} \parallel \overline{HG}$;

$$\text{therefore, } \frac{AH}{GE} = \frac{BH}{GB} \text{ (\#46).} \qquad \text{(II)}$$

From (I) and (II), $\frac{AC}{FC} = \frac{AH}{GE}$.

Since the hypothesis stated that $GE = FC$, it follows that $AC = AH$.

2-10 *In trapezoid* ABCD ($\overline{\text{AB}} \parallel \overline{\text{DC}}$), *with diagonals* $\overline{\text{AC}}$ *and* $\overline{\text{DB}}$ *intersecting at* P, $\overline{\text{AM}}$, *a median of* $\triangle$ADC, *intersects* $\overline{\text{BD}}$ *at* E (*Fig. S2-10*). *Through* E, *a line is drawn parallel to* $\overline{\text{DC}}$ *cutting* $\overline{\text{AD}}$, $\overline{\text{AC}}$, *and* $\overline{\text{BC}}$ *at points* H, F, *and* G, *respectively. Prove that* HE = EF = FG.

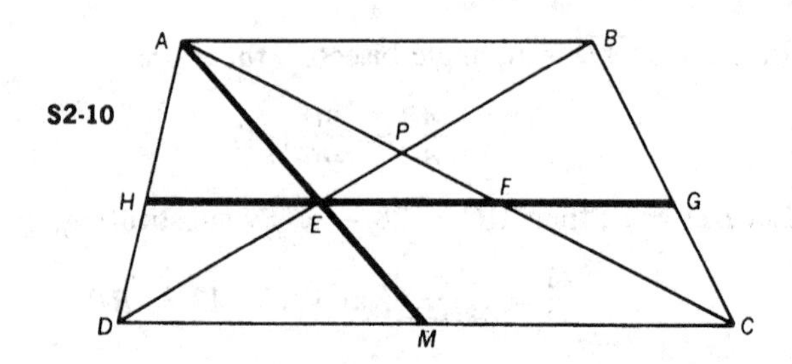

In $\triangle ADM$, since $\overline{HE} \parallel \overline{DM}$, $\triangle AHE \sim \triangle ADM$ (#49).
Therefore, $\frac{HE}{DM} = \frac{AE}{AM}$. In $\triangle AMC$, since $\overline{EF} \parallel \overline{MC}$,
$\triangle AEF \sim \triangle AMC$ (#49). Therefore, $\frac{EF}{MC} = \frac{AE}{AM}$.
In $\triangle DBC$, since $\overline{EG} \parallel \overline{DC}$, $\triangle BEG \sim \triangle BDC$ (#49).
Therefore, $\frac{EG}{DC} = \frac{BG}{BC}$.
But $\frac{BG}{BC} = \frac{AE}{AM}$ (#24); thus, $\frac{EG}{DC} = \frac{AE}{AM}$ (transitivity).
It then follows that $\frac{HE}{DM} = \frac{EF}{MC} = \frac{EG}{DC}$. (I)
But, since M is the midpoint of $\overline{DC}$, $DM = MC$, and
$DC = 2MC$. (II)
Substituting (II) in (I), we find that $HE = EF$, and $\frac{EF}{MC} = \frac{EG}{2(MC)}$.
Thus, $EF = \frac{1}{2}(EG)$ and $EF = FG$.
We therefore get $HE = EF = FG$ (transitivity).

2-11 *A line segment* $\overline{AB}$ *is divided by points* K *and* L *in such a way that* $(AL)^2 = (AK)(AB)$ *(Fig. S2-11). A line segment* $\overline{AP}$ *is drawn congruent to* $\overline{AL}$. *Prove that* $\overline{PL}$ *bisects* $\angle KPB$.

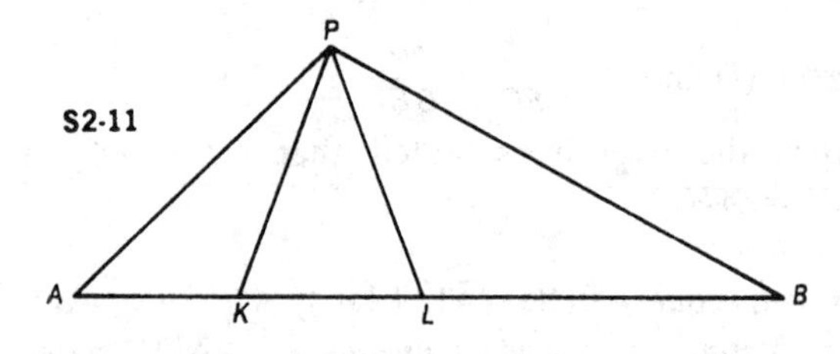

Since $AP = AL$, $(AL)^2 = (AK)(AB)$ may be written $(AP)^2 = (AK)(AB)$, or, as a proportion, $\frac{AK}{AP} = \frac{AP}{AB}$.

$\triangle KAP \sim \triangle PAB$ (#50).
It then follows that $\angle PKA \cong \angle BPA$, and $\angle KPA \cong \angle PBA$.
Since $\angle PKA$ is an exterior angle of $\triangle KPB$,
$m\angle PKA = m\angle KPB + m\angle PBK$ (#12).
$\angle PKA \cong \angle BPA$ may be written as

$$m\angle PKA = m\angle BPL + m\angle KPL + m\angle APK. \qquad \text{(I)}$$

Since $AP = AL$, in $\triangle APL$,

$$m\angle ALP = m\angle KPL + m\angle APK \text{ (\#5)}. \qquad \text{(II)}$$

Considering $\angle PKA$ as an exterior angle of $\triangle KPL$,

$$m\angle PKA = m\angle ALP + m\angle KPL \text{ (\#12)}. \qquad \text{(III)}$$

Combine lines (I) and (III) to get

$$m\angle BPL + m\angle KPL + m\angle APK = m\angle ALP + m\angle KPL.$$

Therefore, $m\angle BPL + m\angle APK = m\angle ALP$ (by subtraction). (IV)

Combine lines (II) and (IV) to get

$$m\angle KPL + m\angle APK = m\angle BPL + m\angle APK.$$

Therefore, $m\angle KPL = m\angle BPL$ (by subtraction), and $\overline{PL}$ bisects $\angle KPB$.

2-12 P *is any point on altitude* $\overline{CD}$ *of* $\triangle$ABC (*Fig. S2-12*). $\overline{AP}$ *and* $\overline{BP}$ *meet sides* $\overline{CB}$ *and* $\overline{CA}$ *at points* Q *and* R, *respectively. Prove that* $\angle$QDC $\cong$ $\angle$RDC.

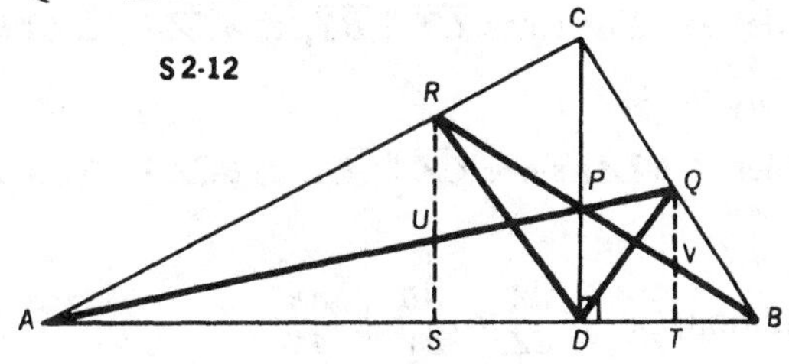

Draw $\overline{RUS} \parallel \overline{CD}$, and $\overline{QVT} \parallel \overline{CD}$.
$\triangle AUR \sim \triangle APC$ (#49), so $\dfrac{RU}{CP} = \dfrac{AU}{AP}$.

$\triangle ASU \sim \triangle ADP$ (#49), so $\frac{US}{PD} = \frac{AU}{AP}$.

Therefore, $\frac{RU}{CP} = \frac{US}{PD}$, and $\frac{RU}{US} = \frac{CP}{PD}$.

$\triangle BVQ \sim \triangle BPC$ (#49), so $\frac{QV}{CP} = \frac{BV}{BP}$.

$\triangle BTV \sim \triangle BDP$ (#49), so $\frac{VT}{PD} = \frac{BV}{BP}$.

Therefore, $\frac{QV}{CP} = \frac{VT}{PD}$ (transitivity), and $\frac{QV}{VT} = \frac{CP}{PD}$.

$\frac{RU}{US} = \frac{QV}{VT}$ (transitivity), and $1 + \frac{US}{RU} = \frac{VT}{QV} + 1$;

therefore, $\frac{RU + US}{RU} = \frac{QV + VT}{QV}$.

$\frac{RS}{RU} = \frac{QT}{QV}$, $\frac{RS}{QT} = \frac{RU}{QV}$.

Since $\overline{RS} \parallel \overline{CD} \parallel \overline{QT}$, $\frac{UP}{PQ} = \frac{DS}{DT}$ (#24).

$\triangle RPU \sim \triangle VPQ$ (#48), and $\frac{RU}{QV} = \frac{UP}{PQ}$.

Therefore, $\frac{RS}{QT} = \frac{DS}{DT}$ (transitivity).

$\angle RSD \cong \angle QTD \cong$ right angle, $\triangle RSD \sim \triangle QTD$ (#50);
$\angle SDR \cong \angle TDQ$, $\angle RDC \cong \angle QDC$ (subtraction).

2-13 *In* $\triangle$ABC, Z *is any point on base* $\overline{\text{AB}}$ *as shown in Fig. S2-13a.* $\overline{\text{CZ}}$ *is drawn. A line is drawn through* A *parallel to* $\overline{\text{CZ}}$ *meeting* $\overleftrightarrow{\text{BC}}$ *at* X. *A line is drawn through* B *parallel to* $\overline{\text{CZ}}$ *meeting* $\overleftrightarrow{\text{AC}}$ *at* Y. *Prove that* $\frac{1}{\text{AX}} + \frac{1}{\text{BY}} = \frac{1}{\text{CZ}}$.

Consider $\triangle AYB$; since $\overline{CZ} \parallel \overline{BY}$, $\triangle ACZ \sim \triangle AYB$ (#49), and $\frac{AZ}{CZ} = \frac{AB}{BY}$.

Consider $\triangle BXA$; since $\overline{CZ} \parallel \overline{AX}$, $\triangle BCZ \sim \triangle BXA$ (#49), and $\frac{BZ}{CZ} = \frac{AB}{AX}$.

By addition, $\frac{AZ}{CZ} + \frac{BZ}{CZ} = \frac{AB}{BY} + \frac{AB}{AX}$.

But $AZ + BZ = AB$, therefore, $\frac{AB}{CZ} = \frac{AB}{BY} + \frac{AB}{AX}$.

Dividing by (AB) we obtain $\frac{1}{CZ} = \frac{1}{BY} + \frac{1}{AX}$.

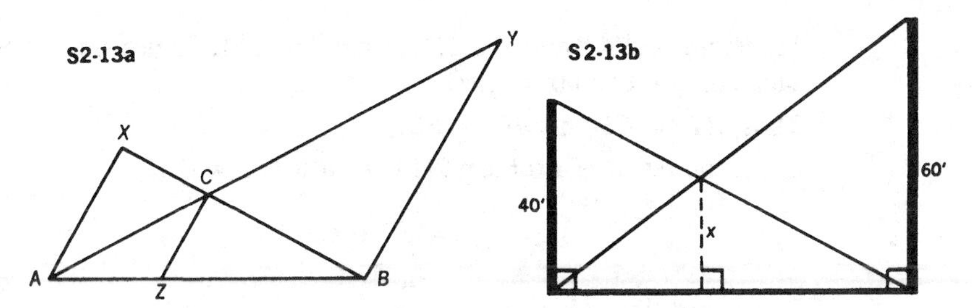

Challenge *Two telephone cable poles,* 40 *feet and* 60 *feet high, respectively, are placed near each other. As partial support, a line runs from the top of each pole to the bottom of the other, as shown in Fig. S2-13b. How high above the ground is the point of intersection of the two support lines?*

Using the result of Problem 2-13, we immediately obtain the following relationship:

$$\frac{1}{X} = \frac{1}{40} + \frac{1}{60};\ \frac{1}{X} = \frac{100}{2400} = \frac{1}{24}.$$

Therefore, $X = 24$. Thus, the point of intersection of the two support lines is 24 feet above the ground.

2-14 *In* $\triangle ABC$, $m\angle A = 120$ *(Fig. S2-14). Express the measure of the internal bisector of* $\angle A$ *in terms of the two adjacent sides.*

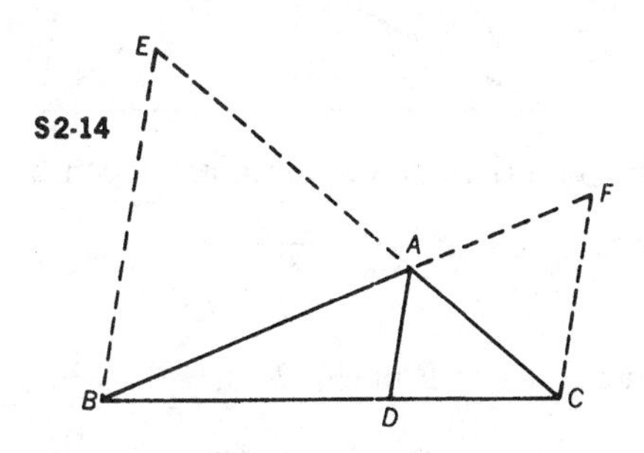

Draw a line through B parallel to $\overline{AD}$ meeting $\overleftrightarrow{CA}$ at E, and a line through C parallel to $\overline{AD}$ meeting $\overleftrightarrow{BA}$ at F.

Since $\angle EAB$ is supplementary to $\angle BAC$, $m\angle EAB = 60$, as does the measure of its vertical angle, $\angle FAC$.

Now, $m\angle BAD = m\angle EBA = 60$ (#8), and $m\angle DAC = m\angle ACF = 60$ (#8).

Therefore, $\triangle EAB$ and $\triangle FAC$ are equilateral triangles since they each contain two 60° angles.

Thus, $AB = EB$ and $AC = FC$.

From the result of Problem 2-13, we also know that

$$\frac{1}{AD} = \frac{1}{EB} + \frac{1}{FC}.$$

By substitution, $\frac{1}{AD} = \frac{1}{AB} + \frac{1}{AC}$.

Combining fractions, $\frac{1}{AD} = \frac{AC + AB}{(AB)(AC)}$.

Therefore, $AD = \frac{(AB)(AC)}{AC + AB}$.

2-15 *Prove that the measure of the segment passing through the point of intersection of the diagonals of a trapezoid and parallel to the bases, with its endpoints on the legs, is the harmonic mean between the measures of the parallel sides. (See Fig. S2-15.) The harmonic mean of two numbers is defined as the reciprocal of the average of the reciprocals of two numbers. The harmonic mean between* a *and* b *is equal to* $\left(\frac{a^{-1} + b^{-1}}{2}\right)^{-1} = \frac{2ab}{a + b}$.

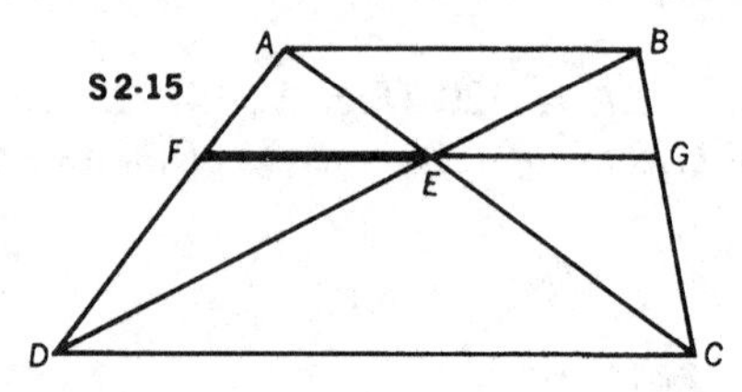

In order for FG to be the harmonic mean between AB and DC it must be true that $FG = \frac{2}{\frac{1}{AB} + \frac{1}{DC}}$.

From the result of Problem 2-13, $\frac{1}{FE} = \frac{1}{AB} + \frac{1}{DC}$, and

$FE = \frac{1}{\frac{1}{AB} + \frac{1}{DC}}$. Similarly, $EG = \frac{1}{\frac{1}{AB} + \frac{1}{DC}}$.

Therefore, $FE = EG$. Thus, since $FG = 2FE$,

$FG = \frac{2}{\frac{1}{AB} + \frac{1}{CD}}$, and FG is the harmonic mean between AB and CD.

2-16 *In parallelogram* ABCD, E *is on* $\overline{BC}$. $\overline{AE}$ *cuts diagonal* $\overline{BD}$ *at* G *and* $\overleftrightarrow{DC}$ *at* F, *as shown in Fig. S2-16. If* AG = 6 *and* GE = 4, *find* EF.

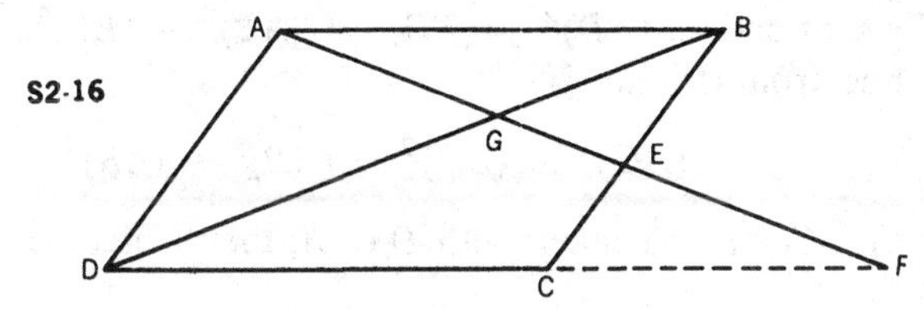

$\triangle FDG \sim \triangle ABG$ (#48), and $\dfrac{GF}{AG} = \dfrac{GD}{GB}$.

$\triangle BGE \sim \triangle DGA$ (#48), and $\dfrac{GD}{GB} = \dfrac{AG}{GE}$.

Therefore, by transitivity, $\dfrac{GF}{AG} = \dfrac{AG}{GE}$.

By substitution, $\dfrac{4 + EF}{6} = \dfrac{6}{4}$ and $EF = 5$.

NOTE: AG is the mean proportional between GF and GE.

3. The Pythagorean Theorem

3-1 *In any* $\triangle$ABC, E *is any point on altitude* $\overline{AD}$ *(Fig. S3-1). Prove that* $(AC)^2 - (CE)^2 = (AB)^2 - (EB)^2$.

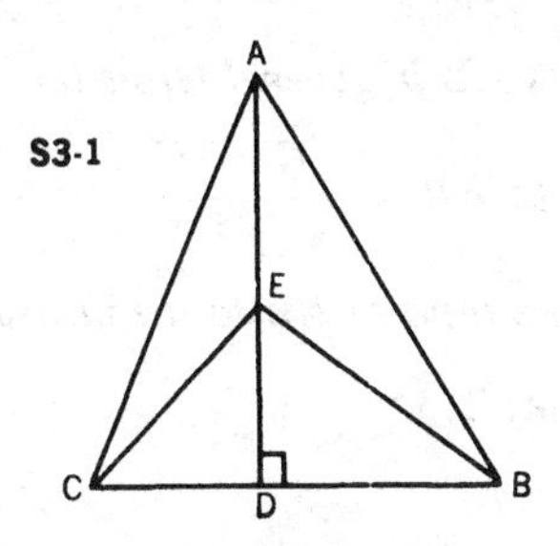

By the Pythagorean Theorem (#55),

$$\text{for } \triangle ADC,\ (CD)^2 + (AD)^2 = (AC)^2;$$
$$\text{for } \triangle EDC,\ (CD)^2 + (ED)^2 = (EC)^2.$$

By subtraction, $(AD)^2 - (ED)^2 = (AC)^2 - (EC)^2$. (I)

By the Pythagorean Theorem (#55),

$$\text{for } \triangle ADB,\ (DB)^2 + (AD)^2 = (AB)^2;$$
$$\text{for } \triangle EDB,\ (DB)^2 + (ED)^2 = (EB)^2.$$

By subtraction, $(AD)^2 - (ED)^2 = (AB)^2 - (EB)^2$. (II)
Thus, from (I) and (II),

$$(AC)^2 - (EC)^2 = (AB)^2 - (EB)^2.$$

NOTE: For E coincident with D or A, the theorem is trivial.

3-2 *In* △ABC, *median* $\overline{AD}$ *is perpendicular to median* $\overline{BE}$ (*Fig.* S3-2). *Find* AB *if* BC = 6 *and* AC = 8.

Let $AD = 3x$; then $AG = 2x$ and $DG = x$ (#29).
Let $BE = 3y$; then $BG = 2y$ and $GE = y$ (#29).
By the Pythagorean Theorem,
for $\triangle DGB$, $x^2 + (2y)^2 = 9$ (#55);
for $\triangle EGA$, $y^2 + (2x)^2 = 16$ (#55).
By addition, $5x^2 + 5y^2 = 25$;
therefore, $x^2 + y^2 = 5$.
However, in $\triangle BGA$, $(2y)^2 + (2x)^2 = (AB)^2$ (#55),

$$\text{or } 4y^2 + 4x^2 = (AB)^2.$$

Since $x^2 + y^2 = 5$, $4x^2 + 4y^2 = 20$.
By transitivity, $(AB)^2 = 20$, and $AB = 2\sqrt{5}$.

Challenge 1 *Express* AB *in general terms for* BC = a, *and* AC = b.

ANSWER: $AB = \sqrt{\dfrac{a^2 + b^2}{5}}$

Challenge 2 *Find the ratio of* AB *to the measure of its median.*

ANSWER: 2:3

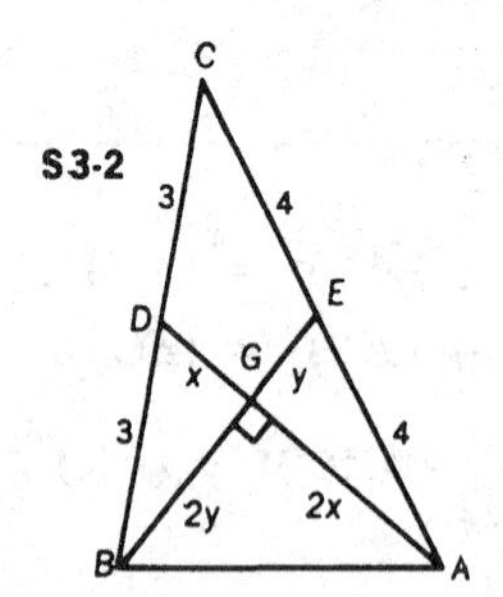

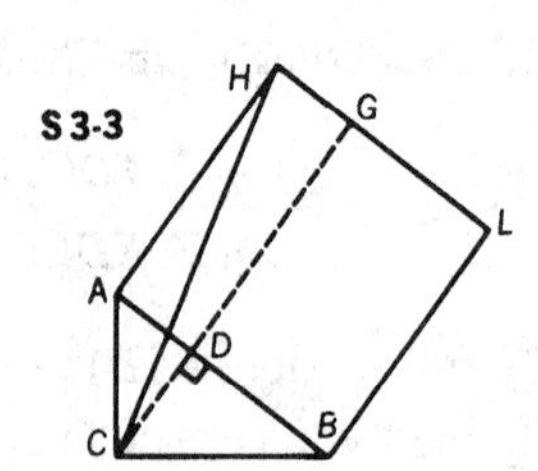

3-3 *On hypotenuse* $\overline{AB}$ *of right* $\triangle ABC$, *draw square* ABLH *externally (Fig. S3-3). If* AC = 6 *and* BC = 8, *find* CH.

Draw $\overline{CDG}$ perpendicular to $\overline{AB}$.

In right $\triangle ABC$, $AB = 10$ (#55), and $\frac{AD}{AC} = \frac{AC}{AB}$ (#51b).

Substituting in this ratio, we find $AD = 3.6$; therefore, $DB = 6.4$.

In right $\triangle ABC$, $\frac{AD}{CD} = \frac{CD}{DB}$ (#51a);

therefore, $CD = 4.8$.

Since $DG = 10$, $CG = 14.8$. $HG = AD = 3.6$.

In right $\triangle HGC$, $(HG)^2 + (CG)^2 = (HC)^2$ (#55), and $HC = 2\sqrt{58}$.

Challenge 1 *Find the area of quadrilateral* HLBC.

ANSWER: 106

Challenge 2 *Solve the problem if square* ABLH *overlaps* $\triangle ABC$.

ANSWER: $2\sqrt{10}$

3-4 *The measures of the sides of a right triangle are* 60, 80, *and* 100 *(Fig. S3-4). Find the measure of a line segment, drawn from the vertex of the right angle to the hypotenuse, that divides the triangle into two triangles of equal perimeters.*

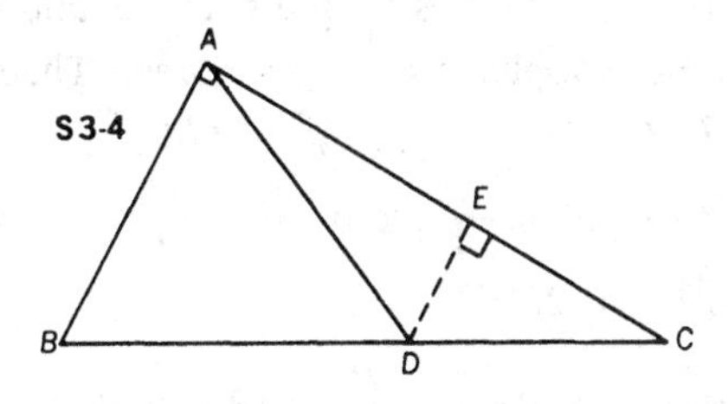

Let $AB = 60$, $AC = 80$, and $BC = 100$. If $\triangle ABD$ is to have the same perimeter as $\triangle ACD$, then $AB + BD$ must equal $AC + DC$, since both triangles share AD; that is, $60 + BD = 80 + 100 - BD$. Therefore, $BD = 60$ and $DC = 40$.

Draw $\overline{DE}$ perpendicular to $\overline{AC}$.

Right $\triangle EDC \sim$ right $\triangle ABC$ (#49); therefore, $\frac{ED}{AB} = \frac{DC}{BC}$.

By substituting the appropriate values, we have $\frac{ED}{60} = \frac{40}{100}$, and $ED = 24$.

By the Pythagorean Theorem (#55), for $\triangle EDC$, we find $EC = 32$; then, by subtraction, $AE = 48$. Again using the Pythagorean Theorem (#55), in $\triangle AED$, $AD = 24\sqrt{5}$.

3-5 *On sides* $\overline{\text{AB}}$ *and* $\overline{\text{DC}}$ *of rectangle* ABCD, *points* F *and* E *are chosen so that* AFCE *is a rhombus, as in Fig. S3-5a. If* AB = 16 *and* BC = 12, *find* EF.

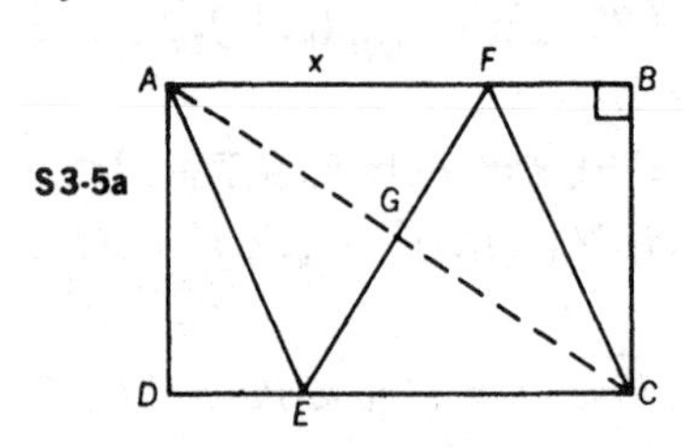

S3-5a

METHOD I: Let $AF = FC = EC = AE = x$ (#21-1).
Since $AF = x$ and $AB = 16$, $BF = 16 - x$.
Since $BC = 12$, in right $\triangle FBC$, $(FB)^2 + (BC)^2 = (FC)^2$ (#55), or $(16 - x)^2 + (12)^2 = x^2$, and $x = \frac{25}{2}$.
Again by applying the Pythagorean Theorem (#55) to $\triangle ABC$, we get $AC = 20$.
Since the diagonals of a rhombus are perpendicular and bisect each other, $\triangle EGC$ is a right triangle, and $GC = 10$.
Once more applying the Pythagorean Theorem (#55),
in $\triangle EGC$, $(EG)^2 + (GC)^2 = (EC)^2$.
$(EG)^2 + 100 = \frac{625}{4}$, and $EG = \frac{15}{2}$.
Thus, $FE = 2(EG) = 15$.

METHOD II: Since $x = \frac{25}{2}$ (see Method I), $EC = \frac{25}{2}$.
Draw a line through B parallel to $\overline{EF}$ meeting $\overline{DC}$ at H (Fig. S3-5b).

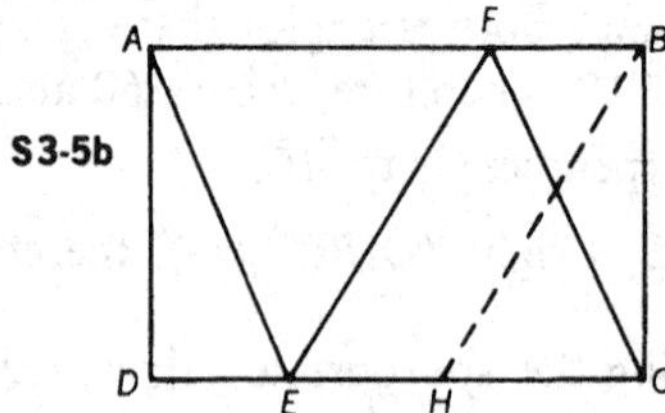

S3-5b

Since quadrilateral $BFEH$ is a parallelogram (#21a), and $FB = AB - AF = \frac{7}{2}$, $EH = \frac{7}{2}$. Therefore, $HC = 9$.

In right $\triangle BCH$, $(BH)^2 = (BC)^2 + (HC)^2$ (#55), so that $BH = 15$.

Therefore, $EF = BH = 15$ (#21b).

Challenge *If* AB = a *and* BC = b, *what general expression will give the measure of* $\overline{EF}$?

ANSWER: $\frac{b}{a}\sqrt{a^2 + b^2}$

3-6 *A man walks one mile east, then one mile northeast, then another mile east (Fig. S3-6). Find the distance, in miles, between the man's initial and final positions.*

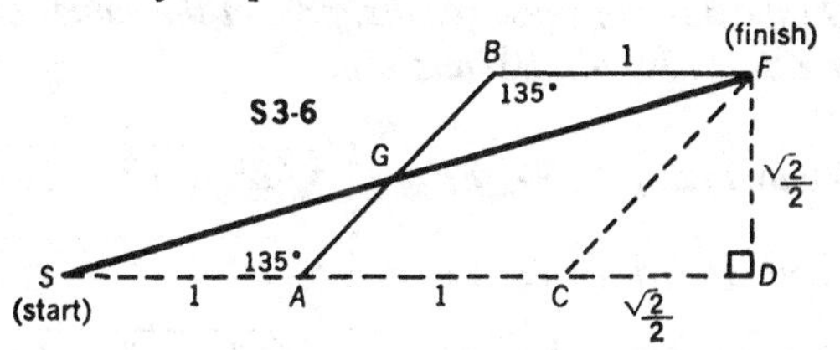

Let S and F be the starting and finishing positions, respectively. Draw $\overline{FD} \perp \overleftrightarrow{SA}$, then draw $\overline{FC} \parallel \overline{AB}$.

In rhombus $ABFC$, $CF = BF = AC = 1$ (#21-l); also $SA = 1$.

In isosceles right $\triangle FDC$, $FD = CD = \frac{\sqrt{2}}{2}$ (#55b).

Applying the Pythagorean Theorem (#55) to right $\triangle DSF$,

$$(FD)^2 + (SD)^2 = (SF)^2$$

$$\left(\frac{\sqrt{2}}{2}\right)^2 + \left(2 + \frac{\sqrt{2}}{2}\right)^2 = (SF)^2$$

$$\sqrt{5 + 2\sqrt{2}} = SF.$$

Challenge *How much shorter (or longer) is the distance if the course is one mile east, one mile north, then one mile east?*

ANSWER: The new course is shorter by $\sqrt{5 + 2\sqrt{2}} - \sqrt{5}$.

3-7 *If the measures of two sides and the included angle of a triangle are* 7, $\sqrt{50}$, *and* 135, *respectively, find the measure of the segment joining the midpoints of the two given sides (Fig. S3-7).*

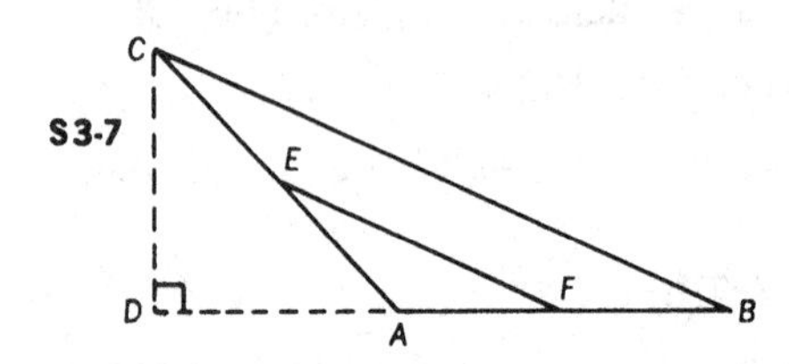

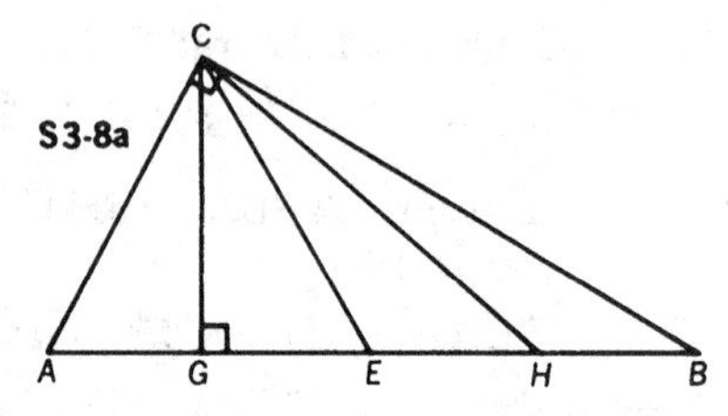

Draw altitude $\overline{CD}$. Since $m\angle CAB = 135$, $m\angle DAC = 45$, therefore, $\triangle ADC$ is an isosceles right triangle. If $AC = \sqrt{50} = 5\sqrt{2}$, then $DA = DC = 5$ (#55b).

In $\triangle DBC$, since $DB = 12$ and $DC = 5$, $BC = 13$ (#55).

Therefore, $EF = \frac{1}{2}(BC) = \frac{13}{2}$ (#26).

Challenge 4 *On the basis of these results, predict the values of* EF *when* m∠A = 30, 45, 60, *and* 90.

$$\text{When } m\angle A = 30,\ EF = \frac{1}{2}\sqrt{b^2 + c^2 - bc\sqrt{3}};$$

$$\text{when } m\angle A = 45,\ EF = \frac{1}{2}\sqrt{b^2 + c^2 - bc\sqrt{2}};$$

$$\text{when } m\angle A = 60,\ EF = \frac{1}{2}\sqrt{b^2 + c^2 - bc\sqrt{1}};$$

$$\text{when } m\angle A = 90,\ EF = \frac{1}{2}\sqrt{b^2 + c^2 - bc\sqrt{0}}.$$

3-8 *Hypotenuse* $\overline{AB}$ *of right* △ABC *is divided into four congruent segments by points* G, E, *and* H, *in the order* A, G, E, H, B (*Fig.* S3-8*a*). *If* AB = 20, *find the sum of the squares of the measures of the line segments from* C *to* G, E, *and* H.

METHOD I: Since $AB = 20$, $AG = GE = EH = HB = 5$. Since the measures of $\overline{AC}$, $\overline{CB}$, and $\overline{CG}$ are not given, $\triangle ABC$ may be constructed so that $\overline{CG}$ is perpendicular to $\overline{AB}$ without affecting the sum required.

Since $\overline{CG}$ is the altitude upon the hypotenuse of right $\triangle ABC$,

$$\frac{5}{CG} = \frac{CG}{15} \text{ (\#51a), and } (CG)^2 = 75.$$

By applying the Pythagorean Theorem to right $\triangle HGC$, we find

$$(CG)^2 + (GH)^2 = (HC)^2 \text{ (\#55),}$$
$$\text{or } 75 + 100 = 175 = (HC)^2.$$

Since $CE = GH$ (#27), $(CE)^2 = (GH)^2 = 100$. Therefore,

$$(CG)^2 + (CE)^2 + (HC)^2 = 75 + 100 + 175 = 350.$$

METHOD II: In Fig. S3-8b, $\overline{CJ}$ is drawn perpendicular to $\overline{AB}$. Since $AB = 20$, $CE = 10$ (#27). Let $GJ = x$, and $JE = 5 - x$. In $\triangle CJG$ and $\triangle CJE$, $(CG)^2 - x^2 = 10^2 - (5 - x)^2$ (#55), or $(CG)^2 = 75 + 10x$. (I)
Similarly, in $\triangle CJH$ and $\triangle CJE$,
$(CH)^2 - (10 - x)^2 = 10^2 - (5 - x)^2$,
or $(CH)^2 = 175 - 10x$. (II)
By addition of (I) and (II), $(CG)^2 + (CH)^2 + (CE)^2 = 75 + 10x + 175 - 10x + 100 = 350$.

Notice that Method II gives a more general proof than Method I.

Challenge *Express the result in general terms when* AB = c.

ANSWER: $\frac{7c^2}{8}$

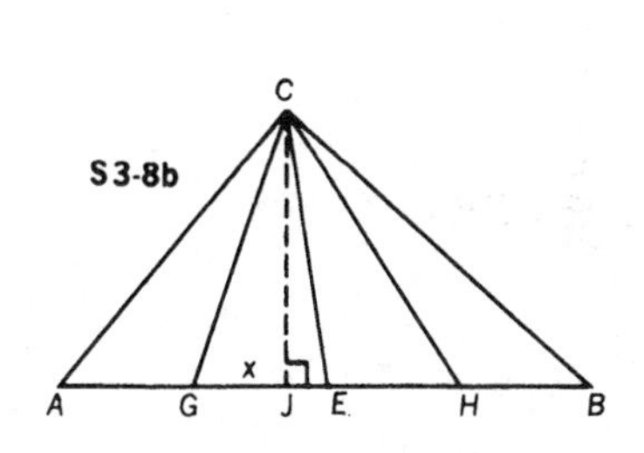

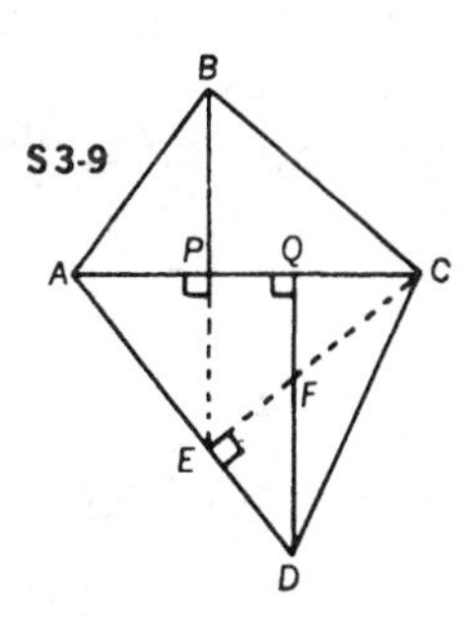

3-9 *In quadrilateral* ABCD, AB = 9, BC = 12, CD = 13, DA = 14, *and diagonal* AC = 15. *Perpendiculars are drawn from* B *and* D *to* $\overline{AC}$, *meeting* $\overline{AC}$ *at points* P *and* Q, *respectively* (*Fig. S*3-9). *Find* PQ.

Consider $\triangle ACD$. If we draw the altitude from C to $\overline{AD}$ we find that $CE = 12$, $AE = 9$, and $ED = 5$ (#55e).

$$\text{Therefore, } \triangle ABC \cong \triangle AEC \text{ (S.S.S.).}$$

Thus, altitude $\overline{BP}$, when extended, passes through E. In $\triangle ABC$, $\frac{AC}{AB} = \frac{AB}{AP}$ (#51b), and $\frac{15}{9} = \frac{9}{AP}$; therefore, $AP = \frac{27}{5}$.

Now consider $\triangle AQD$, where $\overline{PE} \parallel \overline{QD}$ (#9).

$$\frac{AE}{ED} = \frac{AP}{PQ} \text{ (\#46), and } \frac{9}{5} = \frac{\frac{27}{5}}{PQ}\text{; thus, } PQ = 3.$$

3-10 *In $\triangle$ABC, angle C is a right angle (Fig. S3-10). AC and BC are each equal to 1. D is the midpoint of $\overline{\text{AC}}$. $\overline{\text{BD}}$ is drawn, and a line perpendicular to $\overline{\text{BD}}$ at P is drawn from C. Find the distance from P to the intersection of the medians of $\triangle$ABC.*

Applying the Pythagorean Theorem to $\triangle DCB$,

$$(DC)^2 + (CB)^2 = (DB)^2 \text{ (\#55).}$$

$$\frac{1}{4} + 1 = (DB)^2, \; DB = \frac{1}{2}\sqrt{5}$$

Since the centroid of a triangle trisects each of the medians (#29),

$$DG = \frac{1}{3}(DB) = \frac{1}{3}\left(\frac{1}{2}\sqrt{5}\right) = \frac{1}{6}\sqrt{5}$$

Consider right $\triangle DCB$ where $\overline{CP}$ is the altitude drawn upon the hypotenuse.

Therefore, $\frac{DB}{DC} = \frac{DC}{DP}$ (#51b).

$$\frac{\frac{1}{2}\sqrt{5}}{\frac{1}{2}} = \frac{\frac{1}{2}}{DP}, \; DP = \frac{\sqrt{5}}{10}$$

Thus, $PG = DG - DP$, and

$$PG = \frac{1}{6}\sqrt{5} - \frac{1}{10}\sqrt{5} = \frac{1}{15}\sqrt{5}.$$

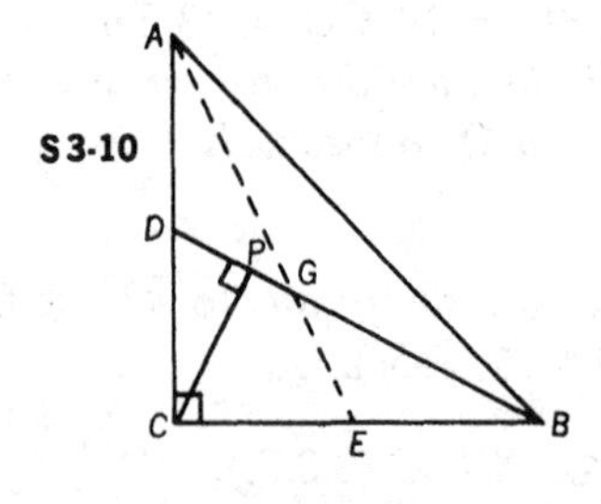

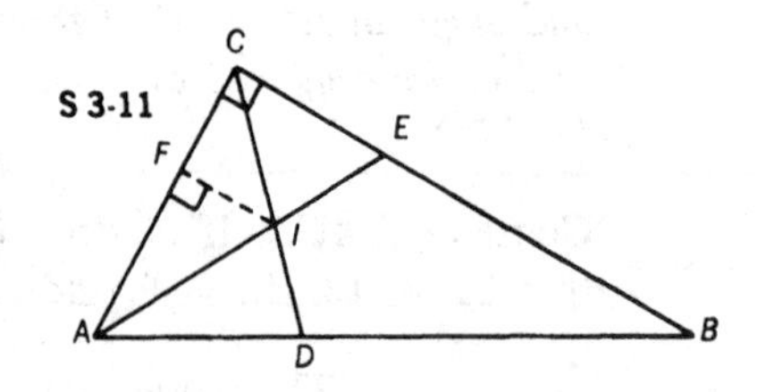

3-11 *A right triangle contains a 60° angle. If the measure of the hypotenuse is 4, find the distance from the point of intersection of the 2 legs of the triangle to the point of intersection of the angle bisectors.*

In Fig. S3-11, $AB = 4$, $m\angle CAB = 60$; therefore $m\angle B = 30$ and $AC = 2$ (#55c). Since $\overline{AE}$ and $\overline{CD}$ are angle bisectors, $m\angle CAE = 30$, and $m\angle ACD = 45$. From the point of intersection, I, of the angle bisectors draw $\overline{FI} \perp \overline{AC}$. Thus, the angles of $\triangle AIF$ measure 30°, 60°, and 90°.

Let $AF = y$. Since $y = \frac{1}{2}(AI)\sqrt{3}$ (#55d), then $AI = \frac{2y}{\sqrt{3}}$, and $FI = \frac{y}{\sqrt{3}} = \frac{y\sqrt{3}}{3}$ (#55c).

Since $m\angle FCI = 45$, $FC = FI = (2 - y)$ (#5).

Therefore, $(2 - y) = \frac{y\sqrt{3}}{3}$, and $y = (3 - \sqrt{3})$.

Hence, $FC = (2 - y) = \sqrt{3} - 1$.

Then $CI = (FC)\sqrt{2} = \sqrt{2}(\sqrt{3} - 1)$, and $CI = \sqrt{6} - \sqrt{2}$ (#55a).

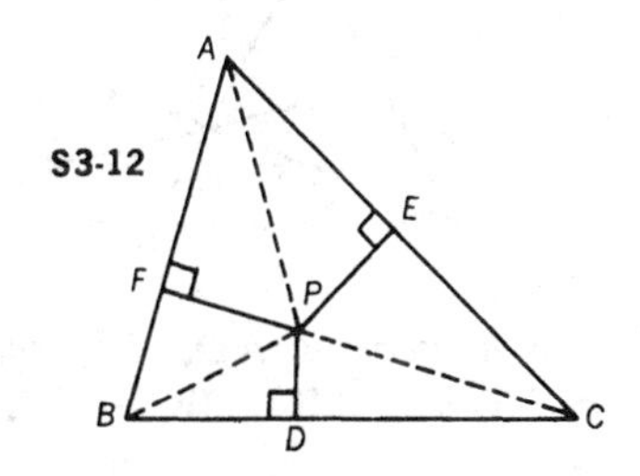

3-12 *From point* P *inside* $\triangle$ABC, *perpendiculars are drawn to the sides meeting* $\overline{\text{BC}}$, $\overline{\text{CA}}$, *and* $\overline{\text{AB}}$, *at points* D, E, *and* F, *respectively* (*Fig.* S3-12). *If* BD = 8, DC = 14, CE = 13, AF = 12, *and* FB = 6, *find* AE. *Derive a general theorem, and then make use of it to solve this problem.*

The Pythagorean Theorem is applied to each of the six right triangles shown in Fig. S3-12.

$(BD)^2 + (PD)^2 = (PB)^2$, $(FB)^2 + (PF)^2 = (PB)^2$;

therefore, $(BD)^2 + (PD)^2 = (FB)^2 + (PF)^2$. (I)

$(DC)^2 + (PD)^2 = (PC)^2$, $(CE)^2 + (PE)^2 = (PC)^2$;

therefore, $(DC)^2 + (PD)^2 = (CE)^2 + (PE)^2$. (II)

$(EA)^2 + (PE)^2 = (PA)^2$, $(AF)^2 + (PF)^2 = (PA)^2$;

therefore, $(EA)^2 + (PE)^2 = (AF)^2 + (PF)^2$. (III)

Subtracting (II) from (I), we have

$(BD)^2 - (DC)^2 = (FB)^2 + (PF)^2 - (CE)^2 - (PE)^2$. (IV)

Rewriting (III) in the form $(EA)^2 = (AF)^2 + (PF)^2 - (PE)^2$, and subtracting it from (IV) we obtain

$$(BD)^2 - (DC)^2 - (EA)^2 = (FB)^2 - (CE)^2 - (AF)^2, \text{ or}$$
$$(BD)^2 + (CE)^2 + (AF)^2 = (DC)^2 + (EA)^2 + (FB)^2.$$

Thus, if, from any point P inside a triangle, perpendiculars are drawn to the sides, the sum of the squares of the measures of every other segment of the sides so formed equals the sum of the squares of the measures of the other three segments.

Applying the theorem to the given problem, we obtain
$8^2 + 13^2 + 12^2 = 6^2 + 14^2 + X^2$, $145 = X^2$, $\sqrt{145} = X$.

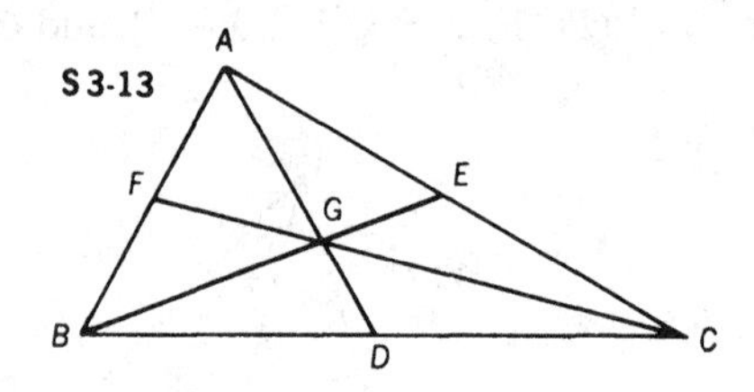

3-13 *For* $\triangle ABC$ *with medians* $\overline{AD}$, $\overline{BE}$, *and* $\overline{CF}$, *let* $m = AD + BE + CF$, *and let* $s = AB + BC + CA$ (*Fig. S3-13*). *Prove that* $\frac{3}{2}s > m > \frac{3}{4}s$.

$BG + GA > AB$, $CG + GA > AC$, and $BG + CG > BC$ (#41).
By addition, $2(BG + GC + AG) > AB + AC + BC$.

Since $BG + GC + AG = \frac{2}{3}(BE + CF + AD)$ (#29),

by substitution, $2\left(\frac{2}{3}m\right) > s$, or $\frac{4}{3}m > s$; therefore, $m > \frac{3}{4}s$.

$\frac{1}{2}AB + FG > BG, \frac{1}{2}BC + GD > CG, \frac{1}{2}AC + GE > AG$ (#41)
By addition, $\frac{1}{2}(AB + BC + AC) + FG + GD + GE > BG + CG + AG$.
Substituting, $\frac{1}{2}s + \frac{1}{3}m > \frac{2}{3}m, \frac{1}{2}s > \frac{1}{3}m, m < \frac{3}{2}s$.

$$\text{Thus, } \frac{3}{2}s > m > \frac{3}{4}s.$$

3-14 *Prove that* $\frac{3}{4}(a^2 + b^2 + c^2) = m_a{}^2 + m_b{}^2 + m_c{}^2$. ($m_c$ *means the measure of the median drawn to side* c.)

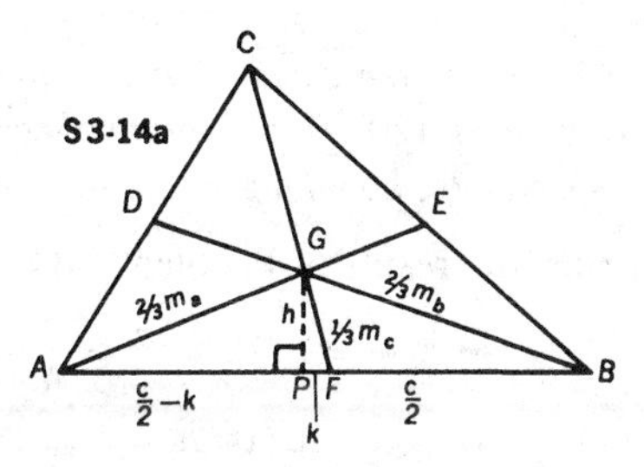

In $\triangle ABC$, medians $\overline{AE}$, $\overline{BD}$, $\overline{CF}$, and $\overline{GP} \perp \overline{AB}$ are drawn, as in Fig. S3-14a. Let $GP = h$, and $PF = k$.

Since $AF = \frac{c}{2}$, then $AP = \frac{c}{2} - k$.

Apply the Pythagorean Theorem (#55),

in $\triangle AGP$, $h^2 + \left(\frac{c}{2} - k\right)^2 = \left(\frac{2}{3} m_a\right)^2$,

$$\text{or } h^2 + \frac{c^2}{4} - ck + k^2 = \frac{4}{9} m_a{}^2. \qquad \text{(I)}$$

In $\triangle BGP$, $h^2 + \left(\frac{c}{2} + k\right)^2 = \left(\frac{2}{3} m_b\right)^2$,

$$\text{or } h^2 + \frac{c^2}{4} + ck + k^2 = \frac{4}{9} m_b{}^2. \qquad \text{(II)}$$

Adding (I) and (II), $2h^2 + \frac{2c^2}{4} + 2k^2 = \frac{4}{9} m_a{}^2 + \frac{4}{9} m_b{}^2$,

$$\text{or } 2h^2 + 2k^2 = \frac{4}{9} m_a{}^2 + \frac{4}{9} m_b{}^2 - \frac{c^2}{2}. \qquad \text{(III)}$$

However, in $\triangle FGP$, $h^2 + k^2 = \left(\frac{1}{3} m_c\right)^2$.

$$\text{Therefore, } 2h^2 + 2k^2 = \frac{2}{9} m_c{}^2. \qquad \text{(IV)}$$

By substitution of (IV) into (III),

$$\frac{2}{9} m_c{}^2 = \frac{4}{9} m_a{}^2 + \frac{4}{9} m_b{}^2 - \frac{c^2}{2}.$$

Therefore, $c^2 = \frac{8}{9} m_a{}^2 + \frac{8}{9} m_b{}^2 - \frac{4}{9} m_c{}^2$.

Similarly, $b^2 = \frac{8}{9} m_a{}^2 + \frac{8}{9} m_c{}^2 - \frac{4}{9} m_b{}^2$,

and $a^2 = \frac{8}{9} m_b{}^2 + \frac{8}{9} m_c{}^2 - \frac{4}{9} m_a{}^2$.

By addition, $a^2 + b^2 + c^2 = \frac{4}{3}(m_a{}^2 + m_b{}^2 + m_c{}^2)$,

$$\text{or } \frac{3}{4}(a^2 + b^2 + c^2) = m_a{}^2 + m_b{}^2 + m_c{}^2.$$

Challenge 2 *The sum of the squares of the measures of the sides of a triangle is* 120. *If two of the medians measure* 4 *and* 5, *respectively, how long is the third median?*

From the result of Problem 3-14, we know that

$$m_a^2 + m_b^2 + m_c^2 = \frac{3}{4}(a^2 + b^2 + c^2).$$

This gives us $5^2 + 4^2 + m^2 = \frac{3}{4}(120)$.

So $m^2 = 49$, and $m = 7$.

Challenge 3 *If* $\overline{AE}$ *and* $\overline{BF}$ *are medians drawn to the legs of right* $\triangle ABC$, *find the numerical value of* $\frac{(AE)^2 + (BF)^2}{(AB)^2}$ (*Fig. S3-14b*).

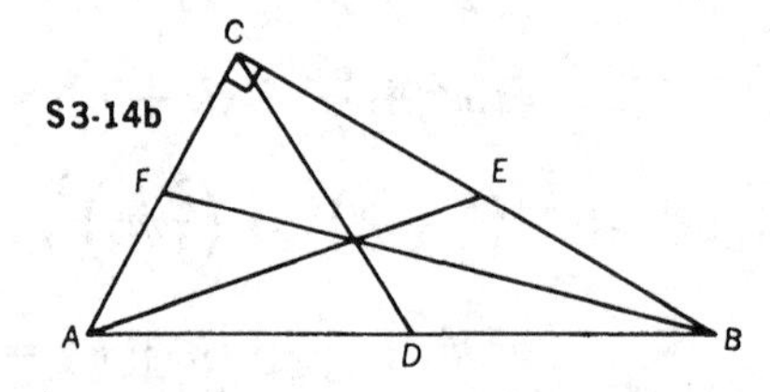

Use the previously proved theorem (Problem 3-14) that the sum of the squares of the measures of the medians equals $\frac{3}{4}$ the sum of the squares of the measures of the sides of the triangle.

$$(AE)^2 + (BF)^2 + (CD)^2$$
$$= \frac{3}{4}[(AC)^2 + (CB)^2 + (AB)^2] \qquad \text{(I)}$$

By the Pythagorean Theorem (#55),

$$(AC)^2 + (CB)^2 = (AB)^2. \qquad \text{(II)}$$

$$\text{Also, } (CD) = \frac{1}{2}(AB) \text{ (\#27).} \qquad \text{(III)}$$

By substituting (II) and (III) into (I),

$$(AE)^2 + (BF)^2 + \left(\frac{1}{2}AB\right)^2 = \frac{3}{4}[(AB)^2 + (AB)^2],$$

$$\text{or } (AE)^2 + (BF)^2 = \frac{3}{2}(AB)^2 - \frac{1}{4}(AB)^2.$$

$$\text{Then } \frac{(AE)^2 + (BE)^2}{(AB)^2} = \frac{5}{4}.$$

4. Circles Revisited

4-1 *Two tangents from an external point* P *are drawn to a circle, meeting the circle at points* A *and* B. *A third tangent meets the circle at* T, *and tangents* $\overrightarrow{PA}$ *and* $\overrightarrow{PB}$ *at points* Q *and* R, *respectively. Find the perimeter* p *of* △PQR.

We first consider the case shown in Fig. S4-1a where $AQ = QT$ and $BR = RT$ (#34).

Therefore, $p = PQ + QT + RT + PR = PQ + QA + BR + PR = PA + PB$.

We next consider the case shown in Fig. S4-1b where $AQ = QT$ and $BR = RT$ (#34).

$$\begin{aligned} \text{Therefore, } p &= PA + AQ + QT + RT + RB + PB \\ &= PA + QR + QR + PB \\ &= PA + PB + 2QR. \end{aligned}$$

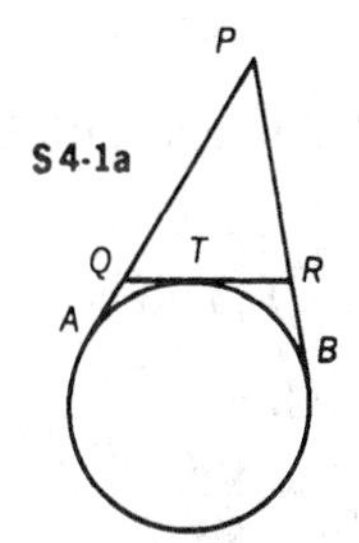

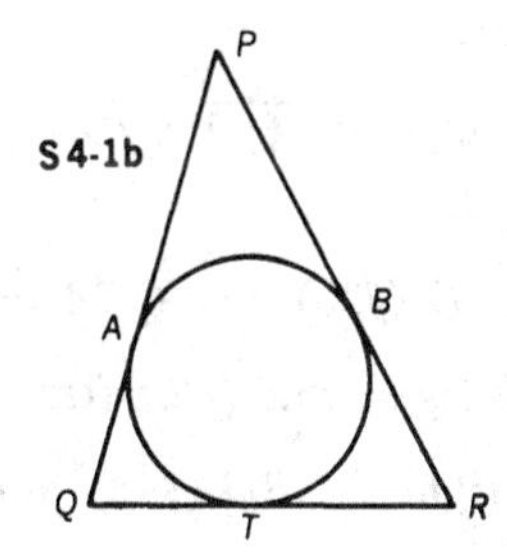

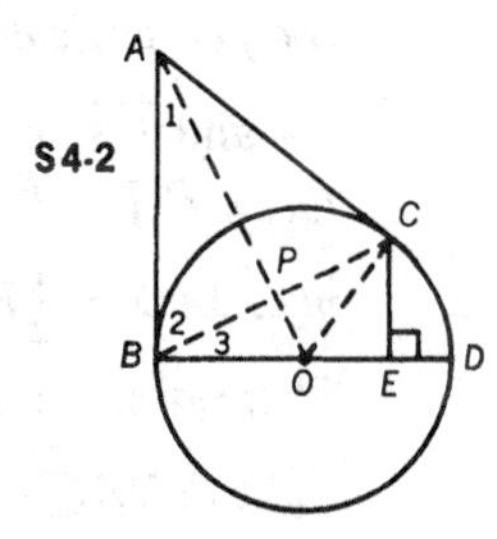

4-2 $\overline{AB}$ *and* $\overline{AC}$ *are tangent to circle* O *at* B *and* C, *respectively, and* $\overline{CE}$ *is perpendicular to diameter* $\overline{BD}$ (*Fig. S4-2*). *Prove* (BE)(BO) = (AB)(CE).

Draw $\overline{AO}$, $\overline{BC}$, and $\overline{OC}$, as in Fig. S4-2. We must first prove $\overline{AO} \perp \overline{BC}$. Since $AB = AC$ (#34), and $BO = OC$ (radii), $\overline{AO}$ is the perpendicular bisector of $\overline{BC}$ (#18). Since $\angle ABD$ is a right angle (#32a), $\angle 3$ is complementary to $\angle 2$. In right$\triangle APB$, $\angle 1$ is complementary to $\angle 2$. Therefore, $\angle 1 \cong \angle 3$.

Thus, right $\triangle BEC \sim$ right $\triangle ABO$, and

$$\frac{AB}{BE} = \frac{BO}{CE}, \text{ or } (BE)(BO) = (AB)(CE).$$

Challenge 3 *Show that* $\frac{AB}{\sqrt{BE}} = \frac{BO}{\sqrt{ED}}$.

We have proved that $\frac{AB}{BE} = \frac{BO}{CE}$. (I)

Since $\triangle BCD$ is a right triangle (#36),

$(CE)^2 = (BE)(ED)$ (#51a). Then $CE = \sqrt{BE}\sqrt{ED}$. (II)

From (I) and (II) we get $\frac{AB}{BE} = \frac{BO}{\sqrt{BE}\sqrt{ED}}$. (III)

By multiplying both sides of (III) by $\frac{1}{\sqrt{BE}}$ we get:

$$\frac{AB}{\sqrt{BE}} = \frac{BO}{\sqrt{ED}}.$$

4-3 *From an external point* P, *tangents* $\overrightarrow{PA}$ *and* $\overrightarrow{PB}$ *are drawn to a circle (Fig. S4-3a). From a point* Q *on the major (or minor) arc* $\widehat{AB}$, *perpendiculars are drawn to* $\overline{AB}$, $\overrightarrow{PA}$, *and* $\overrightarrow{PB}$. *Prove that the perpendicular to* $\overline{AB}$ *is the mean proportional between the other two perpendiculars.*

$\overrightarrow{PA}$ and $\overrightarrow{PB}$ are tangents; $\overline{QD} \perp \overrightarrow{PA}$, $\overline{QE} \perp \overrightarrow{PB}$, and $\overline{QC} \perp \overline{AB}$. Draw $\overline{QA}$ and $\overline{QB}$.

$m\angle DAQ = \frac{1}{2}m\widehat{AQ}$ (#38); $m\angle QBA = \frac{1}{2}m\widehat{AQ}$ (#36)

Therefore, $m\angle DAQ = m\angle QBA$ (transitivity),

right $\triangle DAQ \sim$ right $\triangle CBQ$ (#48), and $\frac{QD}{QC} = \frac{QA}{QB}$.

$m\angle QBE = \frac{1}{2}m\widehat{QB}$ (#38); $m\angle QAB = \frac{1}{2}m\widehat{QB}$ (#36)

Therefore, $m\angle QBE = m\angle QAB$ (transitivity), and

right $\triangle QBE \sim$ right $\triangle QAC$ (#48) so that $\frac{QC}{QE} = \frac{QA}{QB}$.

We therefore obtain $\frac{QD}{QC} = \frac{QC}{QE}$ (transitivity). In Fig. S4-3a, point

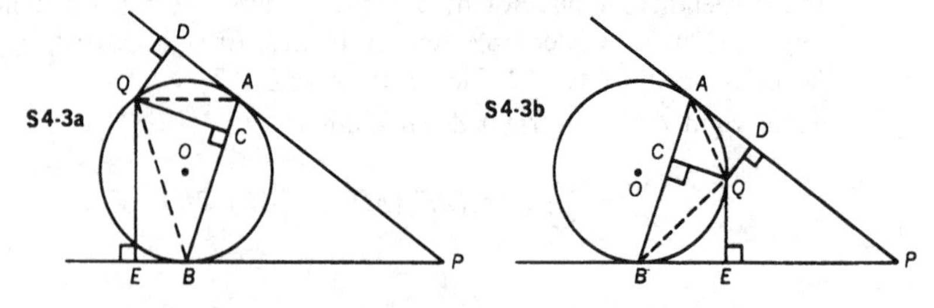

Q is on the major arc of circle O. Fig. S4-3b shows Q on the minor arc. Note that the proof applies equally well in either case.

4-4 *Chords* $\overline{AC}$ *and* $\overline{DB}$ *are perpendicular to each other and intersect at point* G *(Fig. S4-4). In* $\triangle AGD$ *the altitude from* G *meets* $\overline{AD}$ *at* E, *and when extended meets* $\overline{BC}$ *at* P. *Prove that* BP = PC.

In right $\triangle AEG$ $\angle A$ is complementary to $\angle 1$ (#14), and $\angle 2$ is complementary to $\angle 1$. Therefore, $\angle A \cong \angle 2$.
However, $\angle 2 \cong \angle 4$ (#1). Thus, $\angle A \cong \angle 4$.
Since $\angle A$ and $\angle B$ are equal in measure to $\frac{1}{2} m\widehat{DC}$ (#36), they are congruent. Therefore, $\angle 4 \cong \angle B$, and $BP = GP$ (#5).
Similarly, $\angle D \cong \angle 3$ and $\angle D \cong \angle C$ so that $GP = PC$.
Thus, $CP = PB$.

S4-4

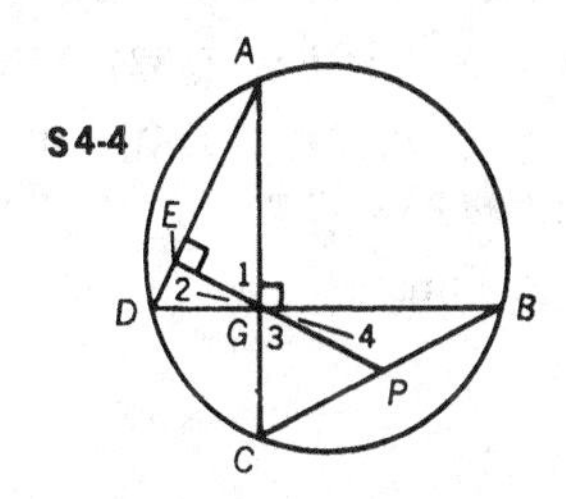

S 4-5

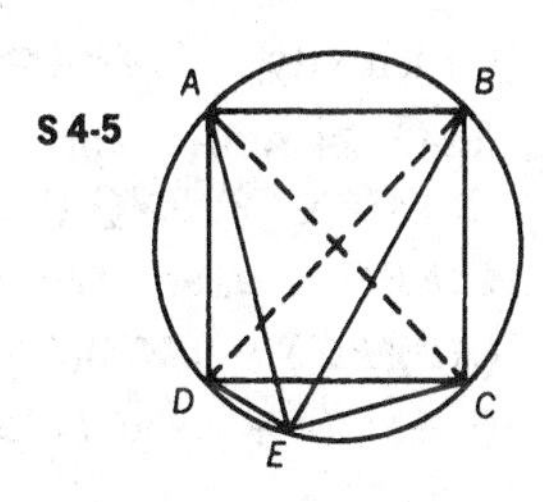

4-5 *Square* ABCD *is inscribed in a circle (Fig. S4-5). Point* E *is on the circle. If* AB = 8, *find the value of* $(AE)^2 + (BE)^2 + (CE)^2 + (DE)^2$.

In this problem we apply the Pythagorean Theorem to various right triangles. $\overline{DB}$ and $\overline{AC}$ are diameters; therefore, $\triangle DEB$, $\triangle DAB$, $\triangle AEC$, and $\triangle ABC$ are right triangles (#36).
In $\triangle DEB$, $(DE)^2 + (BE)^2 = (BD)^2$;
in $\triangle DAB$, $(AD)^2 + (AB)^2 = (BD)^2$ (#55).
Therefore $(DE)^2 + (BE)^2 = (AD)^2 + (AB)^2$.
In $\triangle AEC$, $(AE)^2 + (CE)^2 = (AC)^2$;
in $\triangle ABC$ $(AB)^2 + (BC)^2 = (AC)^2$ (#55).
Therefore, $(AE)^2 + (CE)^2 = (AB)^2 + (BC)^2$.
By addition, $(AE)^2 + (CE)^2 + (DE)^2 + (BE)^2 = (AB)^2 + (BC)^2 + (AD)^2 + (AB)^2$. Since the measures of all sides of square $ABCD$ equal 8,
$(AE)^2 + (CE)^2 + (DE)^2 + (BE)^2 = 4(8^2) = 256.$

To generalize, $(AE)^2 + (CE)^2 + (DE)^2 + (BE)^2 = 4s^2$ where s is the measure of the length of the side of the square. Interpret this result geometrically!

4-6 *In Fig. S4-6, radius* $\overline{AO}$ *is perpendicular to radius* $\overline{OB}$, $\overline{MN}$ *is parallel to* $\overline{AB}$ *meeting* $\overline{AO}$ *at* P *and* $\overline{OB}$ *at* Q, *and the circle at* M *and* N. *If* $MP = \sqrt{56}$, *and* $PN = 12$, *find the measure of the radius of the circle.*

Extend radius $\overline{AO}$ to meet the circle at C. We first prove that $MP = NQ$ by proving $\triangle AMP \cong \triangle BNQ$.
Since $\triangle AOB$ is isosceles, $\angle OAB \cong \angle OBA$ (#5), and trapezoid $APQB$ is isosceles (#23); therefore, $AP = QB$. Since $\overline{MN} \parallel \overline{AB}$, $\angle MPA \cong \angle PAB$ (#8), and $\angle NQB \cong \angle QBA$ (#8). Thus, by transitivity, $\angle MPA \cong \angle BQN$. $\widehat{MA} \cong \widehat{BN}$ (#33). Therefore, $\widehat{MAB} \cong \widehat{NBA}$ and $\angle AMN \cong \angle BNM$ (#36). Therefore, $\triangle AMP \cong \triangle BNQ$ (S.A.A.), and $MP = QN$.
Let $PO = a$, and radius $OA = r$. Thus, $AP = r - a$.
$(AP)(PC) = (MP)(PN)$ (#52)
$(r - a)(r + a) = (\sqrt{56})(12)$

$$r^2 - a^2 = 12\sqrt{56} \qquad \text{(I)}$$

We now find a^2 by applying the Pythagorean Theorem to isosceles right $\triangle POQ$.

$$(PO)^2 + (QO)^2 = (PQ)^2, \text{ so } a^2 + a^2 = (12 - \sqrt{56})^2, \text{ and}$$
$$a^2 = 100 - 12\sqrt{56}.$$

By substituting in equation (I),

$$r^2 = 12\sqrt{56} + 100 - 12\sqrt{56}, \text{ and } r = 10.$$

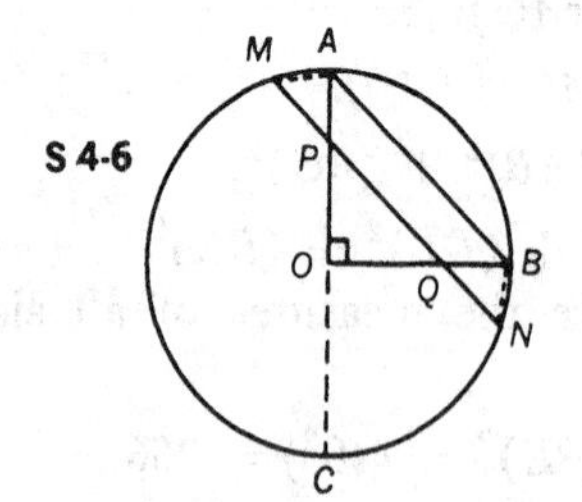

S 4-6

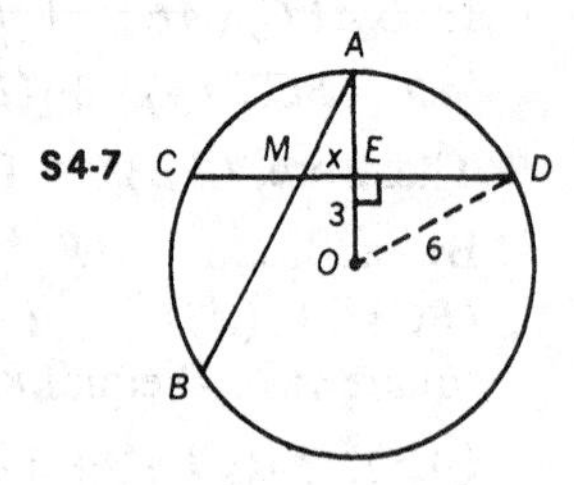

S4-7

4-7 *Chord $\overline{CD}$ is drawn so that its midpoint is 3 inches from the center of a circle with a radius of 6 inches (Fig. S4-7). From A, the midpoint of minor arc $\overset{\frown}{CD}$, any chord $\overline{AB}$ is drawn intersecting $\overline{CD}$ in M. Let v be the range of values of (AB)(AM), as chord $\overline{AB}$ is made to rotate in the circle about the fixed point A. Find v.*

$(AB)(AM) = (AM + MB)(AM) = (AM)^2 + (MB)(AM) = (AM)^2 + (CM)(MD)$ (#52)
E is the midpoint of $\overline{CD}$, and we let $EM = x$. In $\triangle OED$, $ED = \sqrt{27}$ (#55). Therefore, $CE = \sqrt{27}$.
$(CM)(MD) = (\sqrt{27} + x)(\sqrt{27} - x)$, and in $\triangle AEM$, $(AM)^2 = 9 + x^2$, (#55).
$(AB)(AM) = 9 + x^2 + (\sqrt{27} + x)(\sqrt{27} - x) = 36$
Therefore, v has the constant value 36.

QUESTION: Is it permissible to reason to the conclusion that $v = 36$ by considering the two extreme positions of point M, one where M is the midpoint of $\overline{CD}$, the other where M coincides with C (or D)?

4-8 *A circle with diameter $\overline{AC}$ is intersected by a secant at points B and D. The secant and the diameter intersect at point P outside the circle, as shown in Fig. S4-8. Perpendiculars $\overline{AE}$ and $\overline{CF}$ are drawn from the extremities of the diameter to the secant. If EB = 2, and BD = 6, find DF.*

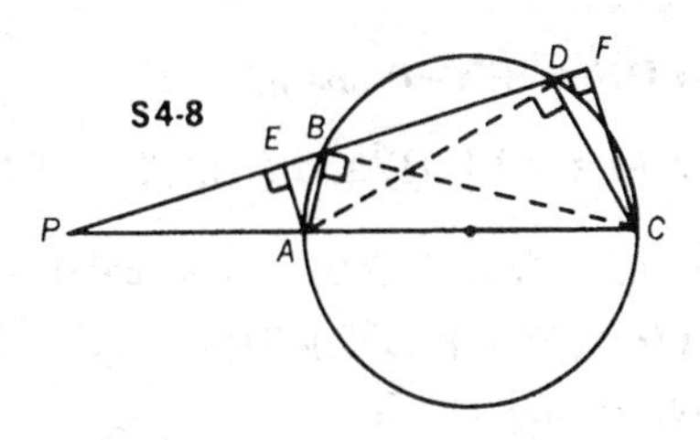

METHOD I: Draw $\overline{BC}$ and $\overline{AD}$. $\angle ABC \cong \angle ADC \cong$ right angle, since they are inscribed in a semicircle. $m\angle FDC + m\angle EDA = 90$ and $m\angle FCD + m\angle FDC = 90$; therefore, $\angle EDA \cong \angle FCD$, since both are complementary to $\angle FDC$.
Thus, right $\triangle CFD \sim$ right $\triangle DEA$ (#48), and $\dfrac{DF}{EA} = \dfrac{FC}{ED}$,

$$\text{or } (EA)(FC) = (DF)(ED). \qquad \text{(I)}$$

Similarly, $m\angle EAB + m\angle EBA = 90$ and $m\angle FBC + m\angle EBA = 90$;
therefore, $\angle EAB \cong \angle FBC$, since both are complementary to $\angle EBA$. Thus, right $\triangle AEB \sim$ right $\triangle BFC$, and $\frac{EB}{FC} = \frac{EA}{FB}$,

$$\text{or } (EA)(FC) = (EB)(FB). \qquad \text{(II)}$$

From (I) and (II) we find $(DF)(ED) = (EB)(FB)$.
Substituting we get $(DF)(8) = (2)(DF + 6)$, and $DF = 2$.

METHOD II: By applying the Pythagorean Theorem (#55),
in $\triangle AED$, $(ED)^2 + (EA)^2 = (AD)^2$;
in $\triangle DFC$, $(DF)^2 + (FC)^2 = (DC)^2$.

$$(ED)^2 + (DF)^2 = (AD)^2 + (DC)^2 - ((EA)^2 + (FC)^2)$$

In $\triangle AEB$, $(EB)^2 + (EA)^2 = (AB)^2$;
in $\triangle BFC$, $(BF)^2 + (FC)^2 = (BC)^2$.

$$(EB)^2 + (BF)^2 = (AB)^2 + (BC)^2 - ((EA)^2 + (FC)^2)$$

Since $(AD)^2 + (DC)^2 = (AC)^2 = (AB)^2 + (BC)^2$,
we get $(ED)^2 + (DF)^2 = (EB)^2 + (BF)^2$. (I)
Let $DF = x$. Substituting, $(8)^2 + x^2 = (2)^2 + (x + 6)^2$,
$64 + x^2 = 4 + x^2 + 12x + 36$, and $x = 2$.

Challenge *Does* DF = EB? *Prove it!*

From Method I, (I) and (II), $(DF)(ED) = (EB)(FB)$.

Then, $(DF)(EB + BD) = (EB)(BD + DF)$,
and $(DF)(EB) + (DF)(BD) = (EB)(BD) + (EB)(DF)$.
Therefore, $DF = EB$.

From Method II, (I),
$(ED)^2 + (DF)^2 = (EB)^2 + (BF)^2$.

Then, $(EB + BD)^2 + (DF)^2 = (EB)^2 + (BD + DF)^2$,
and $(EB)^2 + 2(EB)(BD) + (BD)^2 + (DF)^2 = (EB)^2 + (BD)^2 + 2(BD)(DF) + (DF)^2$.

Therefore, $EB = DF$.

4-9 *A diameter $\overline{CD}$ of a circle is extended through D to external point P. The measure of secant $\overline{CP}$ is 77. From P, another secant is drawn intersecting the circle first at A, then at B. (See Fig. S4-9a.) The measure of secant $\overline{PB}$ is 33. The diameter of the circle measures 74. Find the measure of the angle formed by the secants. (Note that the figure is not drawn to scale.)*

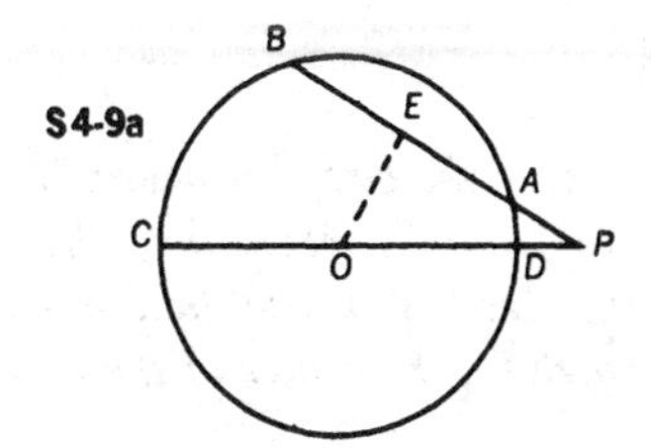

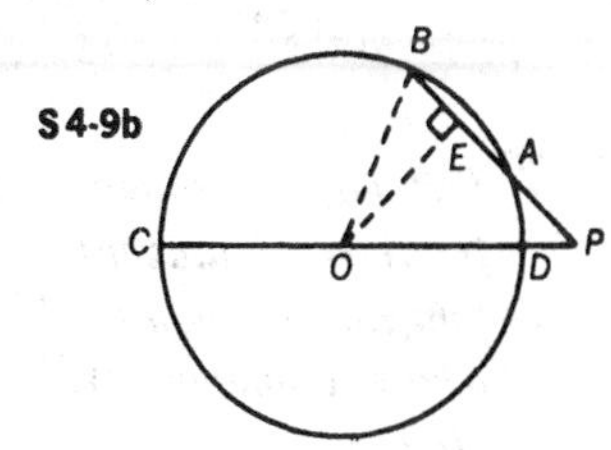

Since $CD = 74$ and $PC = 77$, $PD = 3$. Since $(PA)(PB) = (PD)(PC)$ (#54), $(PA)(33) = (3)(77)$, and $PA = 7$.
Therefore, $BA = 26$. Draw $\overline{OE} \perp \overline{AB}$. Then $AE = BE = 13$ (#30). Since $OD = 37$ and $PD = 3$, $OP = 40$.
In right $\triangle PEO$, $PE = 20$ and $PO = 40$.
Therefore, $m\angle EOP = 30$ and $m\angle P = 60$ (#55c).

Challenge *Find the measure of secant $\overline{PB}$ when $m\angle CPB = 45$ (Fig. S4-9b).*

In right $\triangle PEO$, $OE = 20\sqrt{2}$ (#55b).
Since $OB = 37$, in right $\triangle BEO$, $BE = \sqrt{569}$ (#55).
Therefore, $PB = 20\sqrt{2} + \sqrt{569}$.

4-10 *In $\triangle ABC$, in which $AB = 12$, $BC = 18$, and $AC = 25$, a semicircle is drawn so that its diameter lies on $\overline{AC}$, and so that it is tangent to $\overline{AB}$ and $\overline{BC}$ (Fig. S4-10). If O is the center of the circle, find the measure of $\overline{AO}$.*

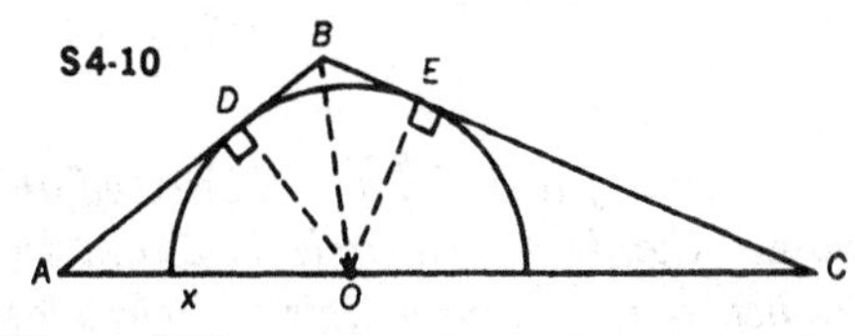

Draw radii $\overline{OD}$ and $\overline{OE}$ to the points of contact of tangents $\overline{AB}$ and $\overline{BC}$, respectively. $OD = OE$ (radii), and $m\angle BDO = m\angle BEO = 90$ (#32a). Since $DB = BE$ (#34), right $\triangle BDO \cong$ right $\triangle BEO$ (#17), and $\angle DBO \cong \angle EBO$.

In $\triangle ABC$, $\overline{BO}$ bisects $\angle B$ so that $\frac{AB}{AO} = \frac{BC}{OC}$ (#47).

Let $AO = x$; then $\frac{12}{x} = \frac{18}{25 - x}$, and $x = 10 = AO$.

Challenge *Find the diameter of the semicircle.*

ANSWER: $\frac{\sqrt{6479}}{6} \approx 13.4$

4-11 *Two parallel tangents to circle* O *meet the circle at points* M *and* N. *A third tangent to circle* O, *at point* P, *meets the other two tangents at points* K *and* L (*Fig. S4-11*). *Prove that a circle, whose diameter is* $\overline{KL}$, *passes through* O, *the center of the original circle.*

Draw $\overline{KO}$ and $\overline{LO}$. If $\overline{KL}$ is to be a diameter of a circle passing through O, then $\angle KOL$ will be an angle inscribed in a semicircle, or a right angle (#36).

Thus, we must prove that $\angle KOL$ is a right angle. It may easily be proved that $\overline{OK}$ bisects $\angle MKP$ and $\overline{OL}$ bisects $\angle PLN$ (Problem 4-10).

Since $m\angle MKP + m\angle NLP = 180$ (#11), we determine that $m\angle OKL + m\angle OLK = 90$ and $m\angle KOL = 90$ (#13).

It then follows that $\angle KOL$ is an inscribed angle in a circle whose diameter is $\overline{KL}$; thus, O lies on the new circle.

S4-11

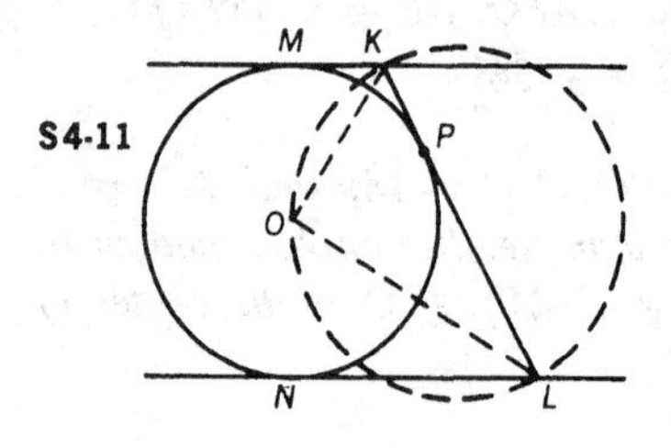

S4-12

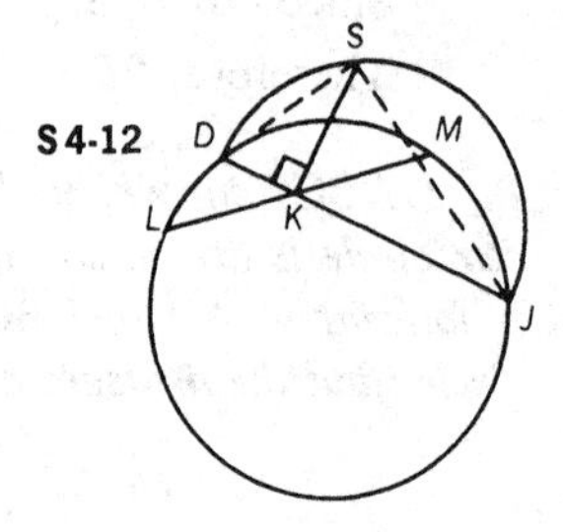

4-12 $\overline{LM}$ *is a chord of a circle, and is bisected at* K (*Fig. S4-12*). $\overline{DKJ}$ *is another chord. A semicircle is drawn with diameter* $\overline{DJ}$. $\overline{KS}$, *perpendicular to* $\overline{DJ}$, *meets this semicircle at* S. *Prove* KS = KL.

Draw $\overline{DS}$ and $\overline{SJ}$.

$\angle DSJ$ is a right angle since it is inscribed in a semicircle. Since $\overline{SK}$ is an altitude drawn to the hypotenuse of a right triangle,

$$\frac{DK}{SK} = \frac{SK}{KJ} \text{ (\#51a), or}$$

$$(SK)^2 = (DK)(KJ). \qquad \text{(I)}$$

However, in the circle containing chord $\overline{LM}$, $\overline{DJ}$ is also a chord and $(DK)(KJ) = (LK)(KM)$ (#52).

$$\text{Since } LK = KM, (DK)(KJ) = (KL)^2. \qquad \text{(II)}$$

From lines (I) and (II), $(SK)^2 = (KL)^2$, or $SK = KL$.

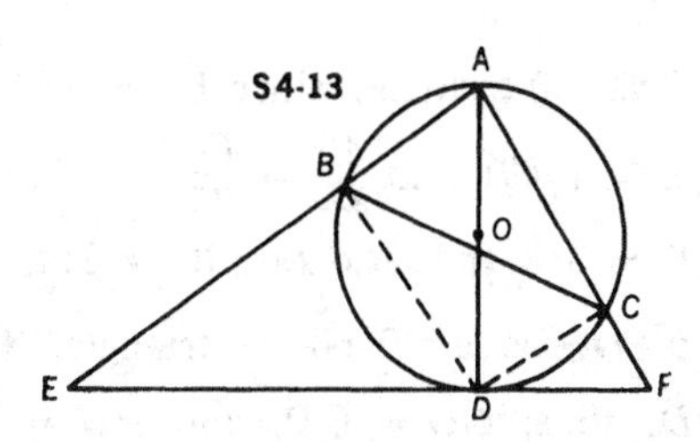

4-13 *Triangle* ABC *is inscribed in a circle with diameter* $\overline{\text{AD}}$, *as shown in Fig. S4-13. A tangent to the circle at* D *cuts* $\overline{\text{AB}}$ *extended at* E *and* $\overline{\text{AC}}$ *extended at* F. *If* AB = 4, AC = 6, *and* BE = 8, *find* CF.

Draw $\overline{DC}$ and $\overline{BD}$.

$\angle ABD \cong \angle ACD \cong$ right angle, since they are inscribed in semicircles (#36).

In right $\triangle ADE$, $\frac{AE}{AD} = \frac{AD}{AB}$ (#51b);

$$\text{thus, } (AD)^2 = (AE)(AB).$$

In right $\triangle ADF$, $\frac{AF}{AD} = \frac{AD}{AC}$ (#51b);

$$\text{thus, } (AD)^2 = (AF)(AC).$$

By transitivity, $(AE)(AB) = (AF)(AC)$.
By substitution, $(12)(4) = (6 + CF)(6)$, and $CF = 2$.

Challenge 1 *Find* m$\angle$DAF.

ANSWER: 30

Challenge 2 *Find* BC.

ANSWER: $2(\sqrt{6} + 1)$

4-14 *Altitude* $\overline{AD}$ *of equilateral* $\triangle ABC$ *is a diameter of circle* O. *If the circle intersects* $\overline{AB}$ *and* $\overline{AC}$ *at* E *and* F, *respectively, as in Fig. S4-14, find the ratio of* EF:BC.

METHOD I: Let $GD = 1$, and draw $\overline{ED}$.

$\angle AED$ is a right angle (#36). $m\angle ABD = 60$ and $\overline{AD} \perp \overline{BC}$; therefore, $m\angle BAD = 30$, and $m\angle ADE = 60$ (#14).

Because of symmetry, $\overline{AD} \perp \overline{EF}$. Therefore, $m\angle GED = 30$.

In $\triangle GED$, since $GD = 1$, we get $ED = 2$ (#55c), and $EG = \sqrt{3}$ (#55d).

In $\triangle AEG$ (30–60–90 triangle), since $EG = \sqrt{3}$, we get $AG = 3$.

$\triangle AEF \sim \triangle ABC$ (#49), and $\dfrac{AG}{AD} = \dfrac{EF}{BC}$.

Since $AG:AD = 3:4$, the ratio $EF:BC = 3:4$.

METHOD II: $\triangle EOG$ is a 30–60–90 triangle. Therefore, $OG = \frac{1}{2}OE = \frac{1}{2}OD$; thus, $OG = GD$, and $AG = \frac{3}{4}AD$. However, $\triangle AEF \sim \triangle ABC$ (#49). Therefore, $EF = \frac{3}{4}BC$, or $EF:BC = 3:4$.

Challenge *Find the ratio of* EB:BD.

ANSWER: 1:2

S4-14

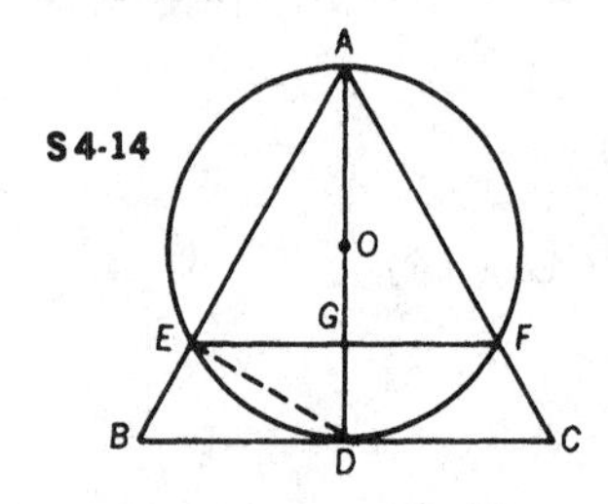

S4-15a

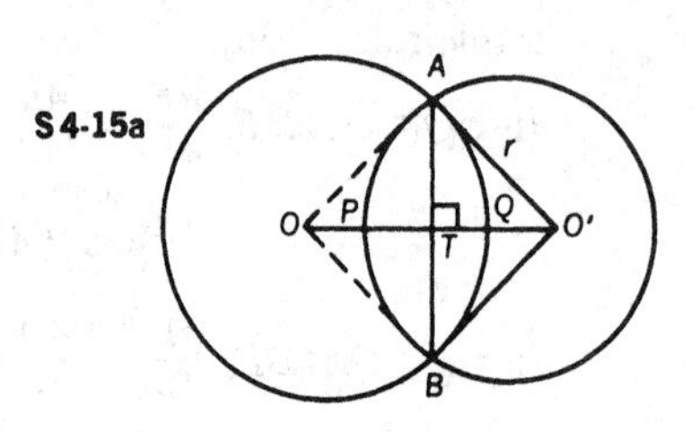

4-15 *Two circles intersect in* A *and* B, *and the measure of the common chord* $\overline{AB} = 10$. *The line joining the centers cuts the circles in* P *and* Q *(Fig. S4-15a). If* PQ = 3 *and the measure of the radius of one circle is* 13, *find the radius of the other circle. (Note that the illustration is not drawn to scale.)*

Since $O'A = O'B$ and $OA = OB$, $\overline{OO'}$ is the perpendicular bisector of $\overline{AB}$ (#18).

Therefore, in right $\triangle ATO$, since $AO = 13$ and $AT = 5$, we find $OT = 12$ (#55).

Since $OQ = 13$ (also a radius of circle O), and $OT = 12, TQ = 1$.
We know that $PQ = 3$.
In Fig. S4-15a, $PT = PQ - TQ$; therefore, $PT = 2$. Let $O'A = O'P = r$, and $PT = 2$, $TO' = r - 2$.
Applying the Pythagorean Theorem in right $\triangle ATO'$,

$$(AT)^2 + (TO')^2 = (AO')^2.$$

Substituting, $5^2 + (r - 2)^2 = r^2$, and $r = \frac{29}{4}$.
In Fig. S4-15b, $PT = PQ + TQ$; therefore, $PT = 4$. Again, let $O'A = O'P = r$; then $TO' = r - 4$.
Applying the Pythagorean Theorem in right $\triangle ATO'$,

$$(AT)^2 + (TO')^2 = (AO')^2.$$

Substituting, $5^2 + (r - 4)^2 = r^2$, and $r = \frac{41}{8}$.

Challenge *Find the second radius if* PQ = 2.

ANSWER: 13

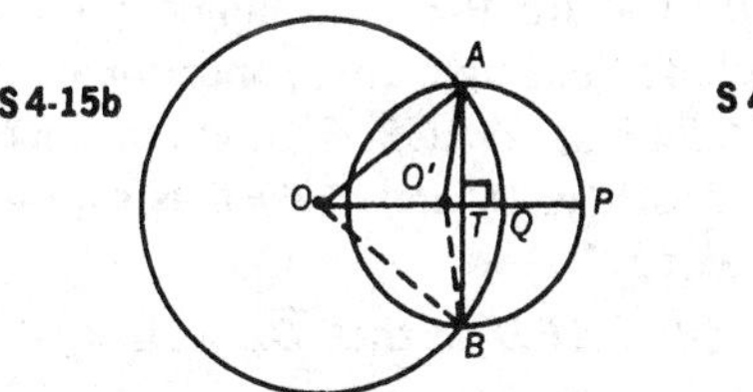

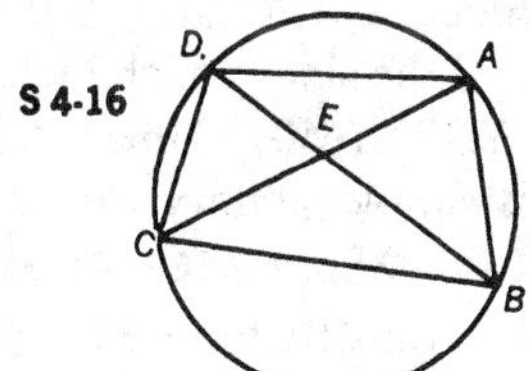

4-16 ABCD *is a quadrilateral inscribed in a circle. Diagonal* $\overline{BD}$ *bisects* $\overline{AC}$, *as in Fig. S4-16. If* AB = 10, AD = 12, *and* DC = 11, *find* BC.

$m\angle DBC = \frac{1}{2}m(\widehat{DC})$; $m\angle DAC = \frac{1}{2}m(\widehat{DC})$ (#36).
Therefore, $\angle DBC \cong \angle DAC$, and $\angle CEB \cong \angle DEA$ (#1).
Thus, $\triangle BEC \sim \triangle AED$ (#48), and

$$\frac{AD}{CB} = \frac{DE}{CE}. \qquad \text{(I)}$$

Similarly, $m\angle CAB = \frac{1}{2}m(\widehat{CB})$, and $m\angle CDB = \frac{1}{2}m(\widehat{CB})$ (#36).
Therefore, $\angle CAB \cong \angle CDB$, and $\angle AEB \cong \angle DEC$ (#1).

Thus, $\triangle AEB \sim \triangle DEC$ (#48), and $\frac{DC}{AB} = \frac{DE}{AE}$. (II)

But $AE = CE$; hence, from (I) and (II), $\frac{AD}{CB} = \frac{DC}{AB}$.

Substituting, $\frac{12}{CB} = \frac{11}{10}$, and $CB = \frac{120}{11}$.

Challenge *Solve the problem when diagonal* $\overline{BD}$ *divides* $\overline{AC}$ *into two segments, one of which is twice as long as the other.*

ANSWER: $CB = \frac{240}{11}$, if $AE = \frac{1}{3}AC$

$CB = \frac{60}{11}$, if $AE = \frac{2}{3}AC$

4-17 A *is a point exterior to circle* O. $\overline{PT}$ *is drawn tangent to the circle so that* PT = PA. *As shown in Fig. S4-17a,* C *is any point on circle* O, *and* $\overline{AC}$ *and* $\overline{PC}$ *intersect the circle at points* D *and* B, *respectively.* $\overline{AB}$ *intersects the circle at* E. *Prove that* $\overline{DE}$ *is parallel to* $\overline{AP}$.

$\frac{PC}{PT} = \frac{PT}{PB}$ (#53). Since $PT = AP$, $\frac{PC}{AP} = \frac{AP}{PB}$.

Since $\triangle APC$ and $\triangle BPA$ share the same angle (i.e., $\angle APC$), and the sides which include this angle are proportional, $\triangle APC \sim \triangle BPA$ (#50). Thus, $\angle BAP \cong \angle ACP$. However, since $\angle ACP$ is supplementary to $\angle DEB$ (#37), and $\angle AED$ is supplementary to $\angle DEB$, $\angle ACP \cong \angle AED$.

By transitivity, $\angle BAP \cong \angle AED$ so that $\overline{DE} \parallel \overline{AP}$ (#8).

Challenge 1 *Prove the theorem for* A *interior to circle* O.

As in the proof just given, we can establish that $\angle BAP \cong \angle ACP$. See Fig. S4-17b. In this case, $\angle DEB \cong \angle ACP$ (#36); therefore, $\angle BAP \cong \angle DEB$, and $\overline{DE}$ is parallel to $\overline{AP}$ (#7).

S4-17a

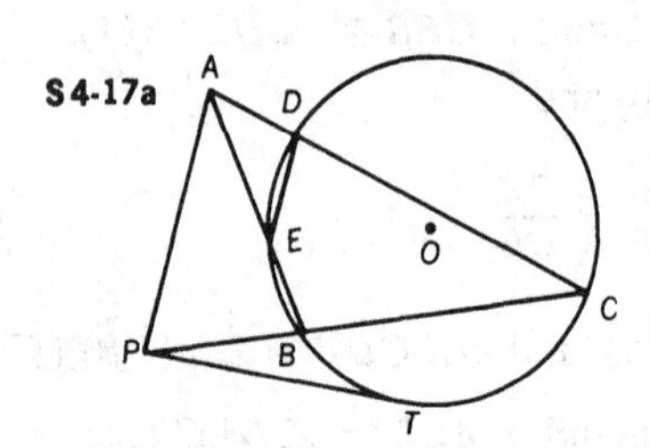

S4-17b

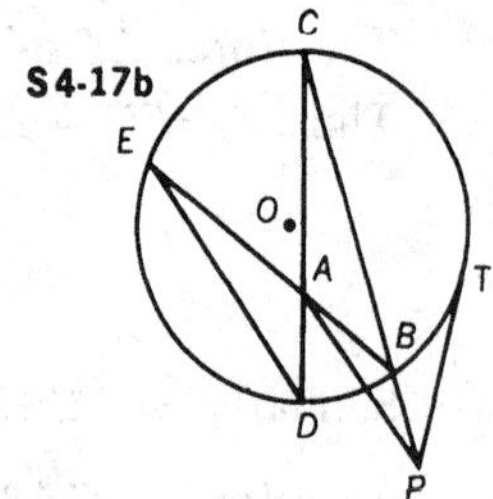

Challenge 2 *Explain the situation when* A *is on circle* O.

ANSWER: $\overline{DE}$ reduces to a point on $\overline{PA}$; thus we have a limiting case of parallel lines.

4-18 $\overline{\text{PA}}$ *and* $\overline{\text{PB}}$ *are tangents to a circle, and* $\overline{\text{PCD}}$ *is a secant. Chords* $\overline{\text{AC}}$, $\overline{\text{BC}}$, $\overline{\text{BD}}$, *and* $\overline{\text{DA}}$ *are drawn, as illustrated in Fig. S4-18. If* AC = 9, AD = 12, *and* BD = 10, *find* BC.

$m\angle PBC = \frac{1}{2}m(\widehat{BC})$ (#38). $m\angle PDB = \frac{1}{2}m(\widehat{BC})$ (#36).

Therefore, $\angle PBC \cong \angle PDB$.

Thus, $\triangle DPB \sim \triangle BPC$ (#48), and $\frac{CB}{DB} = \frac{PB}{PD}$. (I)

Since $PA = PB$ (#34), by substituting in (I) we get $\frac{CB}{DB} = \frac{PA}{PD}$.

Similarly, $\triangle DAP \sim \triangle ACP$ (#48), and $\frac{AC}{AD} = \frac{PA}{PD}$. (II)

From (I) and (II), $\frac{AC}{AD} = \frac{CB}{DB}$, and $\frac{9}{12} = \frac{CB}{10}$, or $CB = 7\frac{1}{2}$.

Challenge *If in addition to the information given above,* PA = 15 *and* PC = 9, *find* AB.

ANSWER: $AB = 11\frac{1}{4}$

S4-18

S4-19a

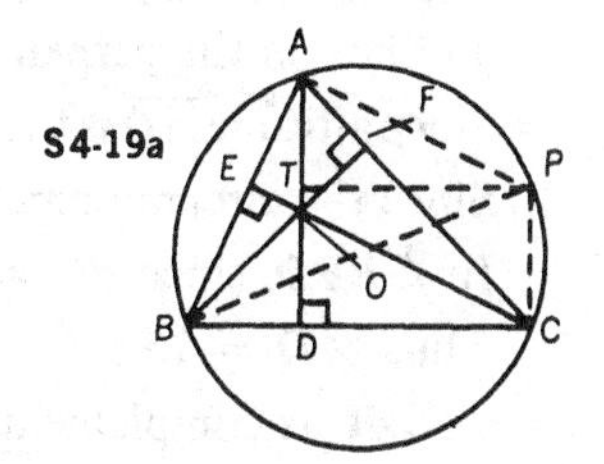

4-19 *The altitudes of* $\triangle$ABC *meet at* O *(Fig. S4-19a).* $\overline{\text{BC}}$, *the base of the triangle, has a measure of* 16. *The circumcircle of* $\triangle$ABC *has a diameter with a measure of* 20. *Find* AO. *(Figure not drawn to scale.)*

METHOD I: Let $\overline{BP}$ be a diameter of the circle circumscribed about $\triangle ABC$. Draw $\overline{PC}$ and $\overline{PA}$. Draw $\overline{PT}$ perpendicular to $\overline{AD}$. Since $\angle BCP$ is inscribed in a semicircle, it is a right angle (#36). Therefore, since $BP = 20$ and $BC = 16$, we get, by the Pythagorean Theorem, $PC = 12$.

$\angle BAP$ is a right angle (#36).

Therefore, $\angle PAO$ is complementary to $\angle EAO$.
Also, in right $\triangle EAO$, $\angle EOA$ is complementary to $\angle EAO$ (#14).
Thus, $\angle PAO \cong \angle EOA$; hence, $\overline{EC} \parallel \overline{AP}$ (#8).
Since $\overline{AD} \perp \overline{BC}$ and $\overline{PC} \perp \overline{BC}$ (#36), $\overline{AD} \parallel \overline{PC}$ (#9).
It then follows that $APCO$ is a parallelogram (#21a), and $AO = PC = 12$.

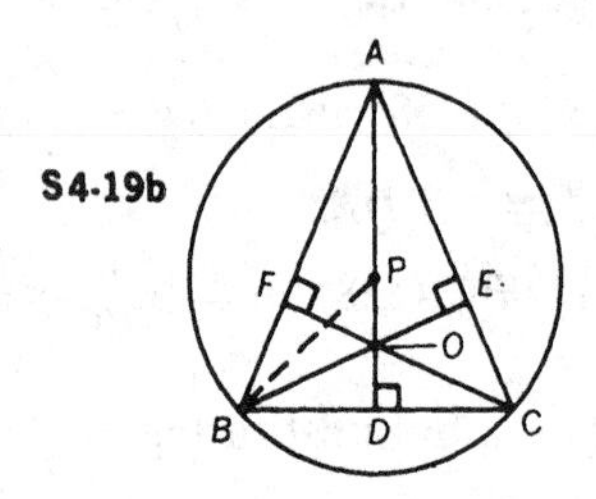

METHOD II: The solution above is independent of the position of point A on the circle. But we may more easily do this problem by letting $\overline{AD}$ be the perpendicular bisector of $\overline{BC}$, in other words letting $\triangle ABC$ be isosceles ($AB = AC$). Our purpose for doing this is to place the circumcenter on altitude $\overline{AD}$ as shown in Fig. S4-19b.
The circumcenter P is equidistant from the vertices ($AP = BP$), and lies on the perpendicular bisectors of the sides (#44).
Since altitude $\overline{AD}$ is the perpendicular bisector of $\overline{BC}$, P lies on $\overline{AD}$.
Since the circumdiameter is 20, $AP = BP = 10$.
In $\triangle PBD$, since $BP = 10$ and $BD = 8$, then $PD = 6$ (#55).
Thus, $AD = 16$.
$\angle DAC$ is complementary to $\angle DCA$, and
$\angle DBO$ is complementary to $\angle BCA$ (#14).
Therefore, $\angle DAC \cong DBO$.
Thus, right $\triangle ACD \sim$ right $\triangle BOD$ (#48), and $\frac{AD}{BD} = \frac{DC}{OD}$.
Substituting, $\frac{16}{8} = \frac{8}{OD}$; then $OD = 4$, and by subtraction, $AO = 12$.

4-20 *Two circles are tangent internally at* P, *and a chord,* $\overline{AB}$, *of the larger circle is tangent to the smaller circle at* C. $\overline{PB}$ *and* $\overline{PA}$ *cut the smaller circle at* E *and* D, *respectively (Fig. S4-20). If* AB = 15, *while* PE = 2 *and* PD = 3, *find* AC.

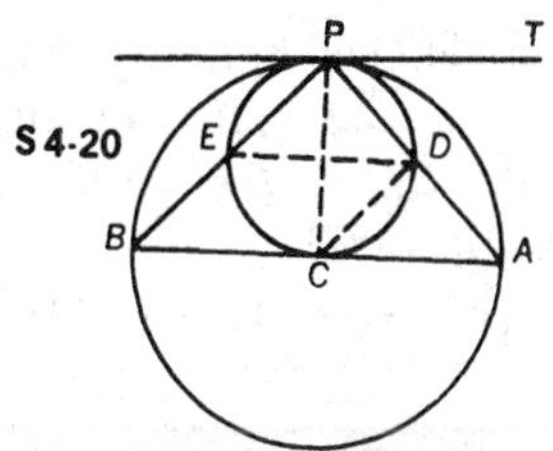

Draw $\overline{ED}$ and the common external tangent through P.

$m\angle PED = \frac{1}{2}m(\widehat{PD})$ (#36), $m\angle TPD = \frac{1}{2}m(\widehat{PD})$ (#38).

Therefore, $\angle PED \cong \angle TPD$. (I)

Similarly, $m\angle PBA = \frac{1}{2}m(\widehat{PA})$ (#36).

$m\angle TPA = \frac{1}{2}m(\widehat{PA})$ (#38).

Therefore, $\angle PBA \cong \angle TPA$. (II)

Thus, from (I) and (II), $\angle PED \cong \angle PBA$, and $\overline{ED} \parallel \overline{BA}$ (#7).

In $\triangle PBA$, $\frac{PB}{PA} = \frac{PE}{PD}$ (#46). Thus, $\frac{PB}{PA} = \frac{2}{3}\cdot$ Now draw $\overline{CD}$. (III)

$m\angle PDC = \frac{1}{2}m(\widehat{PC})$ (#36), while $m\angle PCB = \frac{1}{2}m(\widehat{PC})$ (#38).

Therefore, $\angle PDC \cong \angle PCB$.

Since $m\angle PCD = \frac{1}{2}m(\widehat{PD})$ (#36), and $m\angle TPD = \frac{1}{2}m(\widehat{PD})$ (#38), $\angle PCD \cong \angle TPD$. Since, from (II), $\angle PBA \cong \angle TPD$, $\angle PCD \cong \angle PBA$.

Thus, $\triangle PBC \sim \triangle PCD$ (#48), and $\angle BPC \cong \angle DPC$.

Since, in $\triangle PBA$, $\overline{PC}$ bisects $\angle BPA$, $\frac{PB}{PA} = \frac{BC}{AC}$ (#47).

From (III), $\frac{2}{3} = \frac{BC}{AC}$.

Since $AB = 15$, $BC = AB - AC = 15 - AC$; therefore, $\frac{2}{3} = \frac{15 - AC}{AC}$, and $AC = 9$.

Challenge *Express* AC *in terms of* AB, PE, *and* PD.

ANSWER: $AC = \frac{(AB)(PD)}{PE + PD}$

4-21 *A circle, center* O, *is circumscribed about* $\triangle$ABC, *a triangle in which* $\angle$C *is obtuse (Fig. S4-21). With* $\overline{OC}$ *as diameter, a circle is drawn intersecting* $\overline{AB}$ *in* D *and* D′. *If* AD = 3, *and* DB = 4, *find* CD.

Extend $\overline{CD}$ to meet circle O at E. In the circle with diameter $\overline{OC}$, $\overline{OD}$ is perpendicular to $\overline{CD}$ (#36). In circle O, since $\overline{OD}$ is perpendicular to $\overline{CE}$, $CD = DE$ (#30). Again in circle O, $(CD)(DE) = (AD)(DB)$ (#52).
Since $CD = DE$, $(CD)^2 = (3)(4)$, and $CD = \sqrt{12} = 2\sqrt{3}$.

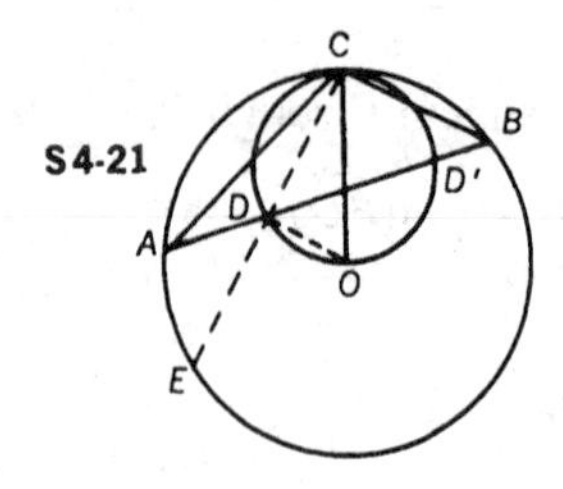

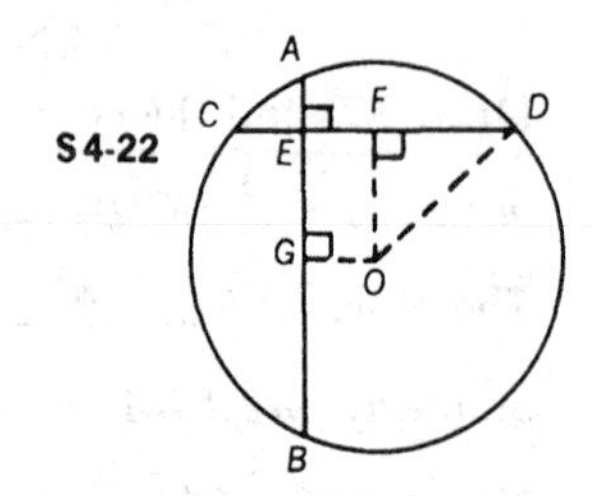

4-22 *In circle* O, *perpendicular chords* $\overline{AB}$ *and* $\overline{CD}$ *intersect at* E *so that* AE = 2, EB = 12, *and* CE = 4 (*Fig.* S4-22). *Find the measure of the radius of circle* O.

From the center O, drop perpendiculars to $\overline{CD}$ and $\overline{AB}$, meeting these chords at points F and G, respectively. Join O and D.
Since $(AE)(EB) = (CE)(ED)$ (#52), $ED = 6$.
$\overline{OF}$ bisects $\overline{CD}$ (#30), and $CD = 10$; therefore, $FD = 5$.
Similarly, since $AB = 14$, then $GB = 7$ and $GE = 5$.
Quadrilateral $EFOG$ is a rectangle (a quadrilateral with three right angles is a rectangle). Therefore, $GE = FO = 5$.
Applying the Pythagorean Theorem to isosceles right $\triangle FOD$, we find $DO = 5\sqrt{2}$, the radius of circle O.

Challenge *Find the shortest distance from* E *to the circle.*

ANSWER: $5\sqrt{2} - \sqrt{26}$

4-23 *Prove that the sum of the squares of the measures of the segments made by two perpendicular chords is equal to the square of the measure of the diameter of the given circle.*

Draw $\overline{AD}$, $\overline{CB}$, diameter $\overline{COF}$, and $\overline{BF}$ as illustrated in Fig. S4-23.
Since $\overline{AB} \perp \overline{CD}$, $\triangle CEB$ is a right triangle, and $c^2 + b^2 = y^2$ (#55). In right $\triangle AED$, $a^2 + d^2 = x^2$. (#55)
By addition, $a^2 + b^2 + c^2 + d^2 = x^2 + y^2$.

In right $\triangle CBF$, $y^2 + z^2 = m^2$.

$\angle 4$ is complementary to $\angle 2$ (#14), and $\angle 3$ is complementary to $\angle 1$.

However, $\angle 2 = \angle 1$ (#36); therefore, $\angle 4 = \angle 3$. Thus, $\widehat{ADF} \cong \widehat{BFD}$, and $\widehat{AD} \cong \widehat{BF}$; hence, $x = z$.

Therefore, $y^2 + x^2 = m^2$, and $a^2 + b^2 + c^2 + d^2 = m^2$.

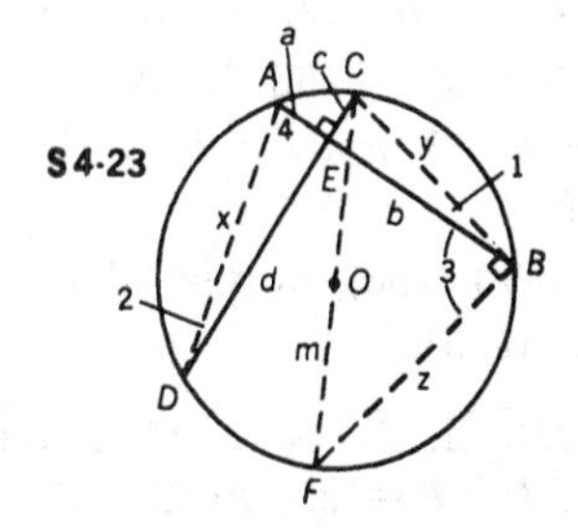

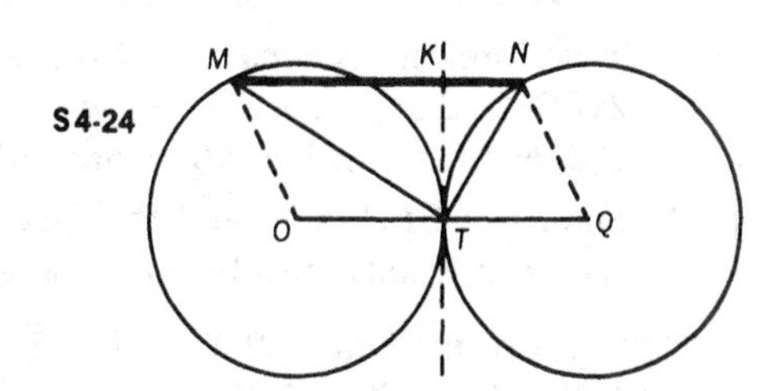

4-24 *Two equal circles are tangent externally at* T. *Chord* $\overline{\text{TM}}$ *in circle* O *is perpendicular to chord* $\overline{\text{TN}}$ *in circle* Q *(Fig. S4-24). Prove that* $\overline{\text{MN}} \parallel \overline{\text{OQ}}$ *and* MN = OQ.

Draw the line of centers $\overline{OQ}$, $\overline{MO}$, and $\overline{NQ}$; then draw the common internal tangent $\overline{KT}$ meeting $\overline{MN}$ at K.

$m\angle KTN = \frac{1}{2} m\widehat{NT}$ and $m\angle KTM = \frac{1}{2} m\widehat{MT}$ (#38).

$m\angle KTN + m\angle KTM = m\angle MTN = 90$.

Therefore, $\frac{1}{2} m\widehat{NT} + \frac{1}{2} m\widehat{MT} = 90$,

or $m\widehat{NT} + m\widehat{MT} = 180$.

Thus, $m\angle MOT + m\angle NQT = 180$ (#35), and $\overline{MO} \parallel \overline{NQ}$ (#11).

Since $MO = NQ$ (radii of equal circles), $MNQO$ is a parallelogram (#22).

It then follows that $MN = OQ$ and $\overline{MN} \parallel \overline{OQ}$.

4-25 *As illustrated, from point* A *on the common internal tangent of tangent circles* O *and* O′, *secants* $\overline{\text{AEB}}$ *and* $\overline{\text{ADC}}$ *are drawn, respectively. If* $\overline{\text{DE}}$ *is the common external tangent, and points* C *and* B *are collinear with the centers of the circles, prove*

(a) m$\angle$1 = m$\angle$2, *and*
(b) $\angle$A *is a right angle.*

(a) Draw common internal tangent $\overline{AP}$ (Fig. S4-25).
For circle O', $(AP)^2 = (AC)(AD)$ (#53);
for circle O, $(AP)^2 = (AB)(AE)$ (#53).
Therefore, $(AC)(AD) = (AB)(AE)$, or $\frac{AC}{AE} = \frac{AB}{AD}$.
Thus, $\triangle ADE \sim \triangle ABC$ (#50), and $m\angle 1 = m\angle 2$.

(b) METHOD I: Draw $\overline{DP}$ and $\overline{PE}$. $GE = GP$ and $DG = GP$ (#34). Therefore, in isosceles $\triangle DGP$, $\angle 3 \cong \angle 4$; and in isosceles $\triangle EGP$, $\angle 5 \cong \angle 6$. Since $m\angle 3 + m\angle 4 + m\angle 5 + m\angle 6 = 180$, $m\angle 4 + m\angle 5 = 90 = m\angle DPE$.
Since $m\angle CDP = 90$ and $m\angle PEB = 90$ (#36), in quadrilateral $ADPE$ $\angle A$ must also be a right angle (#15).

METHOD II: Draw $\overline{DO'}$ and $\overline{OE}$. $\overline{DO'} \perp \overline{DE}$ and $\overline{EO} \perp \overline{DE}$ (#32a). Therefore, $\overline{DO'} \parallel \overline{OE}$, and $m\angle DO'B + m\angle EOC = 180$.
Thus, $m\widehat{DP} + m\widehat{EP} = 180$ (#35).
However, $m\angle DCP = \frac{1}{2} m\widehat{DP}$, and $m\angle EBP = \frac{1}{2} m\widehat{EP}$ (#36).
By addition, $m\angle DCP + m\angle EBP = \frac{1}{2}(m\widehat{DP} + m\widehat{EP}) = \frac{1}{2}(180) = 90$. Therefore, $m\angle BAC = 90$ (#13).

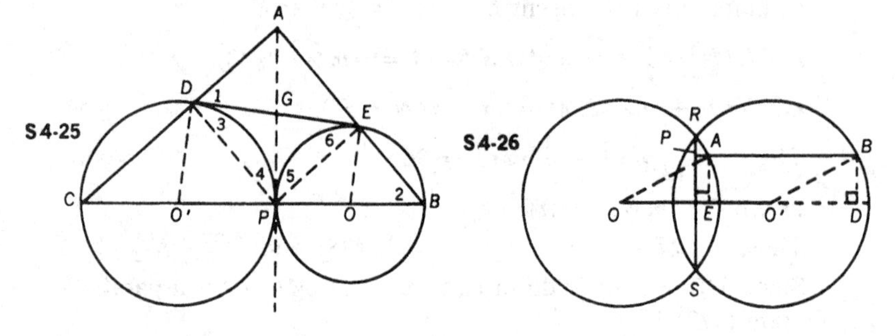

4-26 *Two equal intersecting circles* O *and* O′ *have a common chord* $\overline{RS}$ *(Fig. S4-26). From any point* P *on* $\overline{RS}$ *a ray is drawn perpendicular to* $\overline{RS}$ *cutting circles* O *and* O′ *at* A *and* B, *respectively. Prove that* $\overline{AB}$ *parallel to the line of centers* $\overleftrightarrow{OO'}$, *and that* AB = OO′.

Draw $\overline{OA}$ and $\overline{O'B}$; then draw $\overline{AE} \perp \overline{OO'}$ and $\overline{BD} \perp \overline{OO'}$.
Since $\overline{PAB} \perp \overline{RS}$ and the line of centers $\overline{OO'} \perp \overline{RS}$, $\overline{AB} \parallel \overline{OO'}$ (#9).

Consider $\triangle AOE$ and $\triangle BO'D$. Since $\overline{AB} \parallel \overline{OO'D}$, $AE = BD$ (#20), and $AO = O'B$ since they are radii of equal circles.
Thus, right $\triangle AOE \cong$ right $\triangle BO'D$ (#17).
Therefore, $\angle AOE \cong \angle BO'D$, and $\overline{AO} \parallel \overline{O'B}$ (#7).
It follows that $ABO'O$ is a parallelogram (#22).
Thus, $AB = OO'$ (#21b).

4-27 *A circle is inscribed in a triangle whose sides are* 10, 10, *and* 12 *units in measure (Fig. S4-27). A second, smaller circle is inscribed tangent to the first circle and to the equal sides of the triangle. Find the measure of the radius of the second circle.*

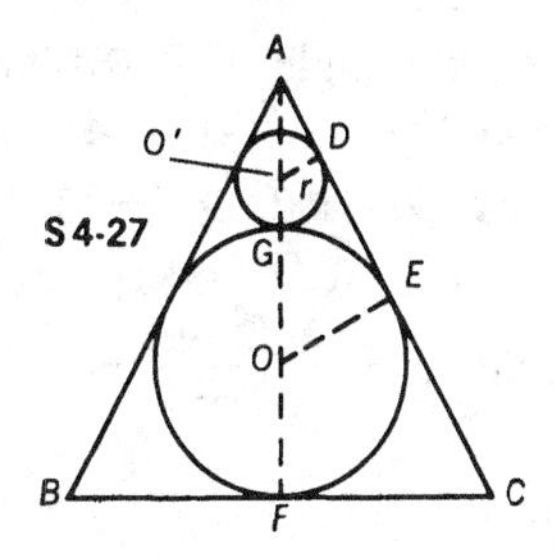

Draw $\overline{AO'OF}$, $\overline{OE}$, and $\overline{O'D}$. $\overline{OE} \perp \overline{AC}$ and $\overline{O'D} \perp \overline{AC}$ (#32a).
$CF = CE = 6$ (#34)
Since $AC = 10$, $AE = 4$. In right $\triangle AFC$, $AF = 8$ (#55).
Right $\triangle AEO \sim$ right $\triangle AFC$ (#48), and $\frac{FC}{OE} = \frac{AF}{AE}$.
Substituting, $\frac{6}{OE} = \frac{8}{4}$, and $OE = 3$.
Therefore, $GF = 6$, and $AG = 2$.
Let $O'D = O'G = r$. Then $O'A = 2 - r$
Since $\overline{O'D} \parallel \overline{OE}$ (#9), right $\triangle ADO' \sim$ right $\triangle AEO$, and $\frac{O'D}{O'A} = \frac{OE}{OA}$.
Since in right $\triangle AEO$, $AE = 4$ and $OE = 3$, $AO = 5$ (#55).
Thus, $\frac{r}{2 - r} = \frac{3}{5}$, and $r = \frac{3}{4}$.

Challenge 1 *Solve the problem in general terms if* AC = a, BC = 2b.

ANSWER: $r = \frac{b(a-b)^{3/2}}{(a+b)^{3/2}}$

Challenge 2 *Inscribe still another, smaller circle, tangent to the second circle and to the equal sides. Find its radius by inspection.*

ANSWER: $\frac{1}{4} \cdot \frac{3}{4} = \frac{3}{16}$

Challenge 3 *Extend the legs of the triangles through* B *and* C, *and draw a circle tangent to the original circle and to the extensions of the legs. What is its radius?*

ANSWER: 12

4-28 *A circle with radius* 3 *is inscribed in a square, as illustrated in Fig. S4-28. Find the radius of the circle that is inscribed between two sides of the square and the original circle.*

Since $\overline{OA}$ bisects right angle A, $\triangle DAO$ and $\triangle EAO'$ are isosceles right triangles. Let $EO' = x$; then $AO' = x\sqrt{2}$ (#55a).
Since $O'F = x$ and $OF = 3$, $OA = 3 + x + x\sqrt{2}$.
But in $\triangle ADO$, $AO = 3\sqrt{2}$ (#55a). It then follows that $3\sqrt{2} = 3 + x + x\sqrt{2}$, and $x = \frac{3\sqrt{2} - 3}{\sqrt{2} + 1} = 3(3 - 2\sqrt{2})$.

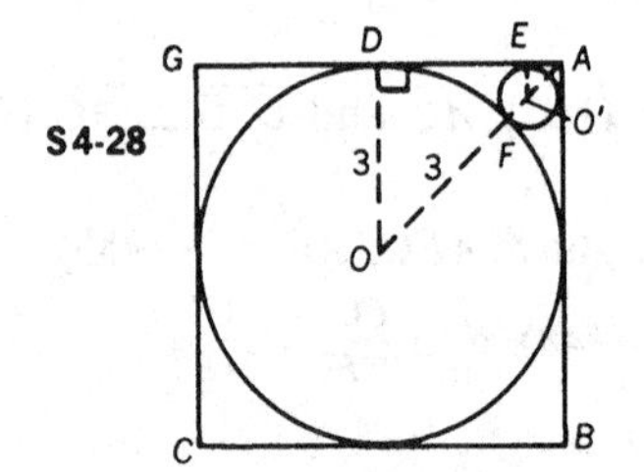

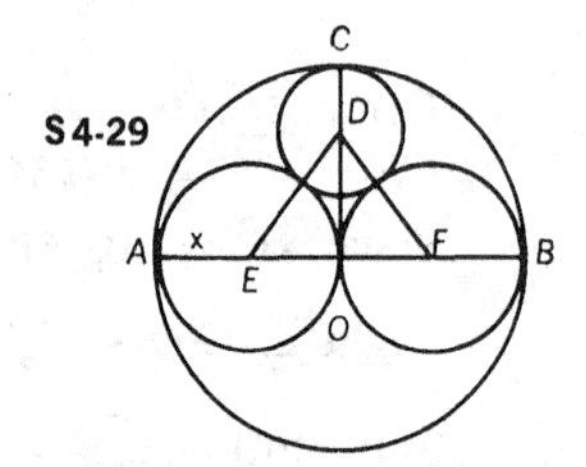

4-29 $\overline{\text{AB}}$ *is a diameter of circle* O *(Fig. S4-29). Two circles are drawn with* $\overline{\text{AO}}$ *and* $\overline{\text{OB}}$ *as diameters. In the region between the circumferences, a circle* D *is inscribed tangent to the three previous circles. If the measure of the radius of circle* D *is* 8, *find* AB.

Let radius $AE = x$. Since $CD = 8$, $DE = AE + CD = x + 8$.
Thus, by applying the Pythagorean Theorem in $\triangle DEO$, $(EO)^2 + (DO)^2 = (DE)^2$, $x^2 + (DO)^2 = (x + 8)^2$, and $DO = 4\sqrt{x + 4}$.
However, $DO + CD = CO = OA = AE + EO$.
Substituting, $4\sqrt{x + 4} + 8 = 2x$, and $x = 12$.
Therefore, $AB = 4x = 48$.

4-30 *A carpenter wishes to cut four equal circles from a circular piece of wood whose area is equal to 9π square feet. He wants these circles of wood to be the largest that can possibly be cut from this piece of wood. Find the measure of the radius of each of the four new circles (Fig. S4-30).*

Let the length of the radius of the four small circles be x. By joining the centers of the four small circles, we get a square whose side equals $2x$ and whose diagonal equals $2x\sqrt{2}$ (#55a). Therefore, the diameter of circle O equals $2x + 2x\sqrt{2}$, and its radius equals $x(1 + \sqrt{2})$. Since the area of circle O is 9π, the radius is 3. Therefore, $x(1 + \sqrt{2}) = 3$, and $x = \dfrac{3}{1 + \sqrt{2}} = 3(\sqrt{2} - 1)$ feet.

Challenge 1 *Find the correct radius if the carpenter decides to cut out three equal circles of maximum size.*

ANSWER: $3(2\sqrt{3} - 3)$

Challenge 2 *Which causes the greater waste of wood, the four circles or the three circles?*

ANSWER: Three circles

S 4-30

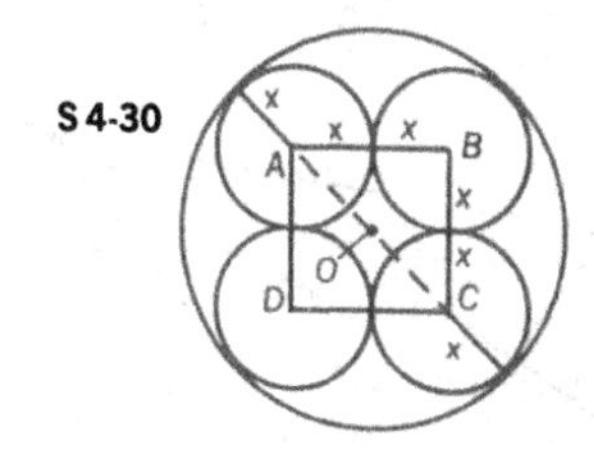

S 4-31

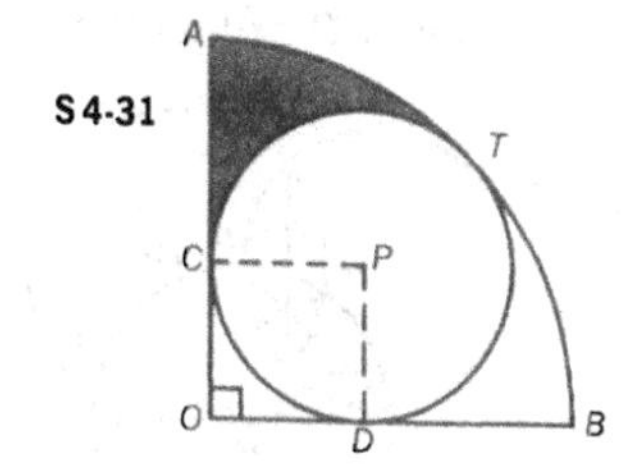

4-31 *A circle is inscribed in a quadrant of a circle of radius 8, as shown in Fig. S4-31. What is the measure of the radius of the inscribed circle?*

Draw radii $\overline{PC}$ and $\overline{PD}$ to points of tangency with $\overline{AO}$ and $\overline{BO}$. Then $\overline{PC} \perp \overline{AO}$, $\overline{PD} \perp \overline{OB}$ (#32a).

Since $\angle AOB$ is a right angle, $PCOD$ is a rectangle (a quadrilateral with three right angles is a rectangle). Moreover, since radius PC = radius PD, $PCOD$ is a square.

Let $PC = PD = r$; then $CO = OD = r$, and $OP = r\sqrt{2}$ (#55a), while $PT = r$. Therefore, $OT = r + r\sqrt{2}$, but $OT = 8$ also; thus, $r + r\sqrt{2} = 8$, and $r = 8(\sqrt{2} - 1)$, $\left(\text{approximately } 3\frac{1}{3}\right)$.

QUESTION: Explain why $\overline{OT}$ goes through P.

Challenge *Find the area of the shaded region.*

ANSWER: $16[4(\sqrt{2} - \pi) + 3(\pi\sqrt{2} - 2)]$

4-32 *Three circles intersect. Each pair of circles has a common chord (Fig. S4-32). Prove that these three chords are concurrent.*

Let chords $\overline{AB}$ and $\overline{CD}$ intersect at P. These are the common chords for circles O and Q, and circles O and R, respectively. Circles R and Q intersect at points E and F. Draw $\overline{EP}$ and extend it.

Assume that $\overleftrightarrow{EP}$ does not pass through F. It therefore meets circles Q and R at points X and Y, respectively.

In circle O, $(AP)(PB) = (CP)(PD)$ (#52).

Similarly, in circle Q, $(AP)(PB) = (EP)(PX)$ (#52).

By transitivity, $(CP)(PD) = (EP)(PX)$.

However, in circle R, $(CP)(PD) = (EP)(PY)$ (#52).

It then follows that X and Y must be the same point and must lie both on circle Q and circle R.

Thus, $\overleftrightarrow{EP}$ will meet the intersection of circles Q and R at F.

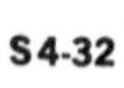
S4-32

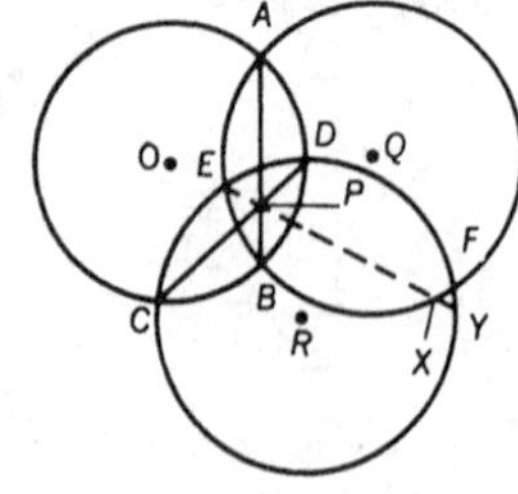

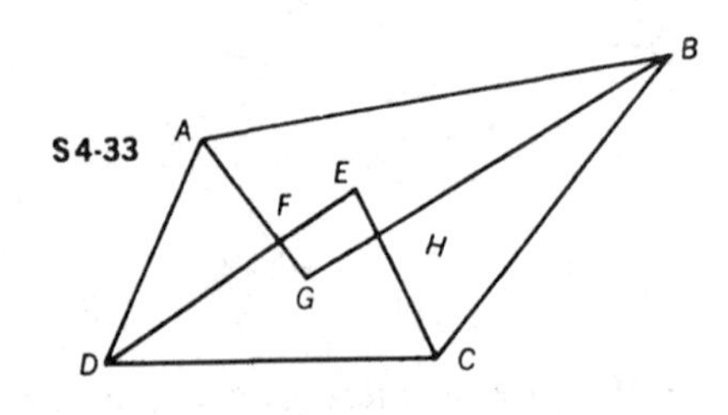

4-33 *The bisectors of the angles of a quadrilateral are drawn. From each pair of adjacent angles, the two bisectors are extended until they intersect, as shown in Fig. S4-33. The line segments connecting the points of intersection form a quadrilateral. Prove that this figure is cyclic (i.e., can be inscribed in a circle).*

$m\angle BAD + m\angle ADC + m\angle DCB + m\angle CBA = 360$ (#15);

therefore, $\frac{1}{2}m\angle BAD + \frac{1}{2}m\angle ADC + \frac{1}{2}m\angle DCB + \frac{1}{2}m\angle CBA = \frac{1}{2}(360) = 180$. Substituting,

$$m\angle EDC + m\angle ECD + m\angle GAB + m\angle ABG = 180. \qquad \text{(I)}$$

Consider $\triangle ABG$ and $\triangle DEC$.

$$m\angle EDC + m\angle ECD + m\angle GAB + m\angle ABG + m\angle AGB + m\angle DEC = 2(180) \qquad \text{(II)}$$

Now, subtracting (I) from (II), we find that

$$m\angle AGB + m\angle DEC = 180.$$

Since one pair of opposite angles of quadrilateral *EFGH* are supplementary, the other pair must also be supplementary, and hence quadrilateral *EFGH* is cyclic (#37).

4-34 *In cyclic quadrilateral* ABCD, *perpendiculars* $\overline{\text{AB}}$ *and* $\overline{\text{CD}}$ *are erected at* B *and* D *and extended until they meet sides* $\overleftrightarrow{\text{CD}}$ *and* $\overleftrightarrow{\text{AB}}$ *at* B′ *and* D′, *respectively (Fig. S4-34). Prove* $\overline{\text{AC}}$ *is parallel to* $\overline{\text{B'D'}}$.

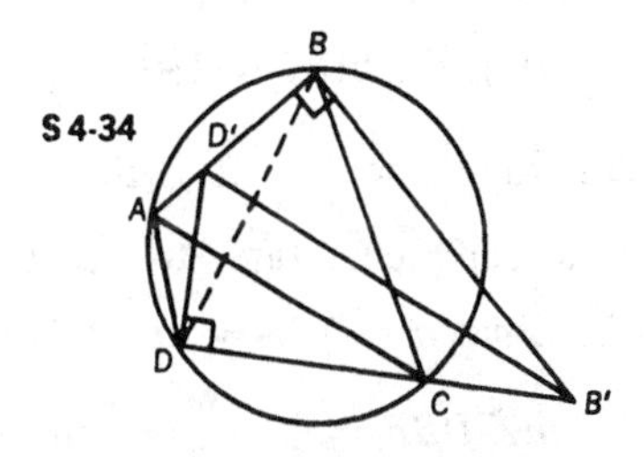

Draw $\overline{BD}$. Consider cyclic quadrilateral *ABCD*.

$$\angle ACD \cong \angle ABD \ (\angle DBD') \ (\#36) \qquad \text{(I)}$$

Since $\angle D'BB' \cong \angle D'DB' \cong$ right angle, quadrilateral $D'BB'D$ is also cyclic (#37).

$$\text{Therefore, } \angle DB'D' \cong \angle DBD' \ (\#36). \qquad \text{(II)}$$

Thus, from (I) and (II), $\angle ACD \cong \angle DB'D'$, and $\overline{AC} \parallel \overline{B'D'}$ (#7).

4-35 *Perpendiculars* $\overline{\text{BD}}$ *and* $\overline{\text{CE}}$ *are drawn from vertices* B *and* C *of* $\triangle$ABC *to the interior bisectors of angles* C *and* B, *meeting them at* D *and* E, *respectively (Fig. S4-35). Prove that* $\overleftrightarrow{\text{DE}}$ *intersects* $\overline{\text{AB}}$ *and* $\overline{\text{AC}}$ *at their respective points of tangency,* F *and* G, *with the circle that is inscribed in* $\triangle$ABC.

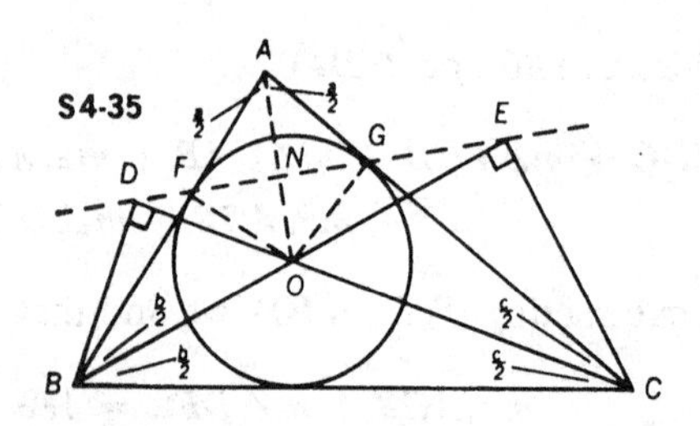

Let a, b, and c respresent the measures of angles A, B, and C, respectively. Draw $\overline{AO}$. Since the angle bisectors of a triangle are concurrent, $\overline{AO}$ bisects $\angle BAC$. Also, $FO = GO$, and $AF = AG$ (#34); therefore, $\overline{AO} \perp \overline{FG}$ at N (#18).

Now, in right $\triangle AFN$, $m\angle GFA = 90 - \frac{a}{2}$ (#14). (I)

Since $\overline{CO}$ and $\overline{BO}$ are angle bisectors, in $\triangle BOC$,

$$m\angle BOC = 180 - \frac{1}{2}(b + c) \text{ (\#13)}. \qquad \text{(II)}$$

However, $b + c = 180 - a$ (#13).

Therefore, from (II), $m\angle BOC = 180 - \frac{1}{2}(180 - a) = 90 + \frac{a}{2}$.

Since $\angle BOD$ is supplementary to $\angle BOC$, $m\angle BOD = 90 - \frac{a}{2}$.

But $\angle DBO$ is complementary to $\angle BOD$ (#14); therefore, $m\angle DBO = \frac{a}{2}$. Since $m\angle DBF = m\angle DBO - m\angle FBO$,

$$m\angle DBF = \frac{a}{2} - \frac{b}{2} = \frac{1}{2}(a - b). \qquad \text{(III)}$$

$\angle BFO \cong \angle BDO \cong$ right angle; therefore, quadrilateral $BDFO$ is cyclic (#36à), and $\angle FDO \cong \angle FBO$ (#36). Thus, $m\angle FDO = \frac{b}{2}$.

It then follows that $m\angle FDB = 90 + \frac{b}{2}$. (IV)

Thus, in $\triangle DFB$, $m\angle DFB = 180 - (m\angle FDB + m\angle DBF)$. (V)

By substituting (III) and (IV) into (V),

$$m\angle DFB = 180 - [90 + \frac{b}{2} + \frac{1}{2}(a - b)] = 90 - \frac{a}{2}. \qquad \text{(VI)}$$

Since $\overline{AFB}$ is a straight line, and $m\angle GFA = 90 - \frac{a}{2} = m\angle DFB$ (See (I) and (VI)), points D, F, and G must be collinear (#1). In a similar manner, points E, G, and F are proved collinear. Thus, points D, F, G, and E are collinear, and $\overleftrightarrow{DE}$ passes through F and G.

4-36 *A line,* $\overline{PQ}$, *parallel to base* $\overline{BC}$ *of* $\triangle ABC$, *cuts* $\overline{AB}$ *and* $\overline{AC}$ *at* P *and* Q, *respectively* (*Fig. S4-36*). *The circle passing through* P *and tangent to* $\overline{AC}$ *at* Q *cuts* $\overline{AB}$ *again at* R. *Prove that the points* R, Q, C, *and* B *lie on a circle.*

Draw RQ. $m\angle 2 = \frac{1}{2}(m\widehat{PQ})$ (#36); also $m\angle 3 = \frac{1}{2}(m\widehat{PQ})$ (#38); therefore, $\angle 3 \cong \angle 2$, and $\angle 3 \cong \angle 5$ (#7).
By transitivity we find that $\angle 2 \cong \angle 5$. But $m\angle 2 + m\angle 4 = 180$. Therefore, $m\angle 4 + m\angle 5 = 180$. Since one pair of opposite angles of a quadrilateral are supplementary, the other pair of opposite angles must also be supplementary, and the quadrilateral is cyclic. Thus, R, Q, C, and B lie on a circle.

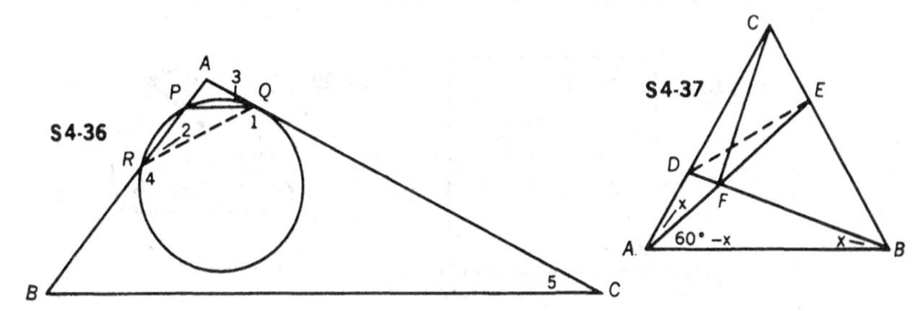

4-37 *In equilateral* $\triangle ABC$, D *is chosen on* $\overline{AC}$ *so that* $AD = \frac{1}{3}(AC)$, *and* E *is chosen on* $\overline{BC}$ *so that* $CE = \frac{1}{3}(BC)$ (*Fig. S4-37*). $\overline{BD}$ *and* $\overline{AE}$ *intersect at* F. *Prove that* $\angle CFB$ *is a right angle.*

Draw $\overline{DE}$. Since $AD = CE$ and $AC = AB$ and $\angle ACB \cong \angle CAB$, $\triangle ACE \cong \triangle BAD$ (S.A.S.), and $m\angle ABD = m\angle CAE = x$.
Since $m\angle FAB = 60 - x$, $m\angle AFB = 120$ (#13).
Then $m\angle DFE = 120$ (#1). Since $m\angle ACB = 60$, quadrilateral $DCEF$ is cyclic because the opposite angles are supplementary.
In $\triangle CED$, $CE = \frac{1}{2}(CD)$ and $m\angle C = 60$; therefore, $\angle CED$ is a right angle (#55c).
Since $\angle DFC$ is inscribed in the same arc as $\angle CED$, $\angle CED \cong \angle DFC \cong$ right angle. Thus, $\angle CFB \cong$ right angle.

4-38 *The measure of the sides of square* ABCD *is* x. F *is the midpoint of* $\overline{BC}$, *and* $\overline{AE} \perp \overline{DF}$ (*Fig. S4-38*). *Find* BE.

Draw $\overline{AF}$. Since $CF = FB$, and $DC = AB$, right $\triangle DCF \cong$ right $\triangle ABF$ (S.A.S.). Then $m\angle CDF = m\angle BAF = \alpha$.

$\angle AEF \cong ABF \cong$ right angle; therefore, quadrilateral $AEFB$ is cyclic (#37).

It follows that $m\angle BAF = m\angle BEF = \alpha$ since both are angles inscribed in the same arc.

Since $\angle DAE$ and $\angle CDF$ are both complementary to $\angle ADE$,

$$m\angle DAE = m\angle CDF = \alpha.$$

Both $\angle BEA$ and $\angle BAE$ are complementary to an angle of measure α; therefore, they are congruent. Thus, $\triangle ABE$ is isosceles, and $AB = BE = x$ (#5).

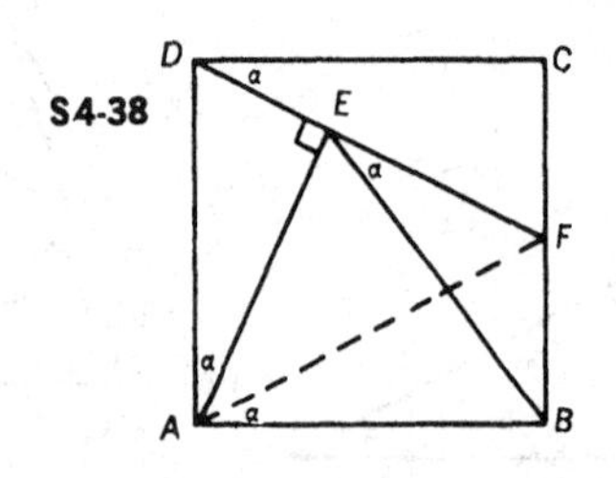

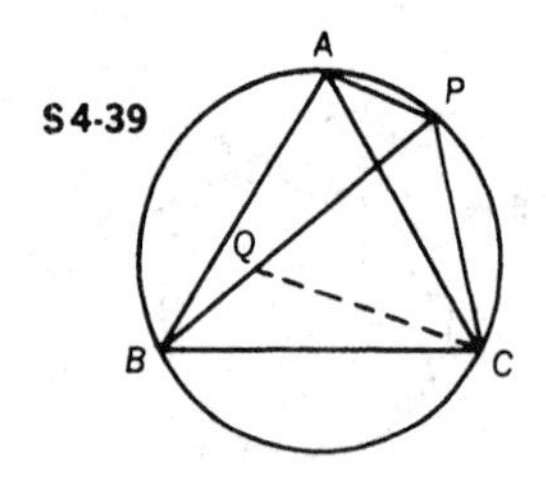

4-39 *If equilateral* $\triangle$ABC *is inscribed in a circle, and a point* P *is chosen on minor arc* $\widehat{AC}$, *prove that* PB = PA + PC (Fig. *S*4-39).

Choose a point Q on $\overline{BP}$ such that $PQ = QC$.

Since $\triangle ABC$ is equilateral, $m\widehat{AB} = m\widehat{BC} = m\widehat{CA} = 120$.

Therefore, $m\angle BPC = \frac{1}{2} m\widehat{BC} = 60$ (#36).

Since in $\triangle PQC$, $PQ = QC$, and $m\angle BPC = 60$, $\triangle QPC$ is equilateral.

$m\angle PQC = 60$, $m\angle BQC = 120$, and $m\angle APC = \frac{1}{2} m\widehat{ABC} = 120$. Therefore, $\angle APC \cong \angle BQC$.

$PC = QC$ and $\angle CAP \cong \angle CBP$ as both are equal in measure to $\frac{1}{2} m\widehat{PC}$ (#36).

Thus, $\triangle BQC \cong \triangle APC$ (S.A.A.), and $BQ = AP$. Since $BQ + QP = BP$, by substitution, $AP + PC = PB$.

4-40 *From point* A, *tangents are drawn to circle* O, *meeting the circle at* B *and* C. *Chord* $\overline{BF} \parallel$ *secant* $\overline{ADE}$, *as in Fig. S*4-40. *Prove that* $\overline{FC}$ *bisects* $\overline{DE}$.

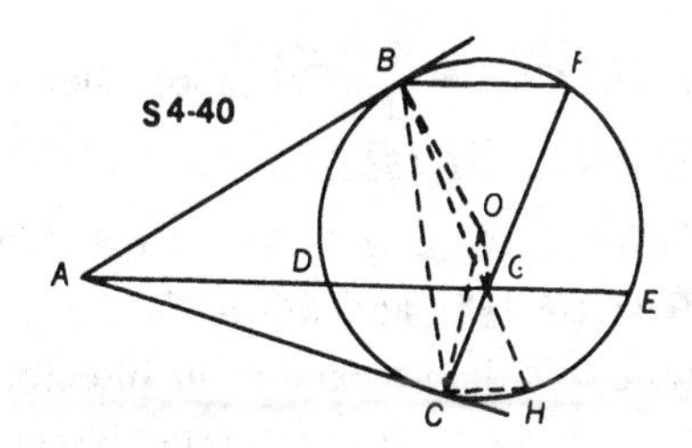

METHOD I: Draw $\overline{BC}$, $\overline{OB}$, and $\overline{OC}$.

$m\angle BAE = \frac{1}{2}(m\widehat{BFE} - m\widehat{BD})$ (#40)

Since $\widehat{BD} \cong \widehat{FE}$ (#33), $m\angle BAE = \frac{1}{2} m\widehat{BF}$.

However, $m\angle BCF = \frac{1}{2} m\widehat{BF}$ (#36).

Therefore, $\angle BAE \cong \angle BCF$, or $\angle BAG \cong \angle BCG$.

It is therefore possible to circumscribe a circle about quadrilateral $ABGC$ since the angles which would be inscribed in the same arc are congruent. Because the opposite angles of quadrilateral $ABOC$ are supplementary, it, too, is cyclic.

We know that three points determine a unique circle, and that points A, B, and C are on both circles; we may therefore conclude that points A, B, O, G, and C lie on the same circle. Since $\angle ACO$ is a right angle (#32a), $\overline{AO}$ must be the diameter of the new circle (#36). $\angle AGO$ is then inscribed in a semicircle and is a right angle (#36). As $\overline{OG} \perp \overline{DE}$, it follows that $DG = EG$ (#30).

METHOD II: Draw $\overline{BG}$ and extend it to meet the circle at H; draw $\overline{CH}$.

$m\angle AGC = \frac{1}{2}(m\widehat{DC} + m\widehat{FE})$ (#39)

Since $\widehat{FE} \cong \widehat{BD}$ (#33), $m\angle AGC = \frac{1}{2}(m\widehat{DC} + m\widehat{BD}) = \frac{1}{2}(m\widehat{BC})$.

$m\angle ABC = \frac{1}{2}(m\widehat{BC})$ (#38), and $m\angle BFC = \frac{1}{2}(m\widehat{BC})$ (#36).

Therefore, $\angle AGC \cong \angle ABC \cong \angle BFC$.

Now we know a circle may be drawn about A, B, G, and C, since $\angle ABC$ and $\angle AGC$ are congruent angles that would be inscribed in the same arc.

It then follows that $\angle CAG \cong \angle CBG$ since they are both inscribed in arc $(\widehat{CG})$.

In circle O, $m\angle CAG$ $(\angle CAE) = \frac{1}{2}(m\widehat{CH} + m\widehat{HE} - m\widehat{CD})$ (#40).

However, $m\angle CBG = \frac{1}{2} m\widehat{CH}$ (#36); therefore, $\widehat{HE} \cong \widehat{CD}$.
Thus, $\overline{CH} \parallel \overline{AE} \parallel \overline{BF}$ (#33).
Since $\widehat{FEH} \cong \widehat{BDC}$, $\angle HBF \cong \angle CFB$ (#36).
Thus, $BG = GF$ (#5), and $BO = FO$.
Therefore, $\overline{OG}$ is the perpendicular bisector of $\overline{BF}$ (#18).
Then $\overline{OG} \perp \overline{DE}$ (#10), and $\overline{OG}$ must bisect $\overline{DE}$ (#30).

5. Area Relationships

5-1 *As shown in Fig. S5-1,* E *is on* $\overline{AB}$ *and* C *is on* $\overline{FG}$. *Prove that parallelogram* ABCD *is equal in area to parallelogram* EFGD.

Draw $\overline{EC}$. Since $\triangle EDC$ and $\square ABCD$ share the same base ($\overline{DC}$) and a common altitude (from E to $\overleftrightarrow{DC}$), the area of $\triangle EDC$ is equal to one-half the area of $\square ABCD$.

Similarly, $\triangle EDC$ and $\square EFGD$ share the same base ($\overline{ED}$), and the same altitude to that base; thus, the area of $\triangle EDC$ is equal to one-half the area of $\square EFDG$.

Since the area of $\triangle EDC$ is equal to one-half the area of each parallelogram, the parallelograms are equal in area..

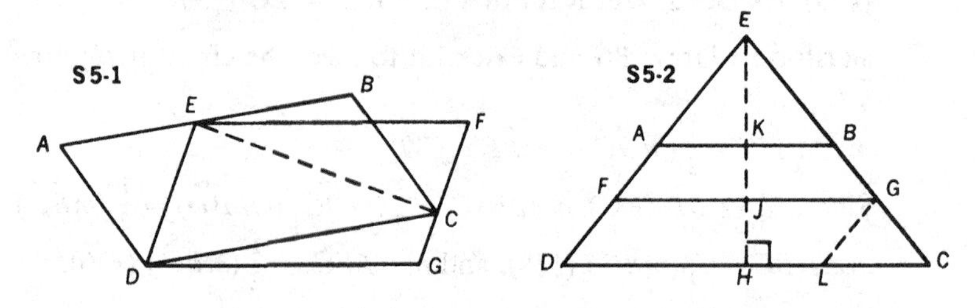

5-2 *The measures of the bases of trapezoid* ABCD *are* 15 *and* 9, *and the measure of the altitude is* 4. *Legs* $\overline{DA}$ *and* $\overline{CB}$ *are extended to meet at* E, *as in Fig. S5-2. If* F *is the midpoint of* $\overline{AD}$, *and* G *is the midpoint of* $\overline{BC}$, *find the area of* $\triangle$FGE. *(The figure is not drawn to scale.)*

METHOD I: $\overline{FG}$ is the median of trapezoid $ABCD$, and

$FG = \frac{15 + 9}{2} = 12$ (#28).

Since $\triangle EFG \sim \triangle EDC$ (#49), $\dfrac{EJ}{EH} = \dfrac{FG}{DC}$. (I)

$KH = 4$ and $HJ = \frac{1}{2} KH = 2$ (#24). Therefore,

$\dfrac{EJ}{EJ + 2} = \dfrac{12}{15}$ and $EJ = 8$.

Hence, the area of $\triangle EFG = \frac{1}{2}(FG)(EJ) = \frac{1}{2}(12)(8) = 48$.

METHOD II: Since $\triangle EFG \sim \triangle EDC$ (#49),

$\dfrac{\text{Area of } \triangle EFG}{\text{Area of } \triangle EDC} = \dfrac{(FG)^2}{(DC)^2} = \dfrac{(12)^2}{(15)^2} = \dfrac{16}{25}$.

Thus, $\dfrac{\left(\frac{1}{2}\right)(FG)(EJ)}{\left(\frac{1}{2}\right)(DC)(EH)} = \dfrac{\left(\frac{1}{2}\right)(12)(EJ)}{\left(\frac{1}{2}\right)(15)(EJ + 2)} = \dfrac{16}{25}$ (Formula #5a).

Therefore, $EJ = 8$, and the area of $\triangle EFG = 48$.

Challenge *Draw $\overline{GL} \parallel \overline{ED}$ and find the ratio of the area of $\triangle$GLC to the area of $\triangle$EDC.*

ANSWER: 1:25

5-3 *The distance from a point A to a line $\overleftrightarrow{BC}$ is 3. Two lines l and l′, parallel to $\overleftrightarrow{BC}$, divide $\triangle$ABC into three parts of equal area, as shown in Fig. S5-3. Find the distance between l and l′.*

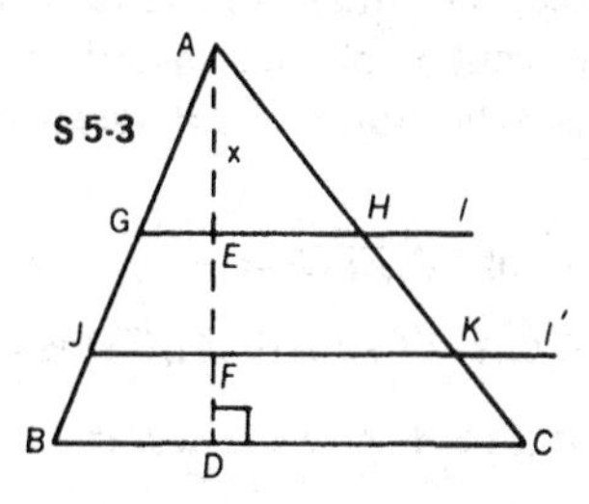

Line l meets $\overline{AB}$ and $\overline{AC}$ at G and H, and line l' meets $\overline{AB}$ and $\overline{AC}$ at J and K. Let $AE = x$.

$\triangle AGH \sim \triangle AJK \sim \triangle ABC$ (#49)

Since l and l' cut off three equal areas,

the area of $\triangle AGH = \frac{1}{2}$ the area of $\triangle AJK$,

and the area of $\triangle AGH = \frac{1}{3}$ the area of $\triangle ABC$.

Since the ratio of the areas is $\triangle AGH{:}\triangle AJK = 1{:}2$,

the ratio of the corresponding altitudes is $AE{:}AF = 1{:}\sqrt{2}$.

Similarly, another ratio of the areas is $\triangle AGH:\triangle ABC = 1:3$. The ratio of the corresponding altitudes is $AE:AD = 1:\sqrt{3}$. Since $AE = x$, $AD = x\sqrt{3}$. However, $AD = 3$, so $x\sqrt{3} = 3$, or $x = \sqrt{3}$.
Similarly, $AF = x\sqrt{2} = \sqrt{6}$. Since $EF = AF - AE$, $EF = \sqrt{6} - \sqrt{3}$.

5-4 *Find the ratio between the area of a square inscribed in a circle, and an equilateral triangle circumscribed about the same circle (Fig. S5-4).*

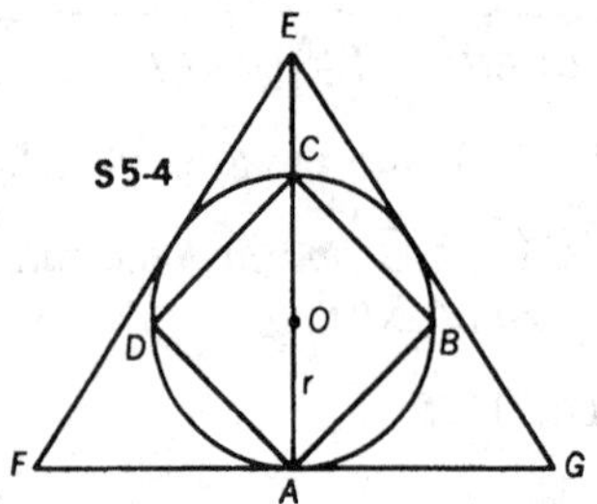

In order to compare the areas of the square and the equilateral triangle we must represent their areas in terms of a common unit, in this instance, the square of the radius r of circle O.

Since the center of the inscribed circle of an equilateral triangle is also the point of intersection of the medians, $EA = 3r$ (#29).

The area of $\triangle EFG = \frac{(3r)^2\sqrt{3}}{3} = 3r^2\sqrt{3}$ (Formula #5f).
Since the diagonal of square $ABCD$ is equal to $2r$,
the area of square $ABCD = \frac{1}{2}(2r)^2 = 2r^2$ (Formula #4b).

Therefore, the ratio of the area of square $ABCD$ to the area of equilateral triangle EFG is

$\frac{2r^2}{3r^2\sqrt{3}} = \frac{2}{3\sqrt{3}} = \frac{2\sqrt{3}}{9}$, approximately 7:18.

Challenge 1 *Using a similar procedure, find the ratio between the area of a square circumscribed about a circle, and an equilateral triangle inscribed in the same circle.*

ANSWER: $\frac{16\sqrt{3}}{9}$

Challenge 2 *Let* D *represent the difference in areas between the circumscribed triangle and the inscribed square. Let* K *represent the area of the circle. Is the ratio* D:K *greater than one, equal to one, or less than one?*

ANSWER: Slightly greater than one

Challenge 3 *Let* D *represent the difference in areas between the circumscribed square and the circle. Let* T *represent the area of the inscribed equilateral triangle. Find the ratio* D:T.

ANSWER: Approximately 2:3

5-5 *A circle* O *is tangent to the hypotenuse* $\overline{BC}$ *of isosceles right* $\triangle ABC$. $\overline{AB}$ *and* $\overline{AC}$ *are extended and are tangent to circle* O *at* E *and* F, *respectively, as shown in Fig. S5-5. The area of the triangle is* X^2. *Find the area of the circle.*

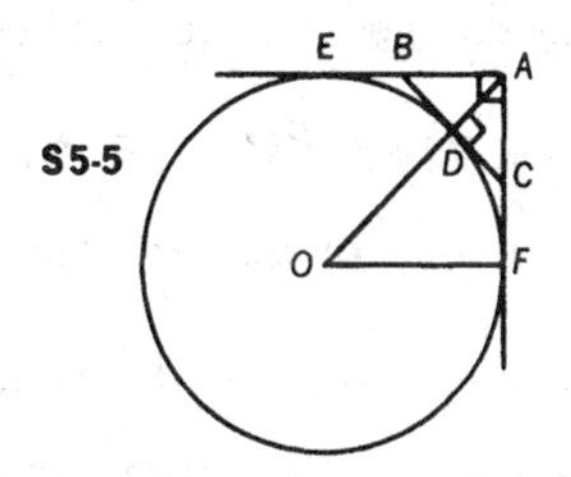

$\overline{OD}$ extended will pass through A (#18).

Since the area of isosceles right $\triangle ABC = X^2$, $AB = AC = X\sqrt{2}$ (Formula #5a).

$BC = 2X$ (#55a). Since $m\angle OAF = 45$, $\triangle ADC$ is also an isosceles right triangle, and $AD = DC = X$.

$\triangle ADC \sim \triangle AFO$ (#48), and $\dfrac{DC}{OF} = \dfrac{AC}{OA}$.

Let radii OF and OD equal r.

Then $\dfrac{X}{r} = \dfrac{X\sqrt{2}}{r + X}$, and $r = X(\sqrt{2} + 1)$.

Hence, the area of the circle $= \pi X^2(3 + 2\sqrt{2})$ (Formula #10).

Challenge *Find the area of trapezoid* EBCF.

ANSWER: $X^2\left(\sqrt{2} + \frac{1}{2}\right)$

5-6 *In Fig. S5-6a,* $\overline{PQ}$ *is the perpendicular bisector of* $\overline{AD}$, $\overline{AB} \perp \overline{BC}$, *and* $\overline{DC} \perp \overline{BC}$. *If* AB = 9, BC = 8, *and* DC = 7, *find the area of quadrilateral* APQB.

METHOD I: To find the area of $APQB$ we must find the sum of the areas of $\triangle ABQ$ and $\triangle PAQ$. Let $BQ = x$.
By the Pythagorean Theorem

$$9^2 + x^2 = AQ^2, \text{ and } 7^2 + (8 - x)^2 = QD^2.$$

But $AQ = QD$ (#18);
therefore, $81 + x^2 = 49 + 64 - 16x + x^2$, and $x = 2$.
Thus, $AQ = \sqrt{85}$.
Draw $\overline{ED} \perp \overline{AB}$. Since $EDBC$ is a rectangle, $DC = EB = 7$, and $AE = 2$.
In $\triangle AED$, $(AE)^2 + (ED)^2 = (AD)^2$, and $AD = 2\sqrt{17}$.
Since $AP = \sqrt{17}$ and $AQ = \sqrt{85}$, we can now find PQ by applying the Pythagorean Theorem to $\triangle APQ$.

$$(\sqrt{85})^2 - (\sqrt{17})^2 = PQ^2, \text{ and } PQ = 2\sqrt{17}.$$

We may now find the area of quadrilateral $APQB$ by adding.

$$\text{The area of } \triangle ABQ = \frac{1}{2}(9)(2) = 9. \quad \text{(Formula \#5a)}$$
$$\text{The area of } \triangle APQ = \frac{1}{2}(\sqrt{17})(2\sqrt{17}) = 17.$$

Therefore, the area of quadrilateral $APQB = 26$.

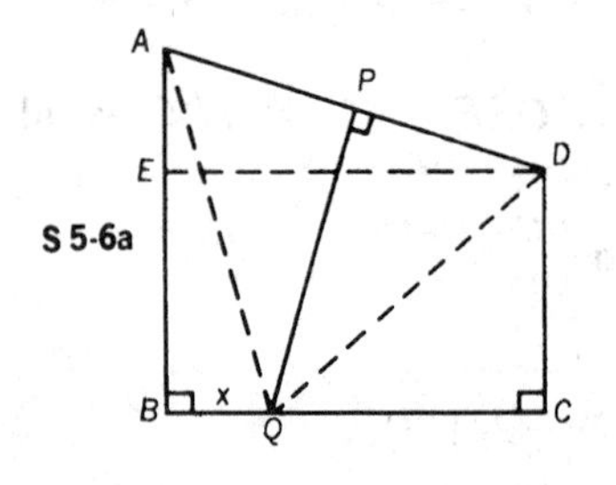

S 5-6a

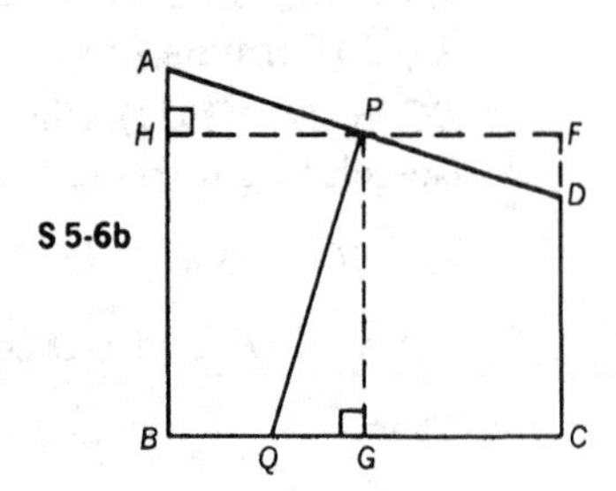

S 5-6b

METHOD II: Draw $\overline{HPF} \parallel \overline{BC}$ (Fig. S5-6b). Then $\triangle APH \cong \triangle DPF$. Since $HF = 8$, $HP = PF = 4$.
Draw $\overline{PG} \perp \overline{BC}$. Since $\overline{PG}$ is the median of trapezoid $ADCB$,

$PG = \frac{1}{2}(AB + DC) = 8.$ Thus, $AH = FD = 1$.

In Method I we found $BQ = 2$; therefore, since $BG = 4$, $QG = 2$. The area of quadrilateral $APQB$ is area of rectangle $HPGB$ − area of $\triangle PGQ$ + area of $\triangle APH$

$= (4)(8) - \left(\frac{1}{2}\right)(2)(8) + \left(\frac{1}{2}\right)(1)(4) = 26.$

5-7 *A triangle has sides that measure 13, 14, and 15. A line perpendicular to the side of measure 14 divides the interior of the triangle into two regions of equal area (Fig. S5-7). Find the measure of the segment of the perpendicular that lies within the triangle.*

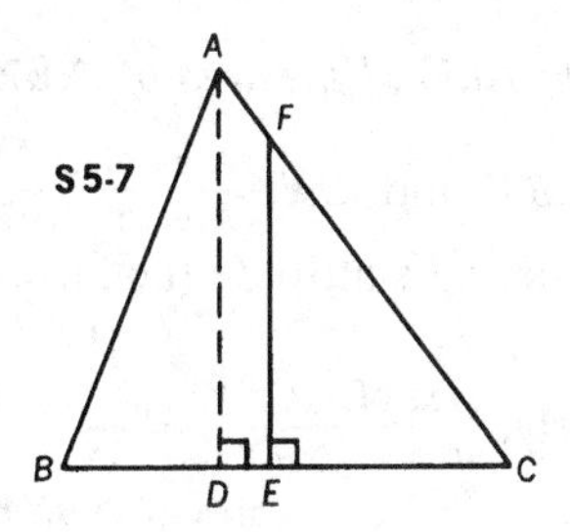

In $\triangle ABC$, $AB = 13$, $AC = 15$, and $BC = 14$; therefore, $AD = 12$ (#55e), $BD = 5$, $DC = 9$.

Since $\overline{FE} \parallel \overline{AD}$ (#9), $\triangle FEC \sim \triangle ADC$ (#49), and $\frac{FE}{EC} = \frac{AD}{DC} = \frac{4}{3}$.

It follows that $EC = \frac{3(FE)}{4}$.

Now the area of $\triangle ABC = \frac{1}{2}(14)(12) = 84$ (Formula #5a).

The area of right $\triangle FEC$ is to be $\frac{1}{2}$ the area of $\triangle ABC$, or 42.

Therefore, the area of right $\triangle FEC = \frac{1}{2}(FE)(EC) = 42$.

Substituting for EC,

$$42 = \frac{1}{2}(FE)\left(\frac{3(FE)}{4}\right), \text{ and } FE = 4\sqrt{7}.$$

Challenge *Find the area of trapezoid* ADEF.

ANSWER: 12

5-8 *Given* $\triangle ABC$ *with* $AB = 20$, $AC = 22\frac{1}{2}$, *and* $BC = 27$. *Points* X *and* Y *are taken on* $\overline{AB}$ *and* $\overline{AC}$, *respectively, so that* $AX = AY$ *(Fig. S5-8). If the area of* $\triangle AXY = \frac{1}{2}$ *area of* $\triangle ABC$, *find* AX.

The area of $\triangle ABC = \frac{1}{2}(AB)(AC)\sin A$ (Formula #5b)
$= \frac{1}{2}(20)\left(22\frac{1}{2}\right)\sin A$
$= (225)\sin A.$

However, the area of $\triangle AXY = \frac{1}{2}(AX)(AY)\sin A.$
Since $AX = AY$, the area of $\triangle AXY = \frac{1}{2}(AX)^2 \sin A.$
Since the area of $\triangle AXY = \frac{1}{2}$ the area of $\triangle ABC$,
$\frac{1}{2}(AX)^2 \sin A = \frac{1}{2}[(225)\sin A]$, and $AX = 15.$

Challenge *Find the ratio of the area of* $\triangle$BXY *to that of* $\triangle$CXY.

Draw BY and CX. $\frac{\text{Area of } \triangle BXY}{\text{Area of } \triangle AXY} = \frac{5}{15} = \frac{1}{3}$, since they share the same altitude (i.e., from Y to AB).

Similarly, $\frac{\text{area of } \triangle CXY}{\text{area of } \triangle AXY} = \frac{7\frac{1}{2}}{15} = \frac{1}{2}.$

Therefore, the ratio $\frac{\text{area of } \triangle BXY}{\text{area of } \triangle CXY} = \frac{2}{3}.$

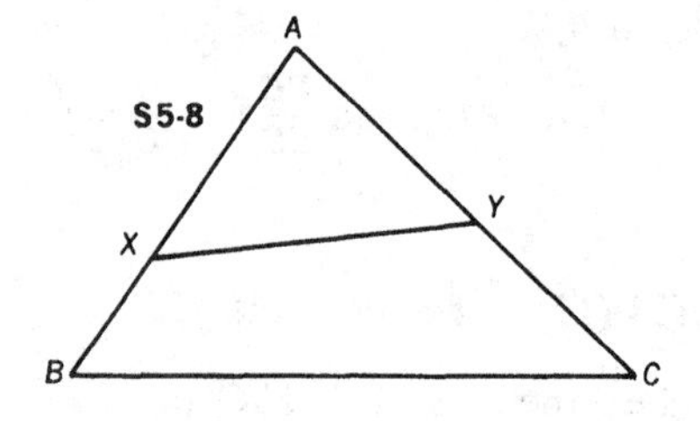

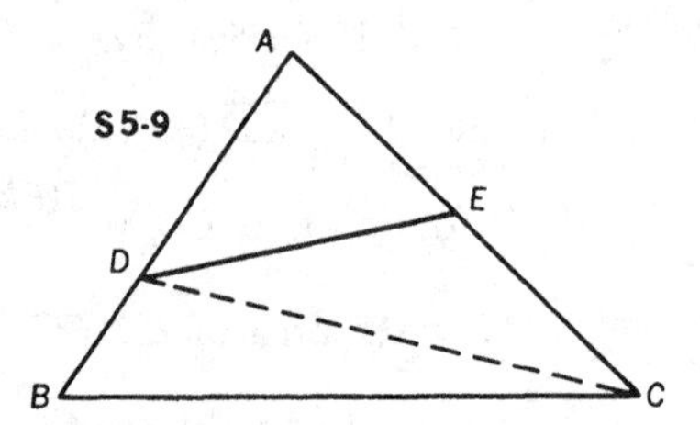

5-9 *In* $\triangle$ABC, AB = 7, AC = 9. *On* $\overline{\text{AB}}$, *point* D *is taken so that* BD = 3. $\overline{\text{DE}}$ *is drawn cutting* $\overline{\text{AC}}$ *in* E *so that quadrilateral* BCED *has* $\frac{5}{7}$ *the area of* $\triangle$ABC. *Find* CE.

In Fig. S5-9, $AD = 4$ while $AB = 7$.
If two triangles share the same altitude, then the ratio of their areas equals the ratio of their bases.
Since $\triangle ADC$ and $\triangle ABC$ share the same altitude (from C to $\overleftrightarrow{AB}$), the area of $\triangle ADC = \frac{4}{7}$ area of $\triangle ABC$.
Since the area of quadrilateral $DECB = \frac{5}{7}$ area of $\triangle ABC$, the area of $\triangle DAE = \frac{2}{7}$ area of $\triangle ABC$.

Thus, the ratio of the areas of $\triangle DAE$ and $\triangle ADC$ equals 1:2. Both triangles DAE and ADC share the same altitude (from D to $\overleftrightarrow{AC}$); therefore, their bases are also in the ratio 1:2.

Thus, $\frac{AE}{AC} = \frac{1}{2}$.

Since $AC = 9$, $AE = 4\frac{1}{2}$, as does CE.

5-10 *An isosceles triangle has a base of measure* 4, *and sides measuring* 3. *A line drawn through the base and one side (but not through any vertex) divides both the perimeter and the area in half, as shown in Fig. S5-10. Find the measures of the segments of the base defined by this line.*

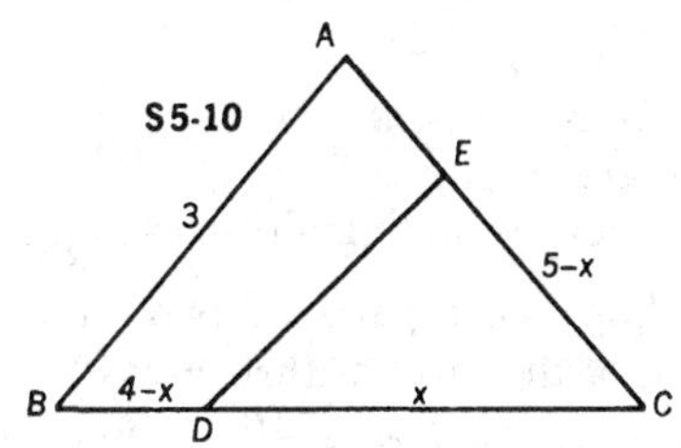

$AB = AC = 3$, and $BC = 4$. If $DC = x$, then $BD = 4 - x$. Since the perimeter of $ABC = 10$, $EC + DC$ must be one-half the perimeter, or 5. Thus, $EC = 5 - x$.

Now the area of $\triangle EDC = \frac{1}{2}(x)(5 - x)\sin C$ (Formula #5b), and the area of $\triangle ABC = \frac{1}{2}(4)(3)\sin C$.

Since the area of $\triangle EDC$ is one-half the area of $\triangle ABC$,

$$\frac{1}{2}(x)(5 - x)\sin C = \frac{1}{2}\left[\frac{1}{2}(4)(3)\sin C\right], \text{ and}$$

$$5x - x^2 = 6.$$

Solving the quadratic equation $x^2 - 5x + 6 = 0$, we find its roots to be $x = 2$ and $x = 3$.

If $x = 2$, then $EC = 3 = AC$, but this cannot be since $\overline{DE}$ may not pass through a vertex.

Therefore, $x = 3$. Thus, $\overline{BC}$ is divided so that $BD = 1$, and $DC = 3$.

Challenge *Find the measure of* DE.

ANSWER: $\sqrt{5}$

5-11 *Through D, a point on base $\overline{BC}$ of $\triangle ABC$, $\overline{DE}$ and $\overline{DF}$ are drawn parallel to sides $\overline{AB}$ and $\overline{AC}$, respectively, meeting $\overline{AC}$ at E and $\overline{AB}$ at F (Fig. S5-11). If the area of $\triangle EDC$ is four times the area of $\triangle BFD$, what is the ratio of the area of $\triangle AFE$ to the area of $\triangle ABC$?*

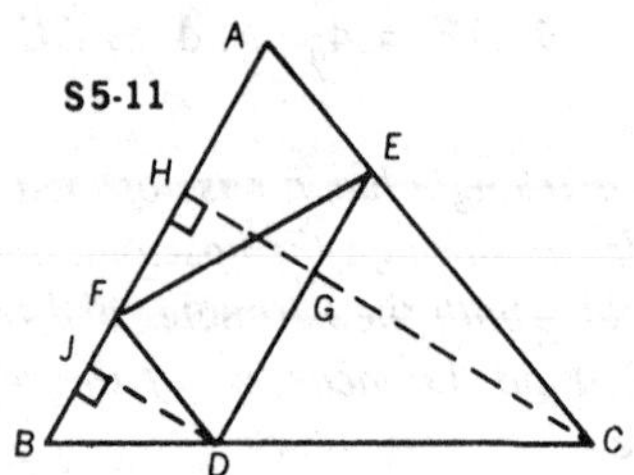

METHOD I: $\dfrac{\text{Area of } \triangle FDB}{\text{Area of } \triangle ECD} = \dfrac{1}{4} \cdot$ $\triangle FDB \sim \triangle ECD$ (#48), and the ratio of the corresponding altitudes is $\dfrac{JD}{CG} = \dfrac{1}{2} \cdot$ (The ratio of the corresponding linear parts of two similar polygons equals the square root of the ratio of their areas.)

$JD = HG$ (#20); therefore, $\dfrac{JD}{GC} = \dfrac{HG}{GC} = \dfrac{1}{2}$,

and $\dfrac{HG + GC}{GC} = \dfrac{1+2}{2}$, or $\dfrac{HC}{GC} = \dfrac{3}{2} \cdot$

Thus, the ratio of $\dfrac{\text{area of } \triangle ABC}{\text{area of } \triangle EDC} = \dfrac{9}{4} \cdot$ (The square of the ratio of corresponding linear parts of two similar polygons equals the ratio of the areas.)

The ratio of area of $\triangle ABC$ to area of $\triangle EDC$ to area of $\triangle FBD$ = 9:4:1.

Area of $\square AEDF$ = area of $\triangle EDC$, and the ratio of area of $\square AEDF$ to area of $\triangle ABC$ = 4:9.

But since area of $\triangle AFE = \dfrac{1}{2}$ area of $\square AEDF$, area of $\triangle AFE$: area of $\triangle ABC$ = 2:9.

METHOD II: Since $\overline{FD} \parallel \overline{AC}$, $\triangle BFD \sim \triangle BAC$ (#49), and since $\overline{ED} \parallel \overline{AB}$, $\triangle DEC \sim \triangle BAC$ (#49).

Therefore, $\triangle BFD \sim \triangle DEC$.

Since the ratio of the areas of $\triangle BFD$ to $\triangle DEC$ is 1:4, the ratio of the corresponding sides is 1:2.

Let $BF = x$, and $FD = y$; then $ED = 2x$ and $EC = 2y$.

Since $AEDF$ is a parallelogram (#21a),
$FD = AE = y$, and $ED = AF = 2x$.

Now, area of $\triangle AFE = \frac{1}{2}(2x)(y)\sin A$, (I)

and area of $\triangle ABC = \frac{1}{2}(3x)(3y)\sin A$ (Formula #5b). (II)

Thus, the ratio of the area of $\triangle AFE$ to the area of $\triangle ABC = \frac{2}{9}$ (from (I) and (II)).

The problem may easily be solved by designating triangle ABC as an equilateral triangle. This approach is left to the student.

5-12 *Two circles, each of which passes through the center of the other, intersect at points* M *and* N. *A line from* M *intersects the circles at* K *and* L, *as illustrated in Fig.* S5-12. *If* KL = 6 *compute the area of* $\triangle$KLN.

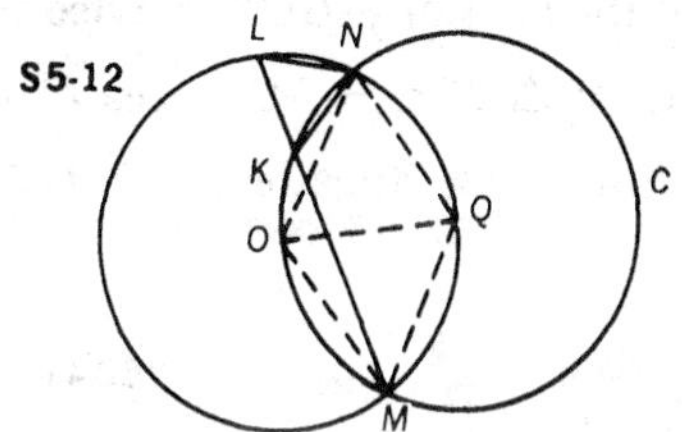

Draw the line of centers $\overline{OQ}$. Then draw $\overline{ON}$, $\overline{OM}$, $\overline{QN}$, and $\overline{QM}$. Since $ON = OQ = QN = OM = QM$, $\triangle NOQ$ and $\triangle MOQ$ are each equilateral.

$m\angle NQO = m\angle MQO = 60$, so $m\angle NQM = 120$.

Therefore, we know that in circle O, $m\angle NLM = 60$ (#36).

Since $m\widehat{NCM} = 240$, in circle Q, $m\angle NKM = 120$ (#36).

Since $\angle NKL$ is supplementary to $\angle NKM$, $m\angle NKL = 60$.

Thus, $\triangle LKN$ is equilateral (#6).

The area of $\triangle KLN = \frac{(KL)^2\sqrt{3}}{4} = 9\sqrt{3}$ (Formula #5e).

Challenge *If* r *is the measure of the radius of each circle, find the least value and the greatest value of the area of* $\triangle$KLN.

ANSWER: The least value is zero, and the greatest value is $\frac{3r^2\sqrt{3}}{4}$.

5-13 *Find the area of a triangle whose medians have measures* 39, 42, 45 (*Fig.* S5-13).

Let $AD = 39$, $CE = 42$, $BF = 45$; then $CG = 28$, $GE = 14$, $AG = 26$, $GD = 13$, $BG = 30$, and $GF = 15$ (#29). Now extend $\overline{AD}$ to K so that $GD = DK$. Quadrilateral $CGBK$ is a parallelogram (#21f).

$CK = BG = 30$ (#21b). $GD = DK = 13$; therefore, $GK = 26$. We may now find the area of $\triangle GCK$ by applying Hero's formula (Formula #5c), or by noting that the altitude to side $\overline{GC}$ must equal 24 (#55e).

In either case, the area of $\triangle GCK = 336$.

Consider the area of $\triangle GCD$ which equals $\frac{1}{3}$ the area of $\triangle ACD$ (#29). However, the area of $\triangle ACD = \frac{1}{2}$ the area of $\triangle ABC$. Therefore, the area of $\triangle GCD$ equals $\frac{1}{6}$ the area of $\triangle ABC$. But the area of $\triangle GCK$ is twice the area of $\triangle GCD$, and thus the area of $\triangle GCK = \frac{1}{3}$ the area of $\triangle ABC$. Then, since the area of $\triangle CGK = 336$, the area of $\triangle ABC = 3(336) = 1008$.

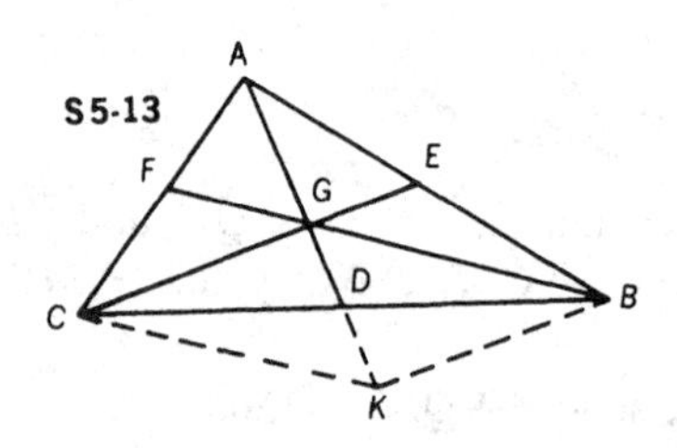

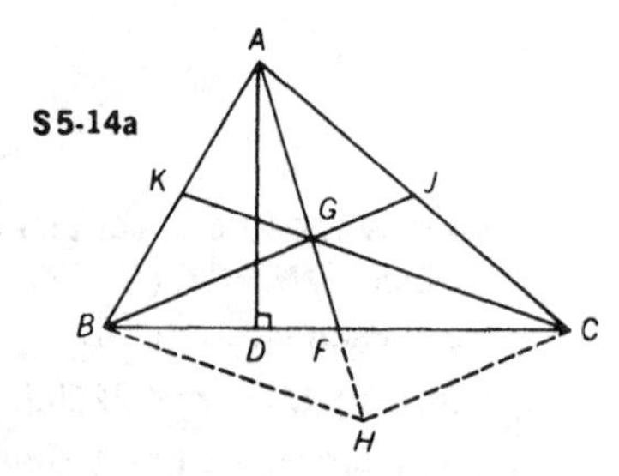

5-14 *The measures of the sides of a triangle are* 13, 14, *and* 15. *A second triangle is formed in which the measures of the three sides are the same as the measures of the medians of the first triangle (Fig.* S5-14*a). What is the area of the second triangle?*

Let $AB = 13$, $AC = 15$, and $BC = 14$.

In $\triangle ABC$, the altitude to side $\overline{BC}$ equals 12 (#55e).

$$\text{Therefore, the area of } \triangle ABC = \frac{1}{2}(BC)(AD)$$
$$= \frac{1}{2}(14)(12) = 84.$$

Another possible method to find the area of $\triangle ABC$ would be to apply Hero's formula (Formula #5c) to obtain $\sqrt{(21)(6)(7)(8)}$. Breaking the expression down into prime factors we have

$\sqrt{7 \cdot 3 \cdot 3 \cdot 2 \cdot 7 \cdot 2 \cdot 2 \cdot 2} = 7 \cdot 3 \cdot 2 \cdot 2 = 84.$

Let us now consider $\triangle ABC$ and its medians, $\overline{AF}$, $\overline{BJ}$, and $\overline{CK}$. Extend $\overline{GF}$ its own length to H.

$GCHB$ is a parallelogram (#21f).

Now consider $\triangle GHC$. $HC = BG = \frac{2}{3}(BJ)$ (#21b, #29).

$GC = \frac{2}{3}CK$, and $GF = \frac{1}{3}AF$ (#29); but $GH = \frac{2}{3}AF$.

Since the measure of each side of $\triangle GHC = \frac{2}{3}$ times the measure of each side of the triangle formed by the lengths of the medians, $\triangle GHC \sim \triangle$ of medians. The ratio of their areas is the square of their ratio of similitude, or $\frac{4}{9}$.

We must now find the area of $\triangle GHC$.

Since $\overline{AF}$ is a median, area of $\triangle AFC = \frac{1}{2}$ area of $\triangle ABC = 42$.

The area of $\triangle GCF = \frac{1}{3}$ area of $\triangle AFC = 14$.

However, the area of $\triangle GHC$ = twice the area of $\triangle GCF = 28$.

Since the ratio of $\dfrac{\text{area of } \triangle GHC}{\text{area of triangle of medians}} = \dfrac{4}{9}$,

$$\frac{28}{\text{area of triangle of medians}} = \frac{4}{9},$$

and the area of triangle of medians $= 63$.

Challenge 1 *Show that* $K(m) = \frac{3}{4}K$ *where* K *represents the area of* $\triangle ABC$, *and* $K(m)$ *represents the area of a triangle with sides* m_a, m_b, m_c, *the medians of* $\triangle ABC$. *(See Fig. S5-14b.)*

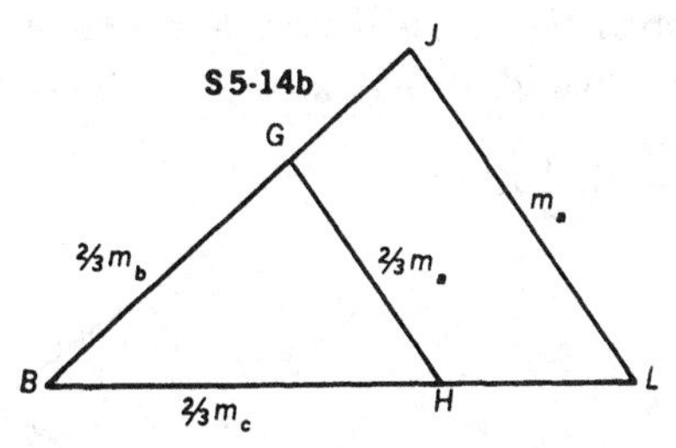

Let $K\left(\frac{2}{3}m\right)$ represent the area of $\triangle BGH$.

We have already shown that $K\left(\frac{2}{3}m\right) = \frac{1}{3}K$.

$$\frac{K(m)}{K\left(\frac{2}{3}m\right)} = \frac{m_a^2}{\left(\frac{2}{3}m_a\right)^2} = \frac{9}{4}$$

Therefore, $K(m) = \frac{9}{4}K\left(\frac{2}{3}m\right) = \left(\frac{9}{4}\right)\left(\frac{1}{3}K\right) = \frac{3}{4}K$.

5-15 *Find the area of a triangle formed by joining the midpoints of the sides of a triangle whose medians have measures* 15, 15, *and* 18 *(Fig. S5-15).*

METHOD I: $AE = 18$, $BD = CF = 15$

In $\triangle ABC$, $\overline{FD} \parallel \overline{BC}$ (#26), and in $\triangle AEC$, $AH = HE$ (#25). Since $AE = 18$, $HE = 9$. Since $GE = 6$ (#29), $GH = 3$. $GD = 5$ (#29). Since $BD = CF$, $\triangle ABC$ is isosceles, and median $\overline{AE} \perp \overline{BC}$, so $\overline{AE} \perp \overline{FD}$ (#10).

Thus, in right $\triangle HGD$, $HD = 4$ (#55).

Since $FH = HD = 4$, $FD = 8$.

Hence, the area of $\triangle FDE = \frac{1}{2}(FD)(HE)$

$= \frac{1}{2}(8)(9) = 36.$

METHOD II: Since the area of the triangle formed by the three medians is $\frac{3}{4}$ the area of $\triangle ABC$ (see Problem 5-14), and the area of the triangle formed by the three medians is equal to 108, the area of $\triangle ABC$ is $\frac{4}{3}(108) = 144$.

Since the area of $\triangle AFD$ = area of $\triangle BFE$ = area of $\triangle CDE$ = area of $\triangle FDE$, area of $\triangle FDE = \frac{1}{4}(144) = 36$.

Challenge *Express the required area in terms of* K(m), *where* K(m) *is the area of the triangle formed from the medians.*

ANSWER: $\frac{1}{3}K(m)$

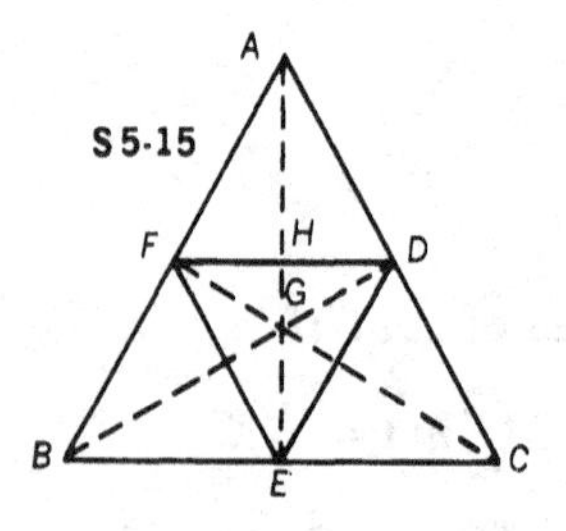

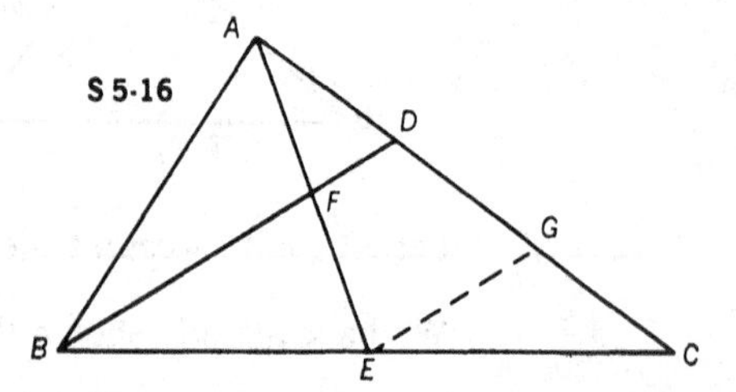

5-16 *In* $\triangle$ABC, E *is the midpoint of* $\overline{BC}$, *while* F *is the midpoint of* $\overline{AE}$, *and* $\overleftrightarrow{BF}$ *meets* AC *at* D, *as shown in Fig. S5-16. If the area of* $\triangle$ABC = 48, *find the area of* $\triangle$AFD.

The area of $\triangle ABE = \frac{1}{2}$ the area of $\triangle ABC$ (#56).
Similarly, the area of $\triangle ABF = \frac{1}{2}$ the area of $\triangle ABE$ (#56).
Therefore, since the area of $\triangle ABC = 48$,
the area of $\triangle ABF = 12$.
Draw $\overline{EG} \parallel \overline{BD}$. In $\triangle EAG$, $AD = DG$ (#25).
Similarly, in $\triangle BCD$, $DG = GC$ (#25).
Therefore, $AD = DG = GC$, or $AD = \frac{1}{3}(AC)$.

Since $\triangle ABD$ and $\triangle ABC$ share the same altitude (from B to $\overleftrightarrow{AC}$) and their bases are in the ratio 1:3,
the area of $\triangle ABD = \frac{1}{3}$ area of $\triangle ABC = 16$.
Thus, the area of $\triangle AFD$ = area of $\triangle ABD$ − area of $\triangle ABF = 4$.

Challenge 2 *Change* AF $= \frac{1}{2}$ AE *to* AF $= \frac{1}{3}$ AE, *and find a general solution.*

ANSWER: The area of $\triangle AFD = \frac{1}{30}$ the area of $\triangle ABC$.

5-17 *In* $\triangle$ABC, D *is the midpoint of side* $\overline{\text{BC}}$, E *is the midpoint of* $\overline{\text{AD}}$, F *is the midpoint of* $\overline{\text{BE}}$, *and* G *is the midpoint of* $\overline{\text{FC}}$. *(See Fig. S5-17.) What part of the area of* $\triangle$ABC *is the area of* $\triangle$EFG?

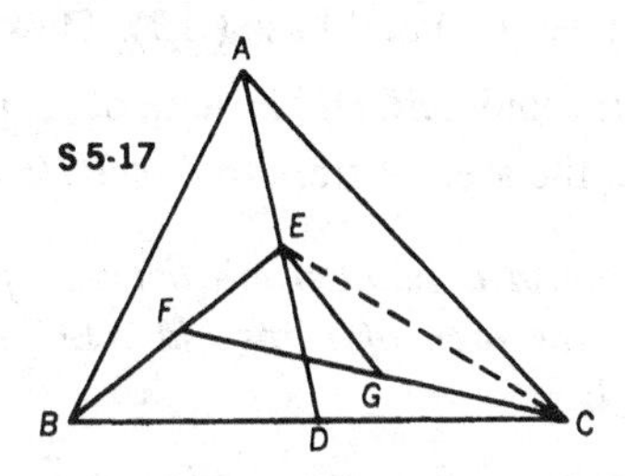

Draw $\overline{EC}$.
Since the altitude of $\triangle BEC$ is $\frac{1}{2}$ the altitude of $\triangle BAC$, and both triangles share the same base, the area of $\triangle BEC = \frac{1}{2}$ area of $\triangle BAC$.
Now, area of $\triangle EFC = \frac{1}{2}$ area of $\triangle BEC$,
and area of $\triangle EGF = \frac{1}{2}$ area of $\triangle EFC$ (#56);
therefore area of $\triangle EGF = \frac{1}{4}$ area of $\triangle BEC$.

Thus, since area of $\triangle BEC = \frac{1}{2}$ area of $\triangle ABC$,
we find that area of $\triangle EGF = \frac{1}{8}$ area of $\triangle ABC$.

Challenge *Solve the problem if* $BD = \frac{1}{3}BC$, $AE = \frac{1}{3}AD$, $BF = \frac{1}{3}BE$, *and* $GC = \frac{1}{3}FC$.

ANSWER: The area of $\triangle EGF = \frac{8}{27}$ the area of $\triangle BAC$.

5-18 *In trapezoid* ABCD *with upper base* $\overline{AD}$, *lower base* $\overline{BC}$, *and legs* $\overline{AB}$ *and* $\overline{CD}$, E *is the midpoint of* $\overline{CD}$ *(Fig. S5-18). A perpendicular,* $\overline{EF}$, *is drawn to* $\overline{BA}$ *(extend* $\overline{BA}$ *if necessary). If* $EF = 24$ *and* $AB = 30$, *find the area of the trapezoid. (Note that the diagram is not drawn to scale.)*

Draw $\overline{AE}$ and $\overline{BE}$. Through E, draw a line parallel to $\overline{AB}$ meeting $\overline{BC}$ at H and $\overline{AD}$, extended at G.
Since $DE = EC$ and $\angle DEG \cong \angle HEC$ (#1) and $\angle DGE \cong \angle CHE$ (#8), $\triangle DEG \cong \triangle CEH$ (A.S.A.).
Since congruent triangles are equal in area, the area of parallelogram $AGHB =$ the area of trapezoid $ABCD$. The area of $\triangle AEB$ is one-half the area of parallelogram $AGHB$, since they share the same altitude ($\overline{EF}$) and base ($\overline{AB}$). Thus, the area of $\triangle AEB = \frac{1}{2}$ area of trapezoid $ABCD$. The area of $\triangle AEB = \frac{1}{2}(30)(24) = 360$. Therefore, the area of trapezoid $ABCD = 720$.

Challenge *Establish a relationship between points* F, A, *and* B *such that the area of trapezoid* ABCD *is equal to the area of* $\triangle$FBH.

ANSWER: A is the midpoint of $\overline{BF}$.

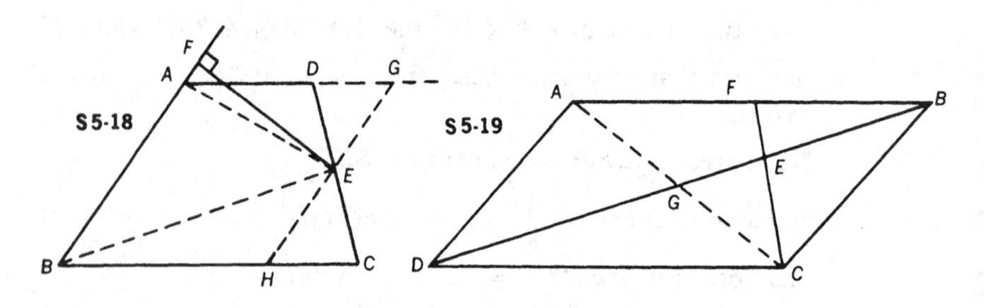

5-19 *In* $\square ABCD$, *a line from* C *cuts diagonal* $\overline{BD}$ *in* E *and* $\overline{AB}$ *in* F, *as shown in Fig.* S5-19. *If* F *is the midpoint of* $\overline{AB}$, *and the area of* $\triangle BEC$ *is* 100, *find the area of quadrilateral* AFED.

Draw $\overline{AC}$ meeting $\overline{DB}$ at G. In $\triangle ABC$, $\overline{BG}$ and $\overline{CF}$ are medians; therefore, $FE = \frac{1}{2}(EC)$ (#29).
If the area of $\triangle BEC = 100$, then the area of $\triangle EFB = 50$, since they share the same altitude.
$\triangle ABD$ and $\triangle FBC$ have equal altitudes (#20), but $AB = 2(FB)$ Therefore, the area of $\triangle ABD$ is twice the area of $\triangle FBC$. Since the area of $\triangle FBC = 150$, the area of $\triangle ABD = 300$. But the area of quadrilateral $AFED$ = the area of $\triangle ABD$ – the area of $\triangle FBE$; therefore, the area of quadrilateral $AFED = 300 - 50 = 250$.

Challenge *Find the area of* $\triangle GEC$.

ANSWER: 50

5-20 P *is any point on side* $\overline{AB}$ *of* $\square ABCD$. $\overline{CP}$ *is drawn through* P *meeting* $\overline{DA}$ *extended at* Q, *as illustrated in Fig.* S5-20. *Prove that the area of* $\triangle DPA$ *is equal to the area of* $\triangle QPB$.

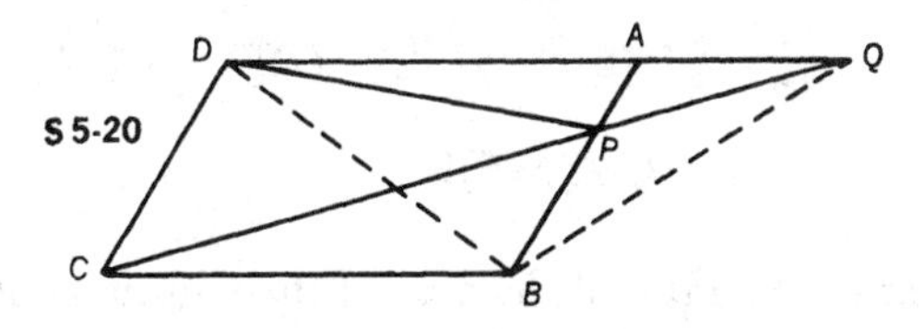

Since $\triangle DPC$ and $\square ABCD$ have the same altitude and share the same base, $\overline{DC}$, the area of $\triangle DPC = \frac{1}{2}$ area of parallelogram $ABCD$.
The remaining half of the area of the parallelogram is equal to the sum of the areas of $\triangle DAP$ and $\triangle PBC$.
However, the area of $\triangle DBC$ is also one-half of the area of the parallelogram.
The area of $\triangle CQB$ = the area of $\triangle CDB$. (They share the same base, $\overline{CB}$, and have equal altitudes since $\overline{DQ} \parallel \overline{CB}$.)
Thus, the area of $\triangle CQB$ equals one-half the area of the parallelogram.

Therefore, the area of $\triangle DAP$ + the area of $\triangle PBC$ = the area of $\triangle CQB$.

Subtracting the area of $\triangle PBC$ from both sides, we find the area of $\triangle DAP$ = the area of $\triangle PQB$.

5-21 *$\overline{RS}$ is the diameter of a semicircle. Two smaller semicircles, $\widehat{RT}$ and $\widehat{TS}$, are drawn on $\overline{RS}$, and their common internal tangent $\overline{AT}$ intersects the large semicircle at A, as shown in Fig. S5-21. Find the ratio of the area of a semicircle with radius $\overline{AT}$ to the area of the shaded region.*

Draw $\overline{RA}$ and $\overline{SA}$. In right $\triangle RAS$ (#36), $\overline{AT} \perp \overline{RS}$ (#32a).

Therefore, $\frac{RT}{AT} = \frac{AT}{ST}$, or $(AT)^2 = (RT)(ST)$ (#51a).

The area of the semicircle, radius

$$\overline{AT} = \frac{\pi}{2}(AT)^2 = \frac{\pi}{2}(RT)(ST).$$

The area of the shaded region

$$= \frac{\pi}{2}\left[\left(\frac{1}{2}RS\right)^2 - \left(\frac{1}{2}RT\right)^2 - \left(\frac{1}{2}ST\right)^2\right]$$

$$= \frac{\pi}{8}[(RS)^2 - (RT)^2 - (ST)^2]$$

$$= \frac{\pi}{8}[(RT + ST)^2 - (RT)^2 - (ST)^2]$$

$$= \frac{\pi}{4}[(RT)(ST)].$$

Therefore, the ratio of the area of the semicircle of radius $\overline{AT}$ to the area of the shaded region is

$$\frac{\frac{\pi}{2}(RT)(ST)}{\frac{\pi}{4}(RT)(ST)} = \frac{2}{1}.$$

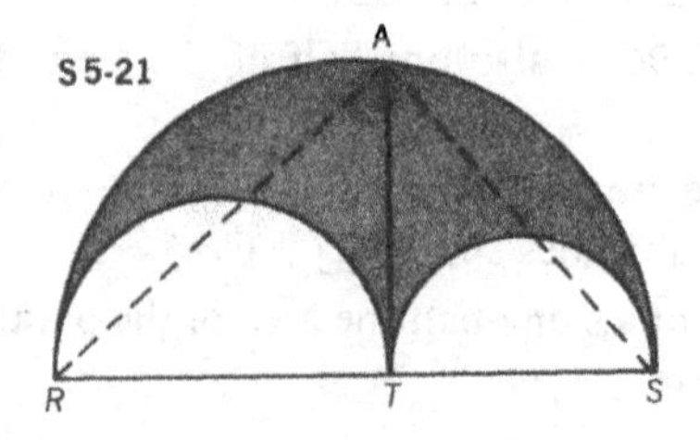

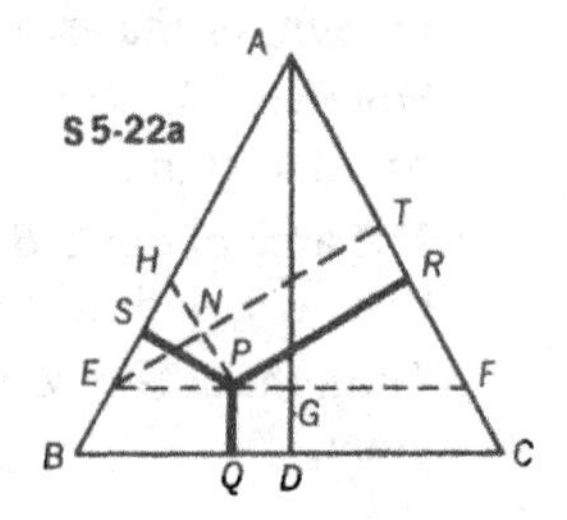

5-22 *Prove that from any point inside an equilateral triangle, the sum of the measures of the distances to the sides of the triangle is constant. (See Fig. S5-22a.)*

METHOD I: In equilateral $\triangle ABC, \overline{PR} \perp \overline{AC}, \overline{PQ} \perp \overline{BC}, \overline{PS} \perp \overline{AB}$, and $\overline{AD} \perp \overline{BC}$.

Draw a line through P parallel to $\overline{BC}$ meeting $\overline{AD}$, $\overline{AB}$, and $\overline{AC}$ at G, E, and F, respectively.

$PQ = GD$ (#20)

Draw $\overline{ET} \perp \overline{AC}$. Since $\triangle AEF$ is equilateral, $AG = ET$ (all the altitudes of an equilateral triangle are congruent).

Draw $\overline{PH} \parallel \overline{AC}$ meeting $\overline{ET}$ at N. $NT = PR$ (#20)

Since $\triangle EHP$ is equilateral, altitudes $\overline{PS}$ and $\overline{EN}$ are congruent.

Therefore, we have shown that $PS + PR = ET = AG$.

Since $PQ = GD$, $PS + PR + PQ = AG + GD = AD$, a constant for the given triangle.

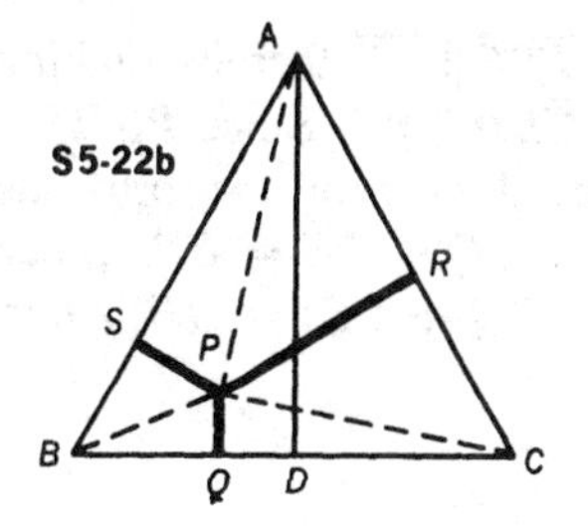

METHOD II: In equilateral $\triangle ABC$, $\overline{PR} \perp \overline{AC}$, $\overline{PQ} \perp \overline{BC}$, $\overline{PS} \perp \overline{AB}$, and $\overline{AD} \perp \overline{BC}$.

Draw $\overline{PA}$, $\overline{PB}$, and $\overline{PC}$ (Fig. S5-22b).

The area of $\triangle ABC$

$$= \text{area of } \triangle APB + \text{area of } \triangle BPC + \text{area of } \triangle CPA$$

$$= \frac{1}{2}(AB)(PS) + \frac{1}{2}(BC)(PQ) + \frac{1}{2}(AC)(PR). \quad \text{(Formula \#5a)}$$

Since $AB = BC = AC$,

the area of $\triangle ABC = \frac{1}{2}(BC)[PS + PQ + PR]$.

However, the area of $\triangle ABC = \frac{1}{2}(BC)(AD)$;

therefore, $PS + PQ + PR = AD$,

a constant for the given triangle.

Challenge *In equilateral* $\triangle$ABC, *legs* $\overline{\text{AB}}$ *and* $\overline{\text{BC}}$ *are extended through* B *so that an angle is formed that is vertical to* $\angle$ABC. *Point* P *lies within this vertical angle. From* P, *perpendiculars are drawn to sides* $\overline{\text{BC}}$, $\overline{\text{AC}}$, *and* $\overline{\text{AB}}$ *at points* Q, R, *and* S, *respectively. See Fig. S5-22c. Prove that* PR − (PQ + PS) *equals a constant for* $\triangle$ABC.

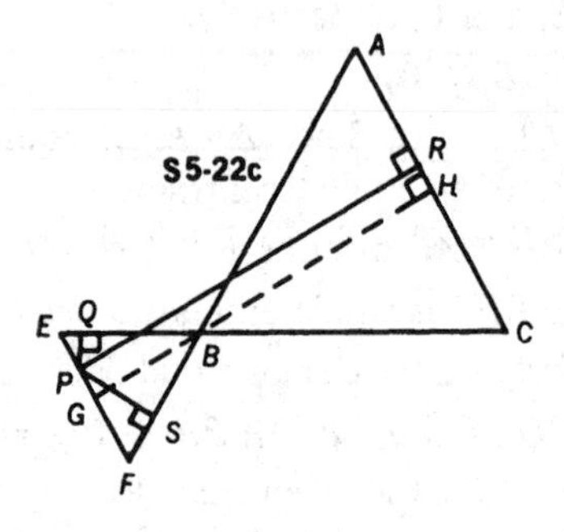

Draw $\overline{EPF} \parallel \overline{AC}$ thereby making $\triangle EBF$ equilateral. Then draw $\overline{GBH} \parallel \overline{PR}$. Since $PGHR$ is a rectangle, $GH = PR$. A special case of the previous problem shows that in $\triangle EBF$, $PQ + PS = GB$. Since $GH - GB = BH$, then $PR - (PQ + PS) = BH$, a constant for $\triangle ABC$.

6. A Geometric Potpourri

6-1 *Heron's Formula is used to find the area of any triangle, given only the measures of the sides of the triangle. Derive this famous formula. The area of any triangle* $= \sqrt{s(s-a)(s-b)(s-c)}$, *where* a, b, c *are measures of the sides of the triangle and* s *is the semiperimeter.*

First inscribe a circle in $\triangle ABC$ and draw the radii $\overline{OD}$, $\overline{OE}$, and $\overline{OF}$ to the points of contact. Then draw $\overline{OB}$, $\overline{OC}$, and $\overline{OA}$. Let a line perpendicular to $\overline{BO}$ at O meet, at point P, the perpendicular to $\overline{BC}$ at C. Extend $\overline{BC}$ to K so that $CK = AD$ (Fig. S6-1).

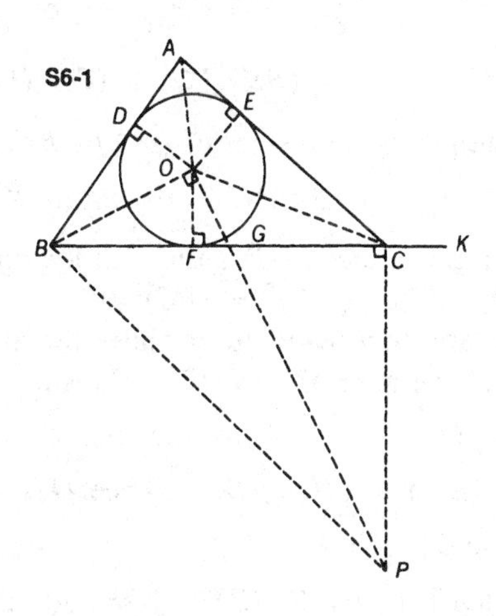

Since $\triangle BOP$ and $\triangle BCP$ are right triangles with $\overline{BP}$ as hypotenuse, it may be said that $\angle BOP$ and $\angle BCP$ are inscribed angles in a circle whose diameter is $\overline{BP}$. Thus, quadrilateral $BOCP$ is cyclic (i.e., may be inscribed in a circle). It follows that $\angle BPC$ is supplementary to $\angle BOC$ (#37).

If we now consider the angles with O as vertex, we note that $\angle DOA \cong \angle AOE$, $\angle COE \cong \angle FOC$, and $\angle BOD \cong \angle BOF$. (This may be proved using congruent triangles.) Therefore, $m\angle BOF + m\angle FOC + m\angle DOA = \frac{1}{2}(360)$, or $\angle DOA$ is supplementary to $\angle BOC$. Thus, $\angle DOA \cong \angle BPC$ because both are supplementary to the same angle. It then follows that right $\triangle DOA \sim$ right $\triangle CPB$ (#48) and

$$\frac{BC}{AD} = \frac{PC}{DO}. \tag{I}$$

Since $\angle OGF \cong \angle PGC$, right $\triangle OGF \sim$ right $\triangle PGC$ (#48) and

$$\frac{GC}{FG} = \frac{PC}{OF}. \tag{II}$$

However $OF = DO$. Therefore, from (I) and (II) it follows that

$$\frac{BC}{AD} = \frac{GC}{FG}. \tag{III}$$

Since $AD = CK$, it follows from (III) that $\frac{BC}{CK} = \frac{GC}{FG}$.

Using a theorem on proportions we get

$$\frac{BC + CK}{CK} = \frac{GC + FG}{FG}, \text{ or } \frac{BK}{CK} = \frac{FC}{FG}.$$

Thus,
$$(BK)(FG) = (CK)(FC). \tag{IV}$$

By multiplying both sides of (IV) by BK, we get

$$(BK)^2(FG) = (BK)(CK)(FC). \tag{V}$$

In right $\triangle BOG$, $\overline{OF}$ is the altitude drawn to the hypotenuse. Thus by (#51a), $(OF)^2 = (FG)(BF)$. (VI)

We are now ready to consider the area of $\triangle ABC$. We may think of the area of $\triangle ABC$ as the sum of the areas of $\triangle AOB$, $\triangle BOC$, and $\triangle AOC$. Thus, the area of $\triangle ABC = \frac{1}{2}(OD)(AB) + \frac{1}{2}(OE)(AC) + \frac{1}{2}(OF)(BC)$. Since $OD = OE = OF$ (the radii of circle O),

$\frac{1}{2}(OF)(AB + AC + BC) = (OF) \cdot$ (semiperimeter of $\triangle ABC$).

Since $BF = BD$, $FC = EC$, and $AD = AE$, $BF + FC + AD =$ half the perimeter of $\triangle ABC$. Since $AD = CK$, $BF + FC + CK = BK$ which equals the semiperimeter of $\triangle ABC$. Hence, the area of $\triangle ABC = (BK)(OF)$.

$$(\text{Area of } \triangle ABC)^2 = (BK)^2(OF)^2.$$
$$(\text{Area of } \triangle ABC)^2 = (BK)^2(FG)(BF). \quad \text{From (VI)}$$
$$(\text{Area of } \triangle ABC)^2 = (BK)(CK)(FC)(BF). \quad \text{From (V)}$$
$$\text{Area of } \triangle ABC = \sqrt{(BK)(CK)(BF)(FC)}.$$

Let s = semiperimeter = BK, $a = BC$, $b = AC$, and $c = AB$. Then $s - a = CK$, $s - b = BF$, and $s - c = FC$. We can now express Heron's Formula for the area of $\triangle ABC$, as it is usually given.

$$\text{Area } \triangle ABC = \sqrt{s(s-a)(s-b)(s-c)}$$

Challenge *Find the area of a triangle whose sides measure* 6, $\sqrt{2}$, $\sqrt{50}$.

$$s = \frac{6 + \sqrt{2} + 5\sqrt{2}}{2} = 3 + 3\sqrt{2}$$

$$K = \sqrt{(3 + 3\sqrt{2})(3\sqrt{2} - 3)(3 + 2\sqrt{2})(3 - 2\sqrt{2})}$$

$$K = \sqrt{[9(2) - 9][9 - 4(2)]}$$

$$K = \sqrt{9} = 3$$

6-2 *An interesting extension of Heron's Formula to the cyclic quadrilateral is credited to Brahmagupta, an Indian mathematician who lived in the early part of the seventh century. Although Brahmagupta's Formula was once thought to hold for all quadrilaterals, it has been proved to be valid only for cyclic quadrilaterals.*

The formula for the area of a cyclic quadrilateral with side measures a, b, c, *and* d *is*

$$K = \sqrt{(s-a)(s-b)(s-c)(s-d)},$$

where s *is the semiperimeter. Derive this formula. (Fig. S6-2.)*

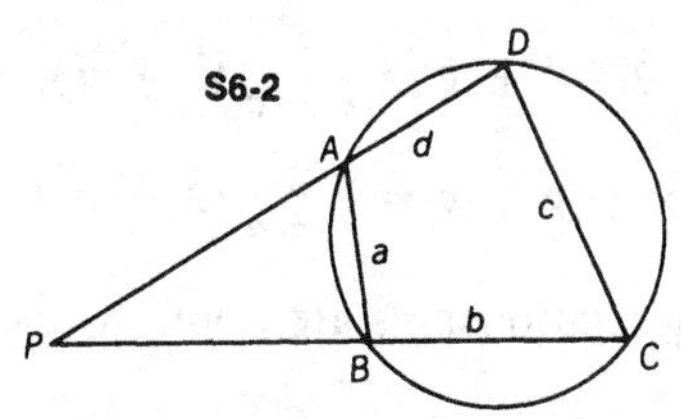

First consider the case where quadrilateral $ABCD$ is a rectangle with $a = c$ and $b = d$. Assuming Brahmagupta's Formula, we have

area of rectangle $ABCD$

$$= \sqrt{(s-a)(s-b)(s-c)(s-d)}$$

$$= \sqrt{(a+b-a)(a+b-b)(a+b-a)(a+b-b)}$$

$$= \sqrt{a^2b^2}$$

$= ab$, which is the area of the rectangle as found by the usual methods.

Now consider any non-rectangular cyclic quadrilateral $ABCD$. Extend $\overline{DA}$ and $\overline{CB}$ to meet at P, forming $\triangle DCP$. Let $PC = x$ and $PD = y$. By Heron's Formula, area of $\triangle DCP$

$$= \frac{1}{4}\sqrt{(x + y + c)(y - x + c)(x + y - c)(x - y + c)} \quad \text{(I)}$$

Since $\angle CDA$ is supplementary to $\angle CBA$ (#37), and $\angle ABP$ is also supplementary to $\angle CBA$, $\angle CDA \cong \angle ABP$. Then by #48,

$$\triangle BAP \sim \triangle DCP. \quad \text{(II)}$$

From (II) we get $\dfrac{\text{area } \triangle BAP}{\text{area } \triangle DCP} = \dfrac{a^2}{c^2}$,

$$\frac{\text{area } \triangle DCP}{\text{area } \triangle DCP} - \frac{\text{area } \triangle BAP}{\text{area } \triangle DCP} = \frac{c^2}{c^2} - \frac{a^2}{c^2},$$

$$\frac{\text{area } \triangle DCP - \text{area } \triangle BAP}{\text{area } \triangle DCP} = \frac{\text{area } ABCD}{\text{area } \triangle DCP} = \frac{c^2 - a^2}{c^2}. \quad \text{(III)}$$

From (II) we also get

$$\frac{x}{c} = \frac{y - d}{a}, \text{ and } \frac{y}{c} = \frac{x - b}{a}. \quad \text{(IV) (V)}$$

By adding (IV) and (V),

$$\frac{x + y}{c} = \frac{x + y - b - d}{a},$$

$$x + y = \frac{c}{c - a}(b + d),$$

$$x + y + c = \frac{c}{c - a}(b + c + d - a). \quad \text{(VI)}$$

The following relationships are found by using similar methods.

$$y - x + c = \frac{c}{c + a}(a + c + d - b) \quad \text{(VII)}$$

$$x + y - c = \frac{c}{c - a}(a + b + d - c) \quad \text{(VIII)}$$

$$x - y + c = \frac{c}{c + a}(a + b + c - d) \quad \text{(IX)}$$

Substitute (VI), (VII), (VIII), and (IX) into (I). Then the area of $\triangle DCP =$

$$\frac{c^2}{4(c^2 - a^2)}\sqrt{(b + c + d - a)(a + c + d - b)} \times \sqrt{(a + b + d - c)(a + b + c - d)}.$$

Since (III) may be read

$$\text{area of } \triangle DCP = \frac{c^2}{c^2 - a^2} (\text{area } ABCD),$$

the area of cyclic quadrilateral $ABCD =$

$$\sqrt{(s-a)(s-b)(s-c)(s-d)}.$$

Challenge 1 *Find the area of a cyclic quadrilateral whose sides measure* 9, 10, 10, *and* 21.

ANSWER: 120

Challenge 2 *Find the area of a cyclic quadrilateral whose sides measure* 15, 24, 7, *and* 20.

ANSWER: 234

6-3 *Sides* $\overline{BA}$ *and* $\overline{CA}$ *of* $\triangle ABC$ *are extended through* A *to form rhombuses* BATR *and* CAKN. (*See Fig.* S6-3.) $\overline{BN}$ *and* $\overline{RC}$, *intersecting at* P, *meet* $\overline{AB}$ *at* S *and* $\overline{AC}$ *at* M. *Draw* $\overline{MQ}$ *parallel to* $\overline{AB}$. (**a**) *Prove* AMQS *is a rhombus and* (**b**) *prove that the area of* $\triangle$BPC *is equal to the area of quadrilateral* ASPM.

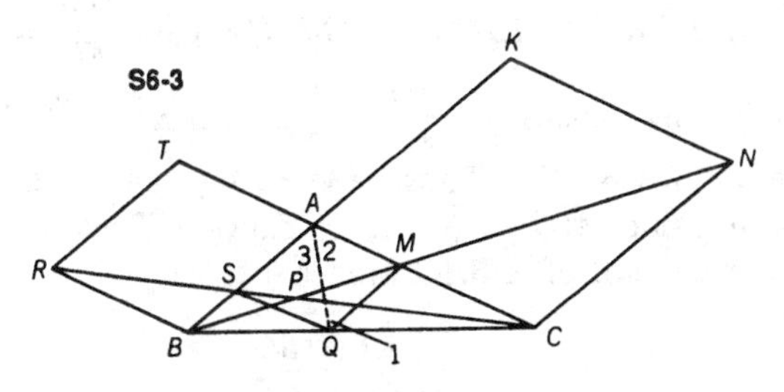

METHOD I: (**a**) Let a side of rhombus $ATRB = a$ and let a side of rhombus $AKNC = b$. Since $\overline{AS} \parallel \overline{RT}$, $\triangle CAS \sim \triangle CTR$ (#49) and $\frac{RT}{AS} = \frac{TC}{AC}$. Since $TC = TA + AC$, we get

$$\frac{a}{AS} = \frac{a+b}{b}; \quad AS = \frac{ab}{a+b}. \tag{I}$$

Similarly, since $\overline{AM} \parallel \overline{KN}$, $\triangle BAM \sim \triangle BKN$ (#49) and $\frac{KN}{AM} = \frac{KB}{AB}$. Since $KB = KA + AB$, we get

$$\frac{b}{AM} = \frac{a+b}{a}; \quad AM = \frac{ab}{a+b}. \tag{II}$$

From (I) and (II) it follows that $AS = AM$.

Since $\overline{QM} \parallel \overline{CN}$, $\triangle BMQ \sim \triangle BNC$ (#49) and

$$\frac{CN}{QM} = \frac{BN}{BM}. \tag{III}$$

Since $\triangle BAM \sim \triangle BKN$ (see above),

$$\frac{KN}{AM} = \frac{BN}{BM}. \tag{IV}$$

Then by transitivity from (III) and (IV),

$$\frac{CN}{QM} = \frac{KN}{AM}.$$

However, since $CN = KN$, it follows that $QM = AM$. Now since $AS = AM = QM$ and $\overline{AS} \parallel \overline{QM}$, $ASQM$ is a parallelogram with adjacent sides $\overline{AS}$ and $\overline{AM}$ congruent. It is, therefore, a rhombus.

METHOD II: Draw $\overline{AQ}$. Since $\overline{MQ} \parallel \overline{AB} \parallel \overline{NC}$, $\triangle MBQ \sim \triangle NBC$ (#49) and $\frac{MQ}{NC} = \frac{MB}{NB}$.

Since $\overline{AM} \parallel \overline{KN}$, $\triangle ABM \sim \triangle KBN$ (#49) and $\frac{AM}{KN} = \frac{MB}{NB}$.

Therefore by transitivity, $\frac{AM}{KN} = \frac{MQ}{NC}$. But $\overline{KN} \cong \overline{NC}$ (#21-l), and therefore $KN = NC$. Thus, $AM = MQ$ and $\angle 1 \cong \angle 2$ (#5). However, since $\overline{MQ} \parallel \overline{AS}$, $\angle 1 \cong \angle 3$ (#8). Thus, $\angle 2 \cong \angle 3$ and $\overline{AQ}$ is a bisector of $\angle BAC$. Hence, by #47,

$$\frac{AB}{AC} = \frac{BQ}{QC}. \tag{I}$$

Since $\triangle RSB \sim \triangle CSA$ (#48), $\frac{BS}{SA} = \frac{RB}{AC}$. But $\overline{RB} \cong \overline{AB}$ (#21-l) and therefore $RB = AB$. By substitution,

$$\frac{BS}{SA} = \frac{AB}{AC}. \tag{II}$$

From (I) and (II), $\frac{BQ}{QC} = \frac{BS}{SA}$. It follows that $\overline{SQ} \parallel \overline{AC}$. Thus, $SQMA$ is a parallelogram (#21a). However, since $AM = MQ$ (previously proved), $SQMA$ is a rhombus.

(b) The area of $\triangle BMQ$ equals the area of $\triangle AMQ$ since they both share the same base $\overline{MQ}$, and their vertices lie on a line parallel to base $\overline{MQ}$. Similarly, the area of $\triangle CSQ$ equals the area of $\triangle ASQ$, since both triangles share base $\overline{SQ}$, and A and C lie on $\overline{AC}$ which is parallel to $\overline{SQ}$. Therefore, by addition,

$$\text{area of } \triangle BMQ + \text{area of } \triangle CSQ = \text{area of } AMQS.$$

By subtracting the area of $SPMQ$ ($\triangle SPQ + \triangle MPQ$) from both of the above, we get,

$$\text{area of } \triangle BPC = \text{area of } ASPM.$$

6-4 *Two circles with centers* A *and* B *intersect at points* M *and* N. *Radii* $\overline{AP}$ *and* $\overline{BQ}$ *are parallel (on opposide sides of* $\overleftrightarrow{AB}$). *If the common external tangents meet* $\overleftrightarrow{AB}$ *at* D, *and* $\overline{PQ}$ *meets* $\overline{AB}$ *at* C, *prove that* $\angle$CND *is a right angle.*

Draw $\overline{AE}$ and $\overline{BF}$, where E and F are the points of tangency of the common external tangent of circles A and B, respectively. Then draw $\overline{BN}$ and extend $\overline{AN}$ through N to K. (See Fig. S6-4.)

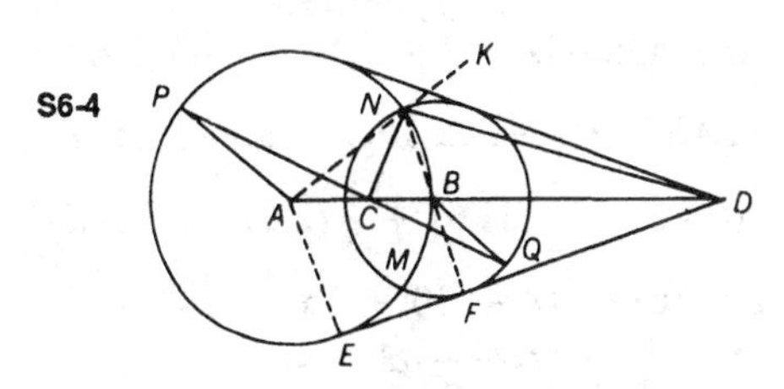

$\triangle APC \sim \triangle BQC$ (#48) and $\frac{CA}{CB} = \frac{AP}{BQ}$. However, $AP = AN$ and $BQ = BN$.

Therefore, $\frac{CA}{CB} = \frac{AN}{BN}$ and, in $\triangle ANB$, $\overline{NC}$ bisects $\angle ANB$ (#47). In $\triangle ADE$, $\overline{BF} \parallel \overline{AE}$ (#9). Therefore, $\triangle DAE \sim \triangle DBF$ (#49) and $\frac{DA}{DB} = \frac{AE}{BF}$. However $AE = AN$ and $BF = BN$. Therefore $\frac{DA}{DB} = \frac{AN}{BN}$ and, in $\triangle ANB$,

$\overline{ND}$ bisects the exterior angle at N ($\angle BNK$) (#47).

Since $\overline{NC}$ and $\overline{ND}$ are the bisectors of a pair of supplementary adjacent angles, they are perpendicular, and thus $\angle CND$ is a right angle.

6-5 *In a triangle whose sides measure* 5″, 6″, *and* 7″, *point* P *is* 2″ *from the* 5″ *side and* 3″ *from the* 6″ *side. How far is* P *from the* 7″ *side?*

There are four cases to be considered here, depending upon the position of Point P which can be within any of the four angles formed at Vertex A. (See Figs. S6-5, a–d.) In each case the area of $\triangle ABC = 6\sqrt{6}$ (by Heron's Formula), and,

AB = 5	BC = 7	PD = 3
AC = 6	PF = 2	PE = x.

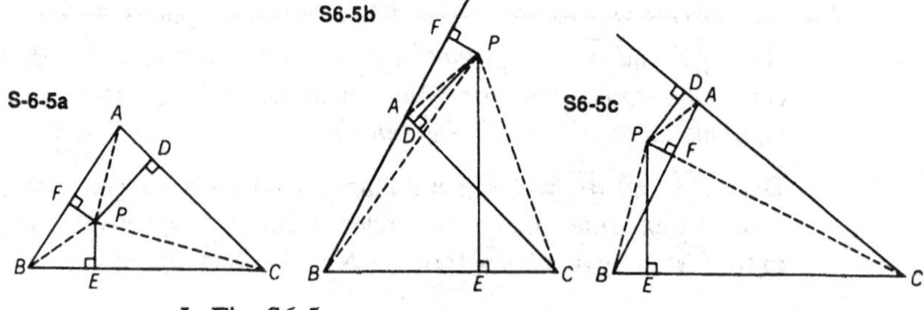

CASE I: In Fig. S6-5a,

$$\text{area } \triangle ABC = \text{area } \triangle APC + \text{area } \triangle APB + \text{area } \triangle BPC.$$

$$(6\sqrt{6}) = \frac{1}{2}(3)(6) + \frac{1}{2}(2)(5) + \frac{1}{2}(7)(x)$$

$$2(6\sqrt{6}) = 18 + 10 + 7x$$

$$x = \frac{12\sqrt{6} - 28}{7}$$

CASE II: In Fig. S6-5b,

$$\text{area } \triangle ABC = \text{area } \triangle APB + \text{area } \triangle BPC - \text{area } \triangle APC.$$

$$6\sqrt{6} = \frac{1}{2}(2)(5) + \frac{1}{2}(x)(7) - \frac{1}{2}(3)(6)$$

$$12\sqrt{6} = 10 + 7x - 18$$

$$x = \frac{12\sqrt{6} + 8}{7}$$

CASE III: In Fig. S6-5c,

$$\text{area } \triangle ABC = \text{area } \triangle BPC + \text{area } \triangle APC - \text{area } \triangle APB.$$

$$6\sqrt{6} = \frac{1}{2}(x)(7) + \frac{1}{2}(3)(6) - \frac{1}{2}(2)(5)$$

$$12\sqrt{6} = 7x + 18 - 10$$

$$x = \frac{12\sqrt{6} - 8}{7}$$

CASE IV: In Fig. S6-5d,

$$\text{area } \triangle ABC = \text{area } \triangle BPC - \text{area } \triangle APC - \text{area } \triangle APB.$$

$$6\sqrt{6} = \frac{1}{2}(x)(7) - \frac{1}{2}(3)(6) - \frac{1}{2}(2)(5)$$

$$12\sqrt{6} = 7x - 18 - 10$$

$$x = \frac{12\sqrt{6} + 28}{7}$$

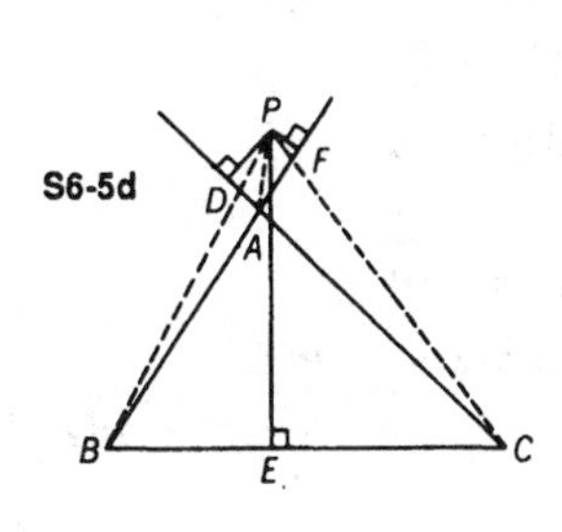

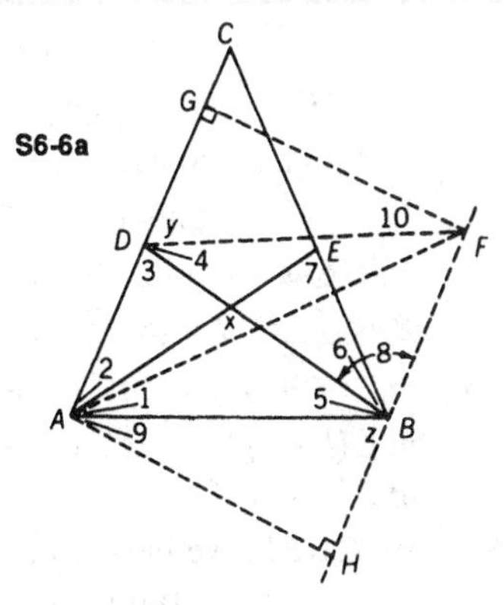

6-6 *Prove that if the measures of the interior bisectors of two angles of a triangle are equal, then the triangle is isosceles.*

METHOD I (DIRECT): $\overline{AE}$ and $\overline{BD}$ are angle bisectors, and $AE = BD$. Draw $\angle DBF \cong \angle AEB$ so that $\overline{BF} \cong \overline{BE}$; draw $\overline{DF}$. Also draw $\overline{FG} \perp \overline{AC}$, and $\overline{AH} \perp \overline{FH}$. (See Fig. S6-6a.) By hypothesis, $\overline{AE} \cong \overline{DB}$, $\overline{FB} \cong \overline{EB}$, and $\angle 8 \cong \angle 7$. Therefore $\triangle AEB \cong \triangle DBF$ (#2), $DF = AB$, and $m\angle 1 = m\angle 4$.

$m\angle x = m\angle 2 + m\angle 3$ (#12)

$m\angle x = m\angle 1 + m\angle 3$ (substitution)

$m\angle x = m\angle 4 + m\angle 3$ (substitution)

$m\angle x = m\angle 7 + m\angle 6$ (#12)

$m\angle x = m\angle 7 + m\angle 5$ (substitution)

$m\angle x = m\angle 8 + m\angle 5$ (substitution)

Therefore, $m\angle 4 + m\angle 3 = m\angle 8 + m\angle 5$ (transitivity).

Thus $m\angle z = m\angle y$.

Right $\triangle FDG \cong$ right $\triangle ABH$ (#16), $DG = BH$, and $FG = AH$.

Right $\triangle AFG \cong$ right $\triangle FAH$ (#17), and $AG = FH$.

Therefore, $GFHA$ is a parallelogram (#21b).

$m\angle 9 = m\angle 10$ (from $\triangle ABH$ and $\triangle FDG$)

$m\angle DAB = m\angle DFB$ (subtraction)
$m\angle DFB = m\angle EBA$ (from $\triangle DBF$ and $\triangle AEB$)
Therefore, $m\angle DAB = m\angle EBA$ (transitivity), and $\triangle ABC$ is isosceles.

METHOD II (INDIRECT): Assume $\triangle ABC$ is *not* isosceles. Let $m\angle ABC > m\angle ACB$. (See Fig. S6-6b.)
$\overline{BF} \cong \overline{CE}$ (hypothesis) $\overline{BC} \cong \overline{BC}$
$m\angle ABC > m\angle ACB$ (assumption) $CF > BE$
Through F, construct $\overline{GF}$ parallel to $\overline{EB}$.
Through E, construct $\overline{GE}$ parallel to $\overline{BF}$.
$BFGE$ is a parallelogram.
$\overline{BF} \cong \overline{EG}$, $\overline{EG} \cong \overline{CE}$, $\triangle GEC$ is isosceles.
$m\angle(g + g') = m\angle(c + c')$ but $m\angle g = m\angle b$
$m\angle(b + g') = m\angle(c + c')$ Therefore, $m\angle g' < m\angle c'$, since $m\angle b > m\angle c$.
In $\triangle GFC$, we have $CF < GF$. But $GF = BE$. Thus $CF < BE$.

The assumption of the inequality of $m\angle ABC$ and $m\angle ACB$ leads to two contradictory results, $CF > BE$ and $CF < BE$. Therefore $\triangle ABC$ is isosceles.

S6-6b

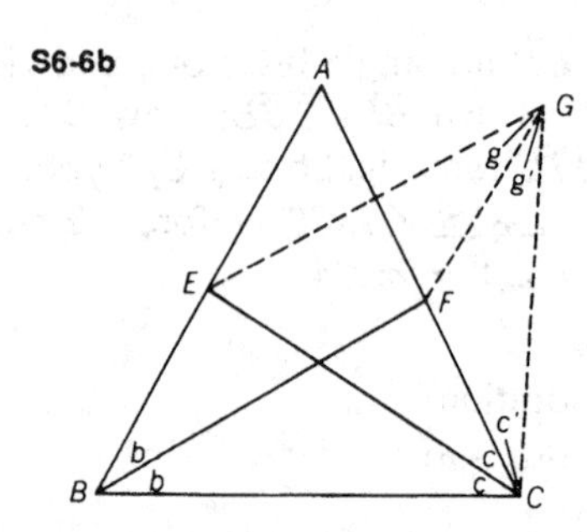

S6-6c

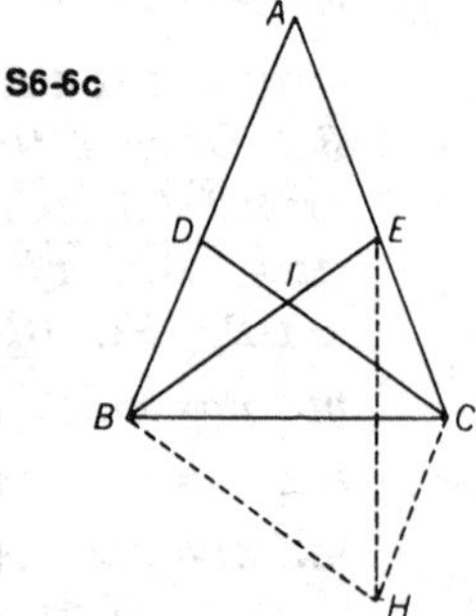

METHOD III (INDIRECT): In $\triangle ABC$, assume $m\angle B > m\angle C$. $\overline{BE}$ and $\overline{DC}$ are the bisectors of $\angle B$ and $\angle C$ respectively, and $BE = DC$. Draw $\overline{BH} \parallel \overline{DC}$ and $\overline{CH} \parallel \overline{DB}$; then draw $\overline{EH}$, as in Fig. S6-6c. $DCHB$ is a parallelogram (#21a).
Therefore, $\overline{BH} \cong \overline{DC} \cong \overline{BE}$, making $\triangle BHE$ isosceles so that, by #5, $$m\angle BEH = m\angle BHE. \quad \text{(I)}$$
From our assumption that $m\angle B > m\angle C$,
$m\angle CBE > m\angle BCD$ and $CE > DB$. Since $CH = DB$, $CE > CH$ which, by #42, leads to $m\angle CHE > m\angle CEH$. (II)

In $\triangle CEH$, by adding (I) and (II), $m\angle BHC > m\angle BEC$.
Since $DCHB$ is a parallelogram, $m\angle BHC = m\angle BDC$.
Thus, by substitution, $m\angle BDC > m\angle BEC$.
In $\triangle DBI$ and $\triangle ECI$, $m\angle DIB = m\angle EIC$.
Since $m\angle BDC > m\angle BEC$, $m\angle DBI < m\angle ECI$.
By doubling this inequality we get $m\angle B < m\angle C$, thereby contradicting the assumption that $m\angle B > m\angle C$.

Since a similar argument, starting with the assumption that $m\angle B < m\angle C$, will also lead to a contradiction, we must conclude that $m\angle B = m\angle C$ and that $\triangle ABC$ is isosceles.

METHOD IV (INDIRECT): In $\triangle ABC$, the bisectors of angles ABC and ACB have equal measures (i.e. $BE = DC$). Assume that $m\angle ABC < m\angle ACB$; then $m\angle ABE < m\angle ACD$.
We then draw $\angle FCD$ congruent to $\angle ABE$. (See Fig. S6-6d.) Note that we may take F between B and A without loss of generality.

In $\triangle FBC$, $FB > FC$ (#42). Choose a point G so that $\overline{BG} \cong \overline{FC}$. Then draw $\overline{GH} \parallel \overline{FC}$. Therefore, $\angle BGH \cong \angle BFC$ (#7) and $\triangle BGH \cong \triangle CFD$ (#3). It then follows that $BH = DC$.

Since $BH < BE$, this contradicts the hypothesis that the angle bisectors are equal. A similar argument will show that it is impossible to have $m\angle ACB < m\angle ABC$. It then follows that $m\angle ACB = m\angle ABC$ and that $\triangle ABC$ is isosceles.

S6-6d

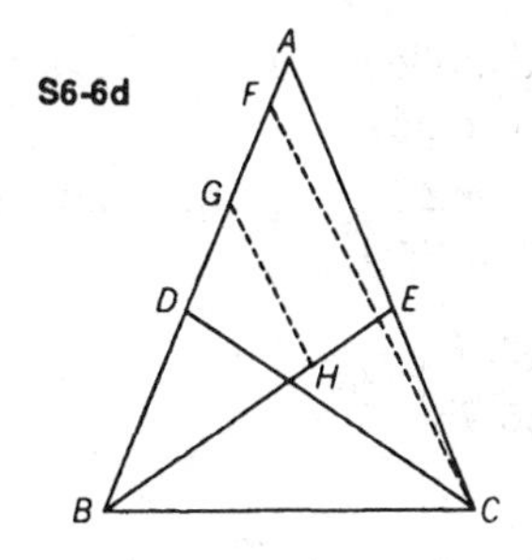

S6-7a

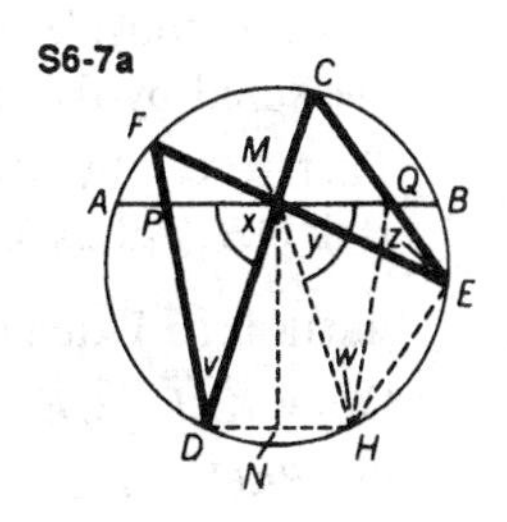

6-7 *In circle O, draw any chord $\overline{AB}$, with midpoint M. Through M two other chords, $\overline{FE}$ and $\overline{CD}$, are drawn. $\overline{CE}$ and $\overline{FD}$ intersect $\overline{AB}$ at Q and P, respectively. Prove that MP = MQ. This problem is often referred to as the butterfly problem.*

METHOD I: With M the midpoint of $\overline{AB}$ and chords $\overline{FME}$ and $\overline{CMD}$ drawn, we now draw $\overline{DH} \parallel \overline{AB}$, $\overline{MN} \perp \overline{DH}$, and lines $\overline{MH}$, $\overline{QH}$, and $\overline{EH}$. (See Fig. S6-7a.) Since $\overline{MN} \perp \overline{DH}$ and $\overline{DH} \parallel \overline{AB}$, $\overline{MN} \perp \overline{AB}$ (#10).

$\overline{MN}$, the perpendicular bisector of $\overline{AB}$, must pass through the center of the circle. Therefore $\overline{MN}$ is the perpendicular bisector of $\overline{DH}$, since a line through the center of the circle and perpendicular to a chord, bisects it.

Thus $MD = MH$ (#18), and $\triangle MND \cong \triangle MNH$ (#17).

$m\angle DMN = m\angle HMN$, so $m\angle x = m\angle y$ (they are the complements of equal angles). Since $\overline{AB} \parallel \overline{DH}$, $m\widehat{AD} = m\widehat{BH}$.

$m\angle x = \frac{1}{2}(m\widehat{AD} + m\widehat{CB})$ (#39)

$m\angle x = \frac{1}{2}(m\widehat{BH} + m\widehat{CB})$ (substitution)

Therefore, $m\angle y = \frac{1}{2}(m\widehat{BH} + m\widehat{CB})$.

But $m\angle CEH = \frac{1}{2}(m\widehat{CAH})$ (#36). Thus, by addition,

$$m\angle y + m\angle CEH = \frac{1}{2}(m\widehat{BH} + m\widehat{CB} + m\widehat{CAH}).$$

Since $m\widehat{BH} + m\widehat{CB} + m\widehat{CAH} = 360, m\angle y + m\angle CEH = 180$. It then follows that quadrilateral $MQEH$ is inscriptible, that is, a circle may be circumscribed about it.

Imagine this circle drawn. $\angle w$ and $\angle z$ are measured by the same arc, $\widehat{MQ}$ (#36), and thus $m\angle w = m\angle z$.

Now consider our original circle $m\angle v = m\angle z$, since they are measured by the same arc, $\widehat{FC}$ (#36).

Therefore, by transitivity, $m\angle v = m\angle w$, and $\triangle MPD \cong \triangle MQH$ (A.S.A.). Thus, $MP = MQ$.

METHOD II: Extend $\overline{EF}$ through F.

Draw $\overline{KPL} \parallel \overline{CE}$, as in Fig. S6-7b.

$m\angle PLC = m\angle ECL$ (#8),

therefore $\triangle PML \sim \triangle QMC$ (#48), and $\frac{PL}{CQ} = \frac{MP}{MQ}$.

$m\angle K = m\angle E$ (#8),

therefore $\triangle KMP \sim \triangle EMQ$ (#48), and $\frac{KP}{QE} = \frac{MP}{MQ}$.

By multiplication, $$\frac{(PL)(KP)}{(CQ)(QE)} = \frac{(MP)^2}{(MQ)^2}. \qquad \text{(I)}$$

Since $m\angle D = m\angle E$ (#36), and $m\angle K = m\angle E$ (#8), $m\angle D = m\angle K$.

Also $m\angle KPF = m\angle DPL$ (#1). Therefore, $\triangle KFP \sim \triangle DLP$ (#48), and $\frac{PL}{DP} = \frac{FP}{KP}$; and so

$$(PL)(KP) = (DP)(FP). \qquad \text{(II)}$$

In equation (I), $\frac{(MP)^2}{(MQ)^2} = \frac{(PL)(KP)}{(CQ)(QE)}$; we substitute from equation (II) to get $\frac{(MP)^2}{(MQ)^2} = \frac{(DP)(FP)}{(CQ)(QE)}$.

Since $(DP)(FP) = (AP)(PB)$ and $(CQ)(QE) = (BQ)(QA)$ (#52),

$$\frac{(MP)^2}{(MQ)^2} = \frac{(AP)(PB)}{(BQ)(QA)} = \frac{(MA - MP)(MA + MP)}{(MB - MQ)(MB + MQ)} = \frac{(MA)^2 - (MP)^2}{(MB)^2 - (MQ)^2}.$$

Then $(MP)^2(MB)^2 = (MQ)^2(MA)^2$.

But $MB = MA$. Therefore $(MP)^2 = (MQ)^2$, or $MP = MQ$.

S6-7b

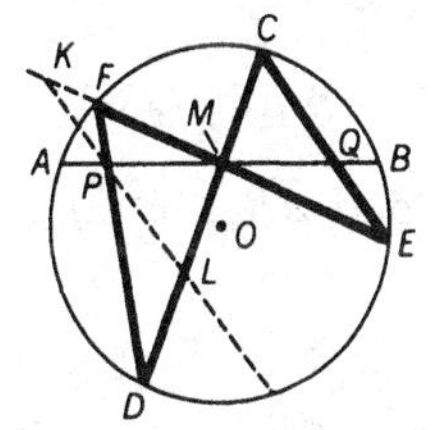

S6-7c

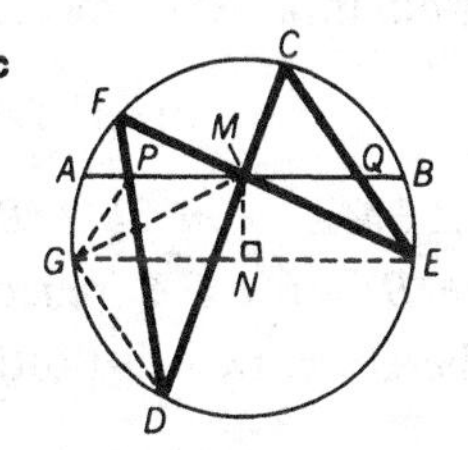

METHOD III: Draw a line through E parallel to $\overline{AB}$ meeting the circle at G, and draw $\overline{MN} \perp \overline{GE}$. Then draw $\overline{PG}$, $\overline{MG}$, and $\overline{DG}$, as in Fig. S6-7c.

$$m\angle GDP(\angle GDF) = m\angle GEF \text{ (\#36).} \qquad \text{(I)}$$

$$m\angle PMG = m\angle MGE \text{ (\#8).} \qquad \text{(II)}$$

Since the perpendicular bisector of $\overline{AB}$ is also the perpendicular bisector of $\overline{GE}$ (#10, #30),

then $GM = ME$ (#18), and $m\angle GEF = m\angle MGE$ (#5). (III)

From (I), (II), and (III), $m\angle GDP = m\angle PMG$. (IV)

Therefore, points P, M, D, and G are concyclic (#36a).

Hence, $m\angle PGM = m\angle PDM$ (#36 in the new circle). (V)

However, $m\angle CEF = m\angle PDM$ $(\angle FDM)$ (#36). (VI)

From (V) and (VI), $m\angle PGM = m\angle QEM$ $(\angle CEF)$.

From (II), we know that $m\angle PMG = m\angle MGE$.

Thus, $m\angle QME = m\angle MEG$ (#8), and $m\angle MGE = m\angle MEG$ (#5).
Therefore, $m\angle PMG = m\angle QME$ and $\triangle PMG \cong \triangle QME$ (A.S.A.). It follows that $PM = QM$.

METHOD IV: A reflection in a line is defined as the replacement of each point by another point (its image), symmetric to the first point with respect to the line of reflection.

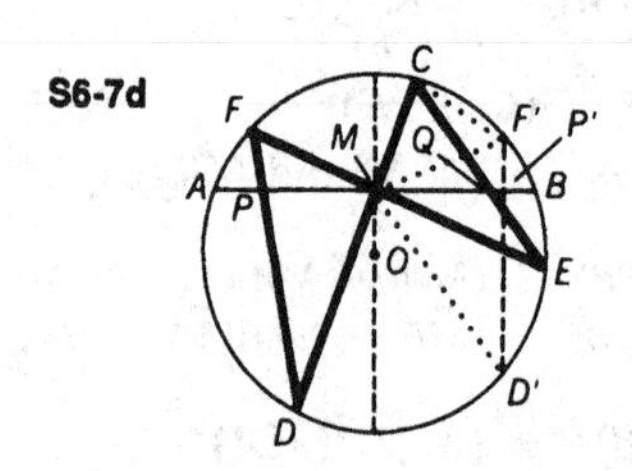

Let $\overline{D'F'}$ be the image of $\overline{DF}$ by reflection in the diameter through M. $\overline{D'F'}$ meets $\overline{AB}$ at P'. (See Fig. S6-7d.)

$$m\angle FMA = \frac{1}{2}(m\widehat{FA} + m\widehat{BE}) \text{ (\#39)} \tag{I}$$

$m\angle FMA = m\angle F'MB$ (reflection and $\overline{AB} \perp \overline{MO}$)
$m\widehat{F'B} = m\angle\widehat{FA}$ (reflection)
Therefore, by substitution in (I),

$$m\angle F'MB = \frac{1}{2}(m\widehat{F'B} + m\widehat{BE}) = \frac{1}{2}m\widehat{F'E}. \tag{II}$$

However, $m\angle F'CE = \frac{1}{2}m\widehat{F'E}$ (#36). (III)

Therefore, from (II) and (III), $m\angle F'MB = m\angle F'CE$.
Thus quadrilateral $F'CMQ$ is cyclic (i.e. may be inscribed in a circle), since if one of two equal angles intercepting the same arc is inscribed in the circle, the other is also inscribed in the circle.

$$m\angle MF'Q(m\angle MF'D') = m\angle MCQ\ (\angle MCE) \text{ (\#36)} \tag{IV}$$

$$m\angle MCE = m\angle DFE \text{ (\#36)} \tag{V}$$

$$m\angle DFE = m\angle D'F'M \text{ (reflection)} \tag{VI}$$

$$m\angle D'F'M = m\angle P'F'M \tag{VII}$$

By transitivity from (IV) through (VII), $m\angle MF'Q = m\angle MF'P'$. Therefore P', the image of P, coincides with Q; and $MP = MQ$, since $\overline{MO}$ must be the perpendicular bisector of $\overline{PQ}$, as dictated by a reflection.

METHOD V (PROJECTIVE GEOMETRY): In Fig. S6-7e, let K be the intersection of $\overleftrightarrow{DF}$ and $\overleftrightarrow{EC}$.

Let I be the intersection of $\overleftrightarrow{FC}$ and $\overleftrightarrow{DE}$.

Let N be the intersection of $\overleftrightarrow{AB}$ and $\overleftrightarrow{KI}$ (not shown).

$\overleftrightarrow{KI}$ is the polar of M with respect to the conic (circle, in this case). Therefore, M, A, B, N form a harmonic range.

Thus, $\frac{MB}{MA} = \frac{BN}{NA}$; and since $MB = MA$, N is at infinity.

Hence $\overleftrightarrow{AB} \parallel \overleftrightarrow{KI}$. Now, $\overleftrightarrow{KE}$, $\overleftrightarrow{KM}$, $\overleftrightarrow{KD}$, $\overleftrightarrow{KI}$ is a harmonic pencil.

Therefore Q, M, P, N is a harmonic range, and $\frac{MQ}{MP} = \frac{QN}{NP}$.

Since N is at infinity, $MQ = MP$.

Note that this method proves that the theorem is true for any conic.

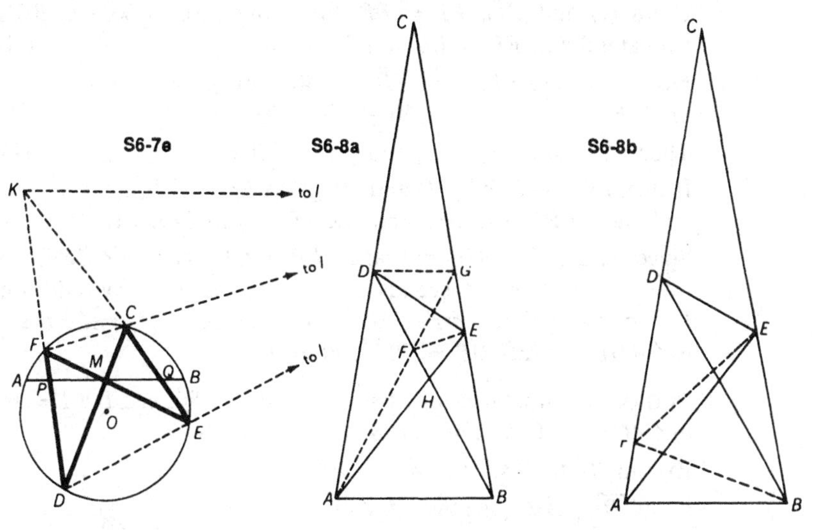

6-8 △ABC *is isosceles, with* CA = CB. m∠ABD = 60, m∠BAE = 50, *and* m∠C = 20. *Find the measure of* ∠EDB.

METHOD I: In isosceles $\triangle ABC$, draw $\overline{DG} \parallel \overline{AB}$, and $\overline{AG}$ meeting $\overline{DB}$ at F. Then draw $\overline{EF}$. (See Fig. S6-8a.)

By hypothesis, $m\angle ABD = 60$, and by theorem #8, $m\angle AGD = m\angle BAG = 60$. Thus $m\angle AFB$ is also 60, and $\triangle AFB$ is equilateral. $AB = FB$ (equilateral triangle), $AB = EB$, and $EB = FB$ (#5). $\triangle EFB$ is therefore isosceles.

Since $m\angle EBF = 20$, $m\angle BEF = m\angle BFE = 80$. As $m\angle DFG = 60$, $m\angle GFE = 40$. $GE = EF$ (equal sides of isosceles triangle), and $DF = DG$ (sides of an equilateral triangle). Thus $DGEF$ is a kite, i.e., two isosceles triangles externally sharing a common base. $\overline{DE}$ bisects $\angle GDF$ (property of a kite), therefore $m\angle EDB = 30$.

METHOD II: In isosceles $\triangle ABC$, $m\angle ACB = 20$, $m\angle CAB = 80$, $m\angle ABD = 60$, and $m\angle EAB = 50$.

Draw $\overline{BF}$ so that $m\angle ABF = 20$; then draw $\overline{FE}$, **Fig. S6-8b.**

In $\triangle ABE$, $m\angle AEB = 50$ (#13);

therefore, $\triangle ABE$ is isosceles and $AB = EB$ (#5). (I)

Similarly, $\triangle FAB$ is isosceles, since $m\angle AFB = m\angle FAB = 80$. Thus, $AB = FB$. (II)

From (I) and (II), $EB = FB$. Since $m\angle FBE = 60$, $\triangle FBE$ is equilateral and $EB = FB = FE$. (III)

Now, in $\triangle DFB$, $m\angle FDB = 40$ (#13), and $m\angle FBD = m\angle ABD - m\angle ABF = 60 - 20 = 40$.

Thus, $\triangle DFB$ is isosceles, and $FD = FB$. (IV)

It then follows from (III) and (IV) that $FE = FD$, making $\triangle FDE$ isosceles, and $m\angle FDE = m\angle FED$ (#5).

Since $m\angle AFB = 80$ and $m\angle EFB = 60$, then $m\angle AFE$, the exterior angle of isosceles $\triangle FDE$, equals 140, by addition. It follows that $m\angle ADE = 70$. Therefore, $m\angle EDB = m\angle ADE - m\angle FDB = 70 - 40 = 30$.

METHOD III: In isosceles $\triangle ABC$, $m\angle CAB = 80$, $m\angle DBA = 60$, $m\angle ACB = 20$, and $m\angle EAB = 50$.

Extend $\overline{BA}$ to G so that $AG = AC$.

Draw $\overline{DF} \parallel \overline{AB}$. (See Fig. S6-8c.)

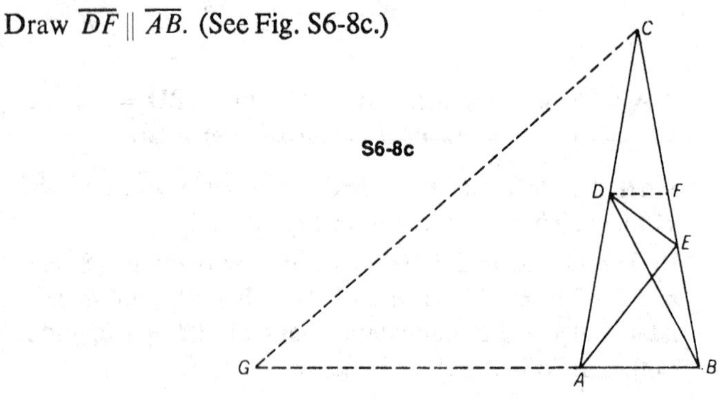

S6-8c

In $\triangle EAB$, $m\angle AEB = 50$; therefore $AB = EB$ (#5).
Since $m\angle CAB$, the exterior angle of $\triangle AGC$, is 80, $m\angle CGA = m\angle GCA = 40$ (#5). The angles of $\triangle BCG$ and $\triangle ABD$ measure 80, 60, and 40 respectively; therefore they are similar, and

$$\frac{AD}{AB} = \frac{BG}{BC}.$$

However, $AD = FB$, $AB = EB$, and $BC = AC = AG$.
By substitution, $\frac{FB}{EB} = \frac{BG}{AG}$. Applying a theorem on proportions, $\frac{FB - EB}{EB} = \frac{BG - AG}{AG}$; or $\frac{FE}{EB} = \frac{AB}{AG}$.
Since $\overline{DF} \parallel \overline{AB}$, in $\triangle ABC$, $\frac{DF}{DC} = \frac{AB}{AC}$.
Since $AG = AC$, $\frac{FE}{EB} = \frac{DF}{DC}$.
In $\triangle CDB$, $m\angle DCB = m\angle DBC = 20$. Therefore $DC = DB$.
It follows that $\frac{FE}{EB} = \frac{DF}{DB}$.
Consider $\triangle FDB$. It can now be established, as a result of the above proportion, that $\overline{DE}$ bisects $\angle FDB$.
Yet $m\angle FDB = m\angle ABD = 60$ (#8).
Therefore, $m\angle EDB = 30$.

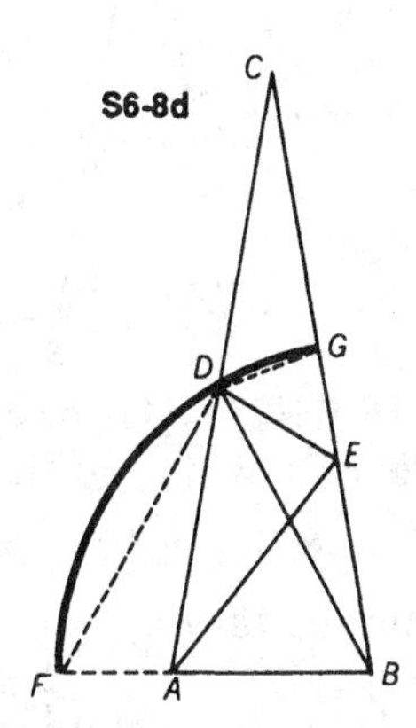

METHOD IV: With B as center, and $\overline{BD}$ as radius, draw a circle meeting $\overleftrightarrow{BA}$ at F and $\overline{BC}$ at G, as in Fig. S6-8d.
$m\angle FAD = 100$, $m\angle ADB = 40$, and $m\angle AEB = 50$ (#13).
Thus, $\triangle FBD$ is equilateral, since it is an isosceles triangle with a 60° angle.

$m\angle F = 60$, and $m\angle FDA = 20$. $BD = CD$ (isosceles triangle), $BD = DF$ (equilateral triangle), and so $CD = DF$. $BA = BE$ (isosceles triangle), $BF = BG$ (radii), and so $FA = GE$ (subtraction). $\triangle DBG$ is isosceles and $m\angle DGB = m\angle BDG = 80$. $m\angle DGC = 100$. Thus we have $\triangle DCG \cong \triangle FDA$ (S.A.A.), and $FA = DG$, since they are corresponding sides. Therefore $DG = GE$, and $m\angle GDE = m\angle GED = 50$.

But we have ascertained earlier that $m\angle BDG = 80$.

Therefore, by subtraction, $m\angle EDB = 30$.

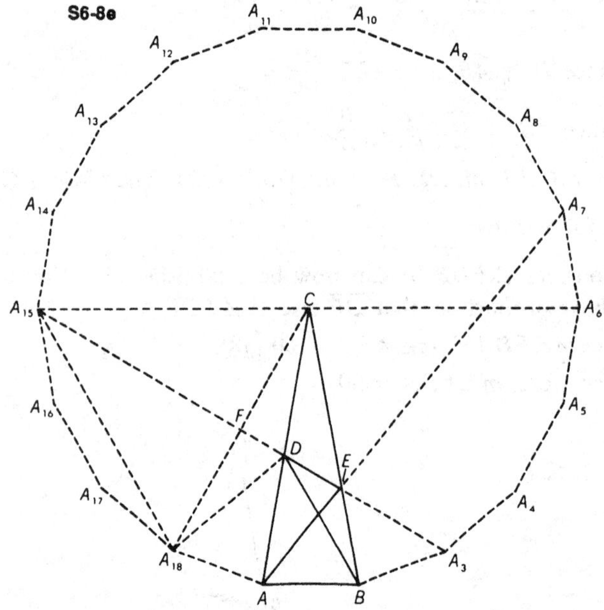

METHOD V: Let $ABA_3A_4 \ldots A_{18}$ be a regular 18-gon with center C. (See Fig. S6-8e.) Draw $\overline{A_3A_{15}}$. By symmetry $\overline{A_3A_{15}}$ and $\overline{AA_7}$ intersect on $\overline{CB}$ at E. $m\angle EAB = 50 = \frac{1}{2} m\widehat{A_7B}$. Consider the circumcircle about the 18-gon.

$$m\angle A_3A_{15}A_6 = \frac{1}{2}(m\widehat{A_3A_6}) = 30 \text{ (\#36)},$$

$$\text{and } m\angle A_{15}CA_{18} = m\widehat{A_{15}A_{18}} = 60 \text{ (\#35)}.$$

$$\text{Therefore } m\angle A_{15}FC = 90 \text{ (\#13)}.$$

However $CA_{15} = CA_{18}$; therefore $\triangle A_{15}CA_{18}$ is equilateral and $CF = FA_{18}$. Thus $\overline{A_3A_{15}}$ is the perpendicular bisector of $\overline{CA_{18}}$.

Since $CA_{18} = CB$, and $A_{18}A = AB$, $\overline{CA}$ is the perpendicular bisector of $\overline{A_{18}B}$ (#18), and $DA_{18} = DB$ (#18). As $m\angle C = m\angle DBC = 20$, $CD = DB$.

It then follows that $DA_{18} = CD$, and thus D must lie on the perpendicular bisector of $\overline{CA_{18}}$. In other words, $\overline{A_3A_{15}}$ passes through D; and A_{15}, D, E, A_3, are collinear.

Once more, consider the circumcircle of the 18-gon.

$m\angle A_{15}A_3B = \frac{1}{2}(m\widehat{A_{15}B}) = 50$ (#36), while

$m\angle CBA_3 = 80$, and $m\angle DBC = 20$.

Thus in $\triangle DBA_3$, $m\angle EDB = 30$ (#13).

METHOD VI (TRIGONOMETRIC SOLUTION I): In isosceles $\triangle ABC$, $m\angle CAB = 80$, $m\angle DBA = 60$, $m\angle ACB = 20$ and $m\angle EAB = 50$. Let $AC = a$, $EB = b$, $BD = c$. (See Fig. S6-8f.)

In $\triangle AEC$ the law of sines yields $\dfrac{CA}{CE} = \dfrac{\sin \angle CEA}{\sin \angle CAE}$ or $\dfrac{a}{a-b} = \dfrac{\sin 130}{\sin 30} = \dfrac{\sin(180 - 130)}{\frac{1}{2}} = 2 \sin 50 = 2 \cos 40.$ (I)

Since $m\angle AEB = 50$ (#13), $\triangle ABE$ is isosceles and $AB = AE$.

In $\triangle ABD$ the law of sines yields $\dfrac{DB}{AB} = \dfrac{\sin \angle DAB}{\sin \angle ADB}$ or $\dfrac{c}{b} = \dfrac{\sin 80}{\sin 40} = \dfrac{\sin 2(40)}{\sin 40} = \dfrac{2 \sin 40 \cos 40}{\sin 40} = 2 \cos 40.$ (II)

Therefore, from (I) and (II), $\dfrac{a}{a-b} = \dfrac{c}{b}$ (transitivity).

$m\angle DBE = m\angle C = 20$. Thus, $\triangle AEC \sim \triangle DEB$, (#50) and $m\angle BDE = m\angle EAC = 30$.

S6-8f

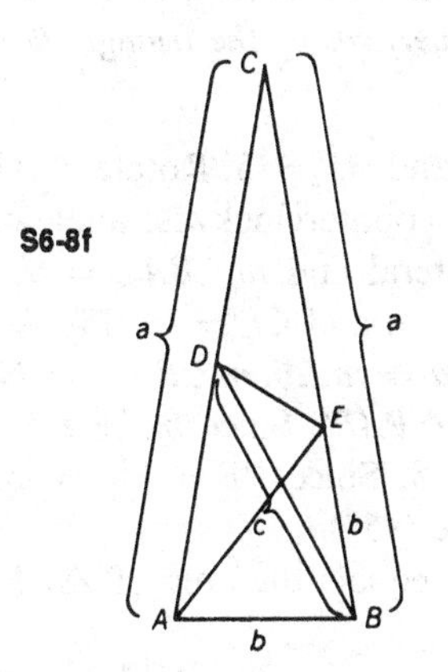

S6-8g

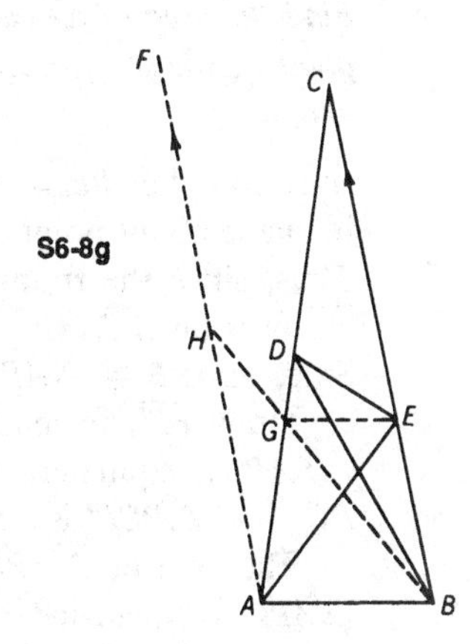

METHOD VII (TRIGONOMETRIC SOLUTION II): In isosceles $\triangle ABC$, $m\angle ABD = 60$, $m\angle BAE = 50$, and $m\angle C = 20$.

Draw $\overrightarrow{AF} \parallel \overline{BC}$, take $AG = BE$, and extend $\overline{BG}$ to intersect $\overrightarrow{AF}$ at H. (See Fig. S6-8g.)

Since $m\angle BAE = 50$, it follows that $m\angle ABG = 50$.

Since $\overrightarrow{AF} \parallel \overline{BC}$, $m\angle CAF = m\angle C = 20$; thus $m\angle BAF = 100$ and $m\angle AHB = 30$.

We know also that $m\angle ADB = 40$. Since $m\angle ABD = 60$, and $m\angle ABC = 80$, $m\angle DBC = 20$. Therefore $\angle GAH \cong \angle DBC$.

By applying the law of sines in $\triangle ADB$, $\frac{BD}{AB} = \frac{\sin \angle BAD}{\sin \angle ADB}$, or $BD =$

$$AB\left(\frac{\sin 80}{\sin 40}\right) = \frac{(AB)\sin 2(40)}{\sin 40} = \frac{(AB)(2)\sin 40 \cos 40}{\sin 40} = 2(AB)\cos 40 \qquad \text{(I)}$$

Now consider $\triangle ABH$. Again, by the law of sines $\frac{AH}{AB} = \frac{\sin \angle ABH}{\sin \angle AHB}$,

$$\text{or } AH = AB\left(\frac{\sin 50}{\sin 30}\right) = \frac{AB \cos 40}{\frac{1}{2}} = 2AB \cos 40. \qquad \text{(II)}$$

From (I) and (II), $BD = AH$ and $\triangle BDE \cong \triangle AHG$ (S.A.S) It thus follows that $m\angle BDE = m\angle GHA = 30$.

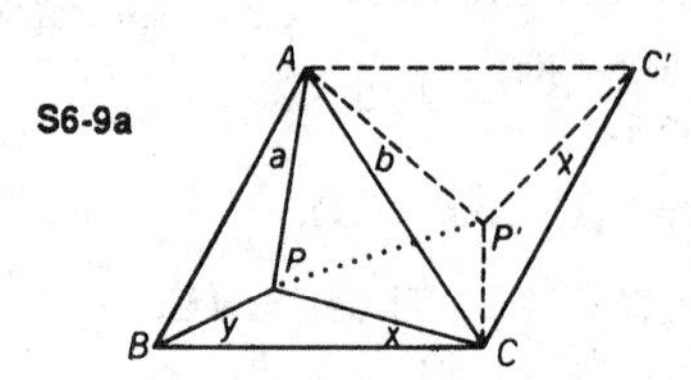

6-9 *Find the area of an equilateral triangle containing in its interior a point* P, *whose distances from the vertices of the triangle are* 3, 4, *and* 5.

METHOD I: Let $BP = 3$, $CP = 4$, and $AP = 5$. Rotate $\triangle ABC$ in its plane about point A through a counterclockwise angle of 60°. Thus, since the triangle is equilateral and $m\angle BAC = 60$ (#6), $\overline{AB}$ falls on $\overline{AC}$, $AP' = 5$, $C'P' = 4$, and $CP' = 3$ (Fig. S6-9a). Since $\triangle APB \cong \triangle AP'C$ and $m\angle a = m\angle b$, $m\angle PAP' = 60$.

Draw $\overline{PP'}$, forming isosceles $\triangle PAP'$. Since $m\angle PAP' = 60$, $\triangle PAP'$ is equilateral and $PP' = 5$. Since $PB = P'C = 3$, and $PC = 4$, $\triangle PCP'$ is a right triangle (#55).

The area of $\triangle APB + \triangle APC$ equals the area of $\triangle AP'C + \triangle APC$, or quadrilateral $APCP'$.

The area of quadrilateral $APCP'$ = the area of equilateral $\triangle APP'$ + the area of right $\triangle PCP'$.

The area of equilateral $\triangle APP' = \frac{25\sqrt{3}}{4}$ (Formula #5e),

and the area of right $\triangle PCP' = \frac{1}{2}(3)(4) = 6$ (Formula #5d).

Thus the area of quadrilateral $APCP' = \frac{25\sqrt{3}}{4} + 6$.

We now find the area of $\triangle BPC$. Since $m\angle BCC' = 2(60) = 120$ and $m\angle PCP' = 90$, $m\angle PCB + m\angle P'CC' = 30$.
Since $m\angle P'CC' = m\angle PBC$, then $m\angle PBC + m\angle PCB = 30$ (by substitution), and $m\angle BPC = 150$.

The proof may be completed in two ways. In the first one, we find that the area of $\triangle BPC = \frac{1}{2}(3)(4)\sin 150° = 3$ (Formula #5b), and the area of $\triangle ABC$ = area of (quadrilateral $APCP'$ +

$$\triangle BPC) = \quad \frac{25\sqrt{3}}{4} + 6 + 3 = \frac{25\sqrt{3}}{4} + 9.$$

Alternatively, we may apply the law of cosines to $\triangle BPC$. Therefore, $(BC)^2 = 3^2 + 4^2 - 2 \cdot 3 \cdot 4 \cos 150° = 25 + 12\sqrt{3}$.
Thus, the area of $\triangle ABC = \frac{1}{4}(BC)^2\sqrt{3} = \frac{1}{4} \cdot 25\sqrt{3} + 9$.

METHOD II: Rotate $\overline{AP}$ through 60° to position $\overline{AP'}$; then draw $\overline{CP'}$. This is equivalent to rotating $\triangle ABP$ into position $\triangle ACP'$. In a similar manner, rotate $\triangle BCP$ into position $\triangle BAP'''$, and rotate $\triangle CAP$ into position $\triangle CBP''$. (See Fig. S6-9b.)
Consider hexagon $AP'CP''BP'''$ as consisting of $\triangle ABC$, $\triangle AP'C$, $\triangle BP''C$, and $\triangle AP'''B$. From the congruence relations,
area $\triangle ABC$ = area $\triangle AP'C$ + area $\triangle BP''C$ + area $\triangle AP'''B$.
Therefore area $\triangle ABC = \frac{1}{2}$ area of hexagon $AP'CP''BP'''$.

Now consider the hexagon as consisting of three quadrilaterals, $PAP'C$, $PCP''B$, and $PBP'''A$, each of which consists of a 3-4-5 right triangle and an equilateral triangle.
Therefore, using formula #5d and #5e, the area of the hexagon =

$$3\left(\frac{1}{2} \cdot 3 \cdot 4\right) + \frac{1}{4} \cdot 5^2\sqrt{3} + \frac{1}{4} \cdot 4^2\sqrt{3} + \frac{1}{4} \cdot 3^2\sqrt{3} = 18 + \frac{1}{2} \cdot 25\sqrt{3}.$$

Therefore, the area of $\triangle ABC = 9 + \frac{1}{4} \cdot 25\sqrt{3}$.

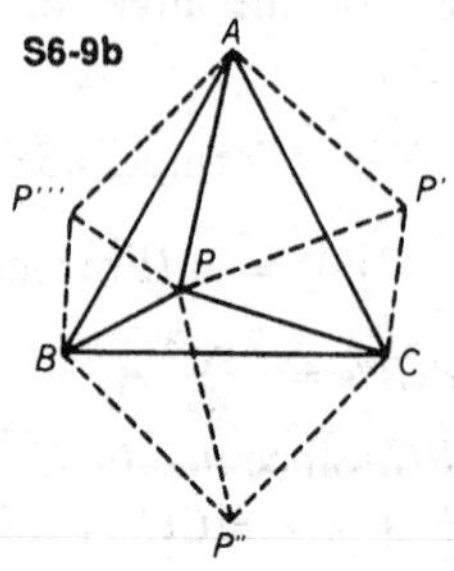

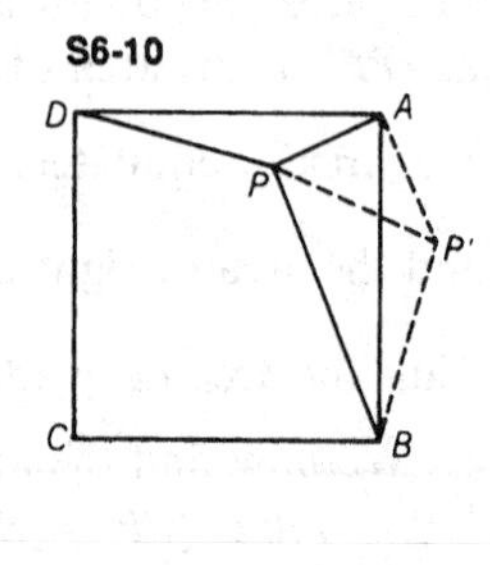

6-10 *Find the area of a square* ABCD *containing a point* P *such that* PA = 3, PB = 7, *and* PD = 5.

Rotate $\triangle DAP$ in its plane 90° about A, so that $\overline{AD}$ falls on $\overline{AB}$ (Fig. S6-10).

$\triangle APD \cong \triangle AP'B$ and $AP' = 3$ and $BP' = 5$. $m\angle PAP' = 90$.

Thus, $\triangle PAP'$ is an isosceles right triangle, and $PP' = 3\sqrt{2}$.

The area of $\triangle PP'B$ by Heron's Formula (Formula #5c) is

$$\sqrt{\left(\frac{3\sqrt{2}+12}{2}\right)\left(\frac{3\sqrt{2}-2}{2}\right)\left(\frac{3\sqrt{2}+2}{2}\right)\left(\frac{12-3\sqrt{2}}{2}\right)} = \frac{21}{2}.$$

Also, the area of $\triangle PP'B = \frac{1}{2}(PB)(PP') \sin \angle BPP'$ (Formula #5b).

Therefore, $\frac{21}{2} = \frac{1}{2}(3\sqrt{2})(7) \sin \angle BPP'$, $\frac{1}{\sqrt{2}} = \sin \angle BPP'$, and $m\angle BPP' = 45$.

In isosceles right $\triangle APP'$, $m\angle APP' = 45$, therefore $m\angle APB = 90$. By applying the Pythagorean Theorem to right $\triangle APB$ we get $(AB)^2 = 58$.

Thus the area of square $ABCD$ is 58 (Formula #4a).

Challenge 1 *Find the measure of* $\overline{PC}$.

ANSWER; $\sqrt{65}$

Challenge 2 *Express* PC *in terms of* PA, PB, *and* PD.

ANSWER: $(PC)^2 = (PD)^2 + (PB)^2 - (PA)^2$.

6-11 *If, on each side of a given triangle, an equilateral triangle is constructed externally, prove that the line segments formed by joining a vertex of the given triangle with the remote vertex of the equilateral triangle drawn on the side opposite it are congruent.*

In $\triangle ADC$, $AD = AC$, and in $\triangle AFB$, $AB = AF$ (equilateral triangles). Also, $m\angle DAC = m\angle FAB$ (Fig. S1-11). $m\angle CAB = m\angle CAB$, and therefore, $m\angle CAF = m\angle DAB$ (addition). By S.A.S., then, $\triangle CAF \cong \triangle DAB$, and thus, $DB = CF$.

Similarly, it can be proved that $\triangle CAE \cong \triangle CDB$, thus yielding $AE = DB$.

Therefore $AE = DB = CF$.

Challenge 1 *Prove that these lines are concurrent.*

Circles K and L meet at point O and A. (Fig. S6-11).

Since $m\widehat{ADC} = 240$, and we know that $m\angle AOC = \frac{1}{2}(m\widehat{ADC})$ (#36), $m\angle AOC = 120$. Similarly, $m\angle AOB = \frac{1}{2}(m\widehat{AFB}) = 120$.

Therefore $m\angle COB = 120$, since a complete revolution $= 360°$.

Since $mCEB = 240$, $\angle COB$ is an inscribed angle and point O must lie on circle M. Therefore, we can see that the three circles are concurrent, intersecting at point O.

Now join point O with points A, B, C, D, E, and F. $m\angle DOA = m\angle AOF = m\angle FOB = 60$, and therefore $\overleftrightarrow{DOB}$. Similarly, $\overleftrightarrow{COF}$ and $\overleftrightarrow{AOE}$.

Thus it has been proved that $\overline{AE}$, $\overline{CF}$, and $\overline{DB}$ are concurrent, intersecting at point O (which is also the point of intersection of circles K, L, and M).

Challenge 2 *Prove that the circumcenters of the three equilateral triangles determine another equilateral triangle.*

Consider equilateral $\triangle DAC$.

Since AK is $\frac{2}{3}$ of the altitude (or median) (#29), we obtain the proportion $\quad AC:AK = \sqrt{3}:1$.

Similarly, in equilateral $\triangle AFB$, $\quad AF:AL = \sqrt{3}:1$. Therefore, $AC:AK = AF:AL$.

$m\angle KAC = m\angle LAF = 30$, $\quad m\angle CAL = m\angle CAL$ (reflexivity), and $m\angle KAL = m\angle CAF$ (addition).

Therefore, $\triangle KAL \sim \triangle CAF$ (#50).

$$\text{Thus, } CF:KL = CA:AK = \sqrt{3}:1.$$

Similarly, we may prove $DB:KM = \sqrt{3}:1$, and $AE:ML = \sqrt{3}:1$.

Therefore, $DB:KM = AE:ML = CF:KL$. But since $DB = AE = CF$, as proved in the solution of Problem 6-11, we obtain $KM = ML = KL$. Therefore, $\triangle KML$ is equilateral.

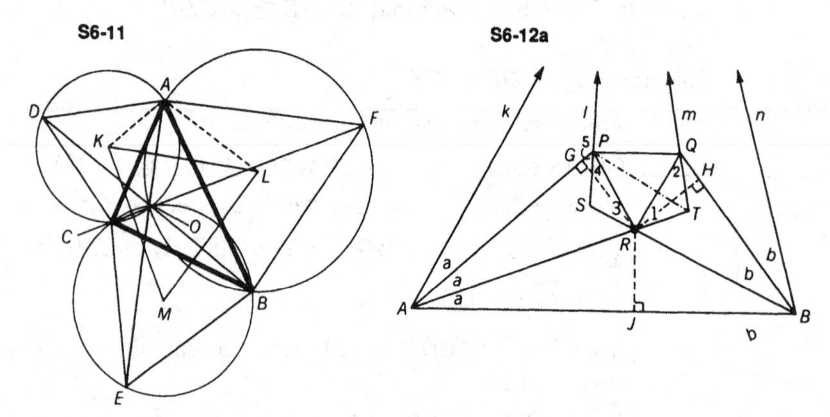

6-12 *Prove that if the angles of a triangle are trisected, the intersections of the pairs of trisectors adjacent to the same side determine an equilateral triangle. (This theorem was first derived by F. Morley about* 1900.)

METHOD I: We begin with the lower part of $\triangle ABC$, with base $\overline{AB}$ and angles $3a$, $3b$, and $3c$, as shown. Let $\overline{AP}$, $\overline{ART}$, $\overline{BQ}$, and $\overline{BRS}$ be angle-trisectors. Point P is determined by making $m\angle ARP = 60 + b$ and point Q is determined by making $m\angle BRQ = 60 + a$. (See Fig. S6-12a.) $m\angle ARB = 180 - b - a$ (#13)

Therefore $m\angle PRQ = 360 - (180 - b - a) - (60 + b) - (60 + a) = 60$.

$m\angle APR = 180 - a - (60 + b)$ (#13)

$m\angle APR = 180 - 60 - a - b = 120 - (a + b)$

However, since $3a + 3b + 3c = 180$, then $a + b + c = 60$, and $a + b = 60 - c$.

Thus $m\angle APR = 120 - (60 - c) = 60 + c$.

Similarly, it can be shown that $m\angle BQR = 60 + c$.

Now, drop perpendiculars from R to $\overline{AP}$, $\overline{BQ}$, and $\overline{AB}$, meeting these sides at points G, H, and J, respectively.

$RG = RJ$, since any point on the bisector of an angle is equidistant from the rays of the angle.

Similarly, $RH = RJ$. Therefore, $RG = RH$ (transitivity).

$\angle RGP$ and $\angle RHQ$ are right angles and are congruent.
From the previous discussion $m\angle APR = m\angle RQB$, since they are both equal to $60 + c$.
Thus $\triangle GPR \cong \triangle HQR$ (S.A.A.), and $RP = RQ$.
This makes $\triangle PRQ$ an equilateral triangle, since it is an isosceles triangle with a 60° vertex angle.

$m\angle ARP = 60 + b$ (it was so drawn at the start). $\angle SRA$ is an exterior angle of $\triangle ARB$ and its measure is equal to $a + b$. Therefore, by subtraction, we obtain $m\angle 3 = 60 + b - (a + b) = 60 - a$. Similarly, $m\angle 1 = 60 - b$.

Through point P, draw line l, making $m\angle 4 = m\angle 3$, and through point Q, draw line m making $m\angle 2 = m\angle 1$. Since $m\angle APR = 60 + c$; and $m\angle 4 = 60 - a$, we now obtain, by subtraction, $m\angle 5 = 60 + c - (60 - a) = a + c$.

By subtracting the measure of one remote interior angle of a triangle from the measure of the exterior angle of the triangle, we obtain the measure of the other remote interior angle. Thus, the measure of the angle formed by lines k and $l = (a + c) - a = c$. Similarly, the measure of the angle formed by lines m and $n = (b + c) - b = c$, while the angle formed by the lines k and $n = 180 - 3a - 3b = 3c$.

If we can now show that lines k, l, m, and n are concurrent, then we have been working properly with $\triangle ABC$. (See Fig. S6-12b.) Since $\triangle QTR$ and $\triangle RPQ$ are each isosceles, it can easily be proved that $\overline{PT}$ bisects $\angle QTR$. Since P is the point of intersection of two of the angle bisectors of $\triangle kATm$, we know that the bisector of $\angle km$ (the angle formed by lines k and m) must travel through P, since the interior angle bisectors of a triangle are concurrent. Consider Fig. S6-12b. Since g is one of the trisectors of $\angle C$, $m\angle kg = c$. g must also pass through P, since all the bisectors of $\triangle kATm$ must pass through P.

It was previously shown that $\angle kl = c$. Therefore, l is parallel to g, and both pass through point P. Thus, l and g are actually the same line. This proves lines k, l, and m to be concurrent.
Similarly, in $\triangle nBSl$, the bisector of $\angle ln$ and m are parallel and pass through point Q.
Thus, n is concurrent with l and m. Since we have proved that lines k, l, m, and n concurrent, it follows that we have properly worked with $\triangle ABC$.

This proof is based upon that given in an article by H. D. Grossman, *American Mathematical Monthly*, 1943, p. 552.

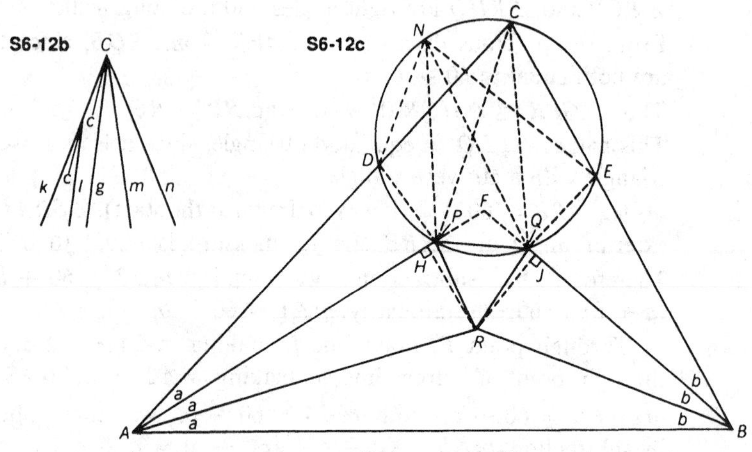

METHOD II: Let $a = \frac{A}{3}$, $b = \frac{B}{3}$, and $c = \frac{C}{3}$. In Fig. S6-12c, trisectors of $\angle A$ and $\angle B$ of $\triangle ABC$ meet at R and F.

$$\text{Construct } m\angle ARP = 60 + b, \quad \text{(I)}$$

$$\text{and } \quad m\angle BRQ = 60 + a, \quad \text{(II)}$$

where P and Q lie on $\overline{AF}$ and $\overline{BF}$, respectively.

$$m\angle APR = 180 - (60 + b) - a = 60 + c \ (\#13) \quad \text{(III)}$$

Similarly, $m\angle BQR = 180 - (60 + a) - b = 60 + c$ (#13). (IV)

Draw $\overline{HR} \perp \overline{AF}$ at H, and $\overline{JR} \perp \overline{BF}$ at J. Since R is the point of intersection of the interior angle bisectors of $\triangle AFB$, R is the center of the inscribed circle, and $HR = JR$. From (III) and (IV), $m\angle APR = m\angle BQR$. Therefore, $\triangle PHR \cong \triangle QJR$ (S.A.A.), and $PR = QR$. (V)

$$m\angle ARB = 180 - (a + b) \ (\#13) \quad \text{(VI)}$$

From (I), (II), and (VI), $m\angle PRQ = 360 - m\angle ARP - m\angle BRQ - m\angle ARB$,

or $m\angle PRQ = 360 - (60 + b) - (60 + a) - [180 - (a + b)] = 60$. Therefore, $\triangle PQR$ is equilateral. (VII)

We must now show that $\overline{PC}$ and $\overline{QC}$ are the trisectors of $\angle C$. Choose points D and E of sides $\overline{AC}$ and $\overline{BC}$ respectively, so that $AD = AR$ and $BE = BR$. It then follows that $\triangle DAP \cong \triangle RAP$ and $\triangle EBQ \cong \triangle RBQ$ (S.A.S.).

Thus, $DP = PR = PQ = RQ = QE,$ (VIII)

and $m\angle DPQ = 360 - m\angle DPA - m\angle APR - m\angle RPQ,$ or $m\angle DPQ = 360 - (60 + c) - (60 + c) - 60$ [from (III) and (VII)]. Therefore $m\angle DPQ = 180 - 2c.$ (IX)
In a like fashion, we may find $m\angle EQP = 180 - 2c.$ (X)
Thus, $m\angle DPQ = m\angle EQP$. It is easily proved that quadrilateral $DPQE$ is an isosceles trapezoid and is thus inscriptible.

In the circle passing through D, P, Q, and E, from (VIII) we know that $m\widehat{DP} = m\widehat{PQ} = m\widehat{QE}$. So from any point N on the circle, $m\angle PNQ = m\angle QNE$ (#36).

Since from (IX), $m\angle DPQ = 180 - 2c$, $m\angle DNQ = 2c$ (#37). Also, since, from (X), $m\angle EQP = 180 - 2c$, $m\angle ENP = 2c$ (#37). Therefore, $m\angle PNQ = c$, as does $m\angle DNP$ and $m\angle ENQ$. Thus, from any point on the circle, line segments issued to points D and E form an angle with measure equal to $3c$. C lies on the circle, and $\overline{PC}$ and $\overline{QC}$ are the trisectors of $\angle C$. We have thus proven that the intersections of angle trisectors adjacent to the same side of a triangle determine an equilateral triangle.

6-13 *Prove that, in any triangle, the centroid trisects the line segment joining the center of the circumcircle and the orthocenter (i.e. the point of intersection of the altitudes). This theorem was first published by Leonhard Euler in 1765.*

Let M be the midpoint of $\overline{BC}$. (See Fig. S6-13.) G, the centroid, lies on $\overline{AM}$ so that $\frac{AG}{GM} = \frac{2}{1}$ (#29). (I)
The center of the circumcircle,
point O, lies on the perpendicular bisector of $\overline{BC}$ (#44). (II)
Extend $\overline{OG}$ to point H so that $\frac{HG}{GO} = \frac{2}{1}$. (III)
From (I) and (III), $\frac{AG}{GM} = \frac{HG}{GO}$.
Therefore, $\triangle AHG \sim \triangle MOG$ (#50), and $m\angle HAG = m\angle OMG$.
Thus, $\overline{AH} \parallel \overline{MO}$, and since $\overline{MO} \perp \overline{BC}$ and $\overleftrightarrow{AH} \perp \overline{BC}$, $\overline{AH}$ extended to $\overline{BC}$ is an altitude.

The same argument will hold if we use a side other than $\overline{BC}$. Each time the point H obtained will lie on an altitude, thus making it the orthocenter of $\triangle ABC$, because, by definition, the point of concurrence of the three altitudes of a triangle is the orthocenter.

Challenge 1 (Vector Geometry)

The result of this theorem leads to an interesting problem first published by James Joseph Sylvester (1814–1897). The problem is to find the resultant of the three vectors $\overrightarrow{OA}$, $\overrightarrow{OB}$, *and* $\overrightarrow{OC}$, *acting on the center of the circumcircle* O *of* $\triangle ABC$.

$\overrightarrow{OM}$ is one-half the resultant of vectors $\overrightarrow{OB}$ and $\overrightarrow{OC}$. Since $\triangle AHG \sim \triangle MOG$, then $\frac{AH}{OM} = \frac{AG}{GM} = \frac{2}{1}$, or $\overrightarrow{AH} = 2(\overrightarrow{OM})$. Thus $\overrightarrow{AH}$ represents the whole resultant of vectors $\overrightarrow{OB}$ and $\overrightarrow{OC}$.

Since $\overrightarrow{OH}$ is the resultant of vectors $\overrightarrow{OA}$ and $\overrightarrow{AH}$, $\overrightarrow{OH}$ is the resultant of vectors $\overrightarrow{OA}$, $\overrightarrow{OB}$, and $\overrightarrow{OC}$.

COMMENT: It follows that $\overrightarrow{OG} = \frac{1}{3}(\overrightarrow{OA} + \overrightarrow{OB} + \overrightarrow{OC})$.

S6-13

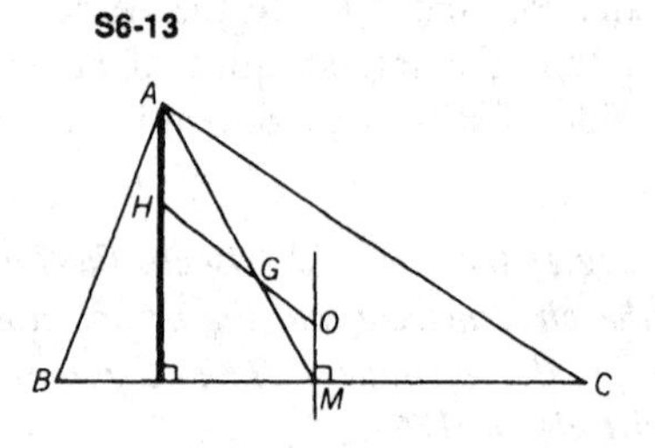

S6-14a

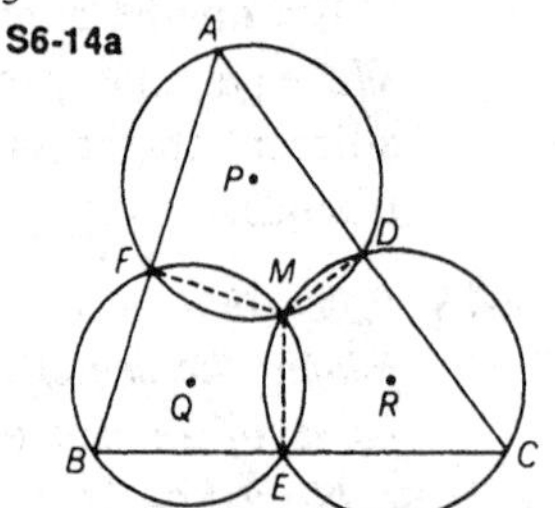

6-14 *Prove that if a point is chosen on each side of a triangle, then the circles determined by each vertex and the points on the adjacent sides, pass through a common point (Figs. 6-14a and 6-14b). This theorem was first published by A. Miquel in 1838.*

CASE I: Consider the problem when M is inside $\triangle ABC$, as shown in Fig. S6-14a. Points D, E, and F are any points on sides $\overline{AC}$, $\overline{BC}$, and $\overline{AB}$, respectively, of $\triangle ABC$. Let circles Q and R, determined by points F, B, E and D, C, E, respectively, meet at M. Draw $\overline{FM}$, $\overline{ME}$, and $\overline{MD}$. In cyclic quadrilateral $BFME$, $m\angle FME = 180 - m\angle B$ (#37). Similarly, in cyclic quadrilateral $CDME$, $m\angle DME = 180 - m\angle C$.

By addition, $m\angle FME + m\angle DME = 360 - (m\angle B + m\angle C)$.

Therefore, $m\angle FMD = m\angle B + m\angle C$.

However, in $\triangle ABC$, $m\angle B + m\angle C = 180 - m\angle A$.

Therefore, $m\angle FMD = 180 - m\angle A$ and quadrilateral $AFMD$ is cyclic. Thus, point M lies on all three circles.

CASE II: Fig. S6-14b illustrates the problem when M is outside $\triangle ABC$.

Again let circles Q and R meet at M. Since quadrilateral $BFME$ is cyclic, $m\angle FME = 180 - m\angle B$ (#37).

Similarly, since quadrilateral $CDME$ is cyclic, $m\angle DME = 180 - m\angle DCE$ (#37).

By subtraction,

$$m\angle FMD = m\angle FME - m\angle DME = m\angle DCE - m\angle B. \quad \text{(I)}$$

$$\text{However, } m\angle DCE = m\angle BAC + m\angle B \text{ (\#12).} \quad \text{(II)}$$

By substituting (II) into (I),

$$m\angle FMD = m\angle BAC = 180 - m\angle FAD.$$

Therefore, quadrilateral $ADMF$ is also cyclic and point M lies on all three circles.

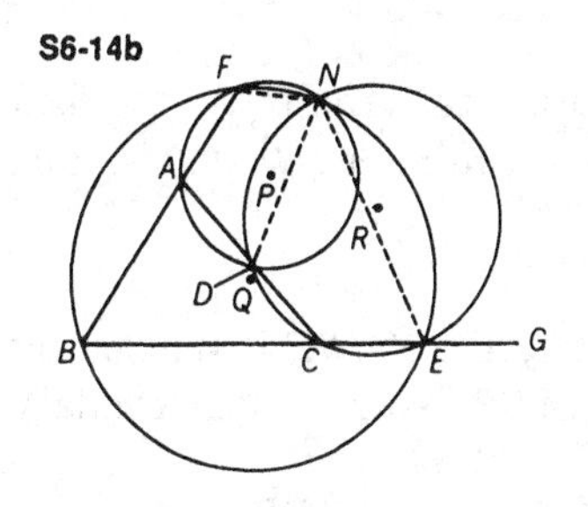

S6-14b

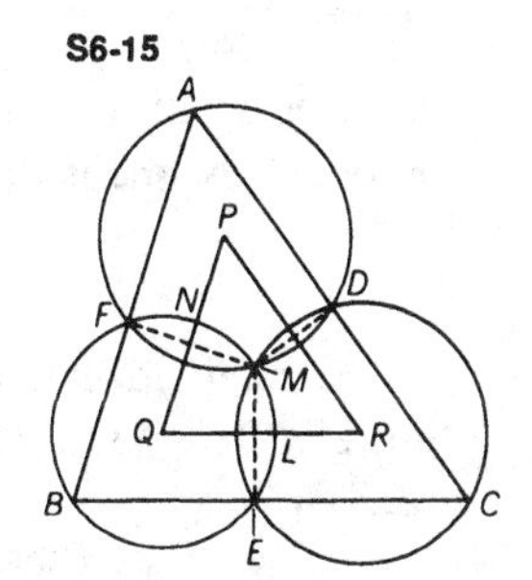

S6-15

6-15 *Prove that the centers of the circles in Problem* 6-14 *determine a triangle similar to the original triangle.*

Draw common chords $\overline{FM}$, $\overline{EM}$, and $\overline{DM}$. $\overline{PQ}$ meets circle Q at N and $\overline{RQ}$ meets circle Q at L. (See Fig. S6-15.) Since the line of centers of two circles is the perpendicular bisector of their common chord, $\overline{PQ}$ is the perpendicular bisector of $\overline{FM}$, and therefore $\overline{PQ}$ also bisects $\overparen{FM}$ (#30), so that $m\overparen{FN} = m\overparen{NM}$. Similarly, $\overline{QR}$ bisects $\overparen{EM}$ so that $m\overparen{ML} = m\overparen{LE}$.

Now $m\angle NQL = (m\overparen{NM} + m\overparen{ML}) = \frac{1}{2}(m\overparen{FE})$ (#35), and

$$m\angle FBE = \frac{1}{2}(m\overparen{FE}) \text{ (\#36).}$$

Therefore, $m\angle NQL = m\angle FBE$.

In a similar fashion it may be proved that $m\angle QPR = m\angle BAC$.

Thus, $\triangle PQR \sim \triangle ABC$ (#48).

7. Ptolemy and the Cyclic Quadrilateral

7-1 *Prove that in a cylic quadrilateral the product of the diagonals is equal to the sum of the products of the pairs of opposite sides (Ptolemy's Theorem).*

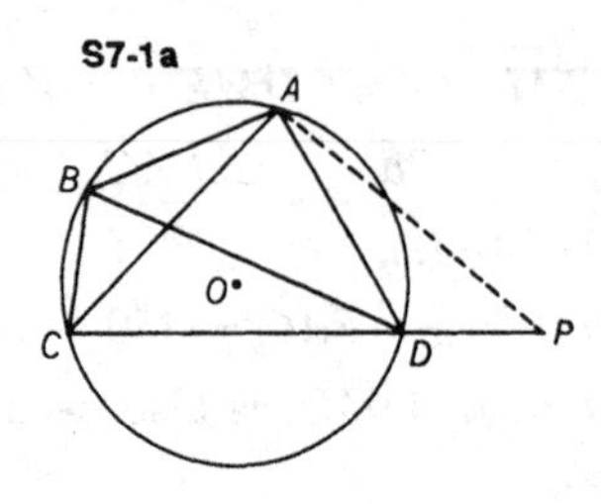

METHOD I: In Fig. S7-1a, quadrilateral $ABCD$ is inscribed in circle O. A line is drawn through A to meet $\overleftrightarrow{CD}$ at P, so that

$$m\angle BAC = m\angle DAP. \qquad \text{(I)}$$

Since quadrilateral $ABCD$ is cyclic, $\angle ABC$ is supplementary to $\angle ADC$ (#37). However, $\angle ADP$ is also supplementary to $\angle ADC$.

$$\text{Therefore, } m\angle ABC = m\angle ADP. \qquad \text{(II)}$$

$$\text{Thus, } \triangle BAC \sim \triangle DAP \text{ (\#48)}, \qquad \text{(III)}$$

$$\text{and } \frac{AB}{AD} = \frac{BC}{DP}, \text{ or } DP = \frac{(AD)(BC)}{AB}. \qquad \text{(IV)}$$

From (I), $m\angle BAD = m\angle CAP$, and from (III), $\frac{AB}{AD} = \frac{AC}{AP}$.

Therefore, $\triangle ABD \sim \triangle ACP$ (#50), and $\frac{BD}{CP} = \frac{AB}{AC}$,

$$\text{or } CP = \frac{(AC)(BD)}{AB}. \qquad \text{(V)}$$

$$CP = CD + DP. \qquad \text{(VI)}$$

Substituting (IV) and (V) into (VI),

$$\frac{(AC)(BD)}{AB} = CD + \frac{(AD)(BC)}{AB}.$$

Thus, $(AC)(BD) = (AB)(CD) + (AD)(BC)$.

METHOD II: In quadrilateral $ABCD$ (Fig. S7-1b), draw $\triangle DAP$ on side $\overline{AD}$ similar to $\triangle CAB$.

$$\text{Thus, } \frac{AB}{AP} = \frac{AC}{AD} = \frac{BC}{PD}, \qquad \text{(I)}$$

$$\text{and } (AC)(PD) = (AD)(BC). \qquad \text{(II)}$$

Since $m\angle BAC = m\angle PAD$, then $m\angle BAP = m\angle CAD$. Therefore, from (I), $\triangle BAP \sim \triangle CAD$ (#50), and $\frac{AB}{AC} = \frac{BP}{CD}$,

$$\text{or } (AC)(BP) = (AB)(CD). \qquad \text{(III)}$$

Adding (II) and (III), we have

$$(AC)(BP + PD) = (AD)(BC) + (AB)(CD). \qquad \text{(IV)}$$

Now $BP + PD > BD$ (#41), unless P is on $\overline{BD}$.
However, P will be on $\overline{BD}$ if and only if $m\angle ADP = m\angle ADB$. But we already know that $m\angle ADP = m\angle ACB$ (similar triangles). And if $ABCD$ were cyclic, then $m\angle ADB$ would equal $m\angle ACB$ (#36a), and $m\angle ADB$ would equal $m\angle ADP$. Therefore, we can state that if and only if $ABCD$ is cyclic, P lies on $\overline{BD}$.
This tells us that $\qquad BP + PD = BD. \qquad$ (V)

Substituting (V) into (IV), $(AC)(BD) = (AD)(BC) + (AB)(CD)$.
Notice we have proved both Ptolemy's Theorem and its converse. For a statement of the converse alone and its proof, see Challenge 1.

Challenge 1 *Prove that if the product of the diagonals of a quadrilateral equals the sum of the products of the pairs of opposite sides, then the quadrilateral is cyclic. This is the converse of Ptolemy's Theorem.*

Assume quadrilateral $ABCD$ is not cyclic.
If $\overline{CDP}$, then $m\angle ADP \neq m\angle ABC$.
If C, D, and P are not collinear then it is possible to have $m\angle ADP = m\angle ABC$. However, then $CP < CD + DP$ (#41) and from steps (IV) and (V), Method I, above.

$$(AC)(BD) < (AB)(CD) + (AD)(BC).$$

But this contradicts the given information that $(AC)(BD) = (AB)(CD) + (AD)(BC)$. Therefore, quadrilateral $ABCD$ is cyclic.

Challenge 2 *To what familiar result does Ptolemy's Theorem lead when the cyclic quadrilateral is a rectangle?*

By Ptolemy's Theorem applied to Fig. S7-1c

$$(AC)(BD) = (AD)(BC) + (AB)(DC).$$

However, since $ABCD$ is a rectangle,

$$AC = BD,\ AD = BC, \text{ and } AB = DC \text{ (\#21g).}$$

Therefore, $(AC)^2 = (AD)^2 + (DC)^2$, which is the Pythagorean Theorem, as applied to any of the right triangles of the given rectangle.

Challenge 3 *Find the diagonal* d *of the trapezoid with bases* a *and* b, *and equal legs* c.

ANSWER: $d = \sqrt{ab + c^2}$

S7-1b

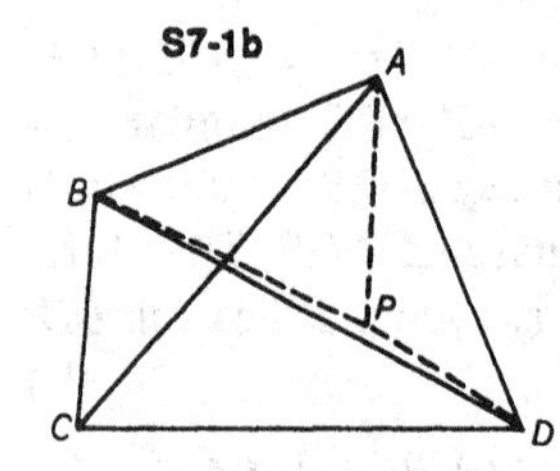

S7-1c

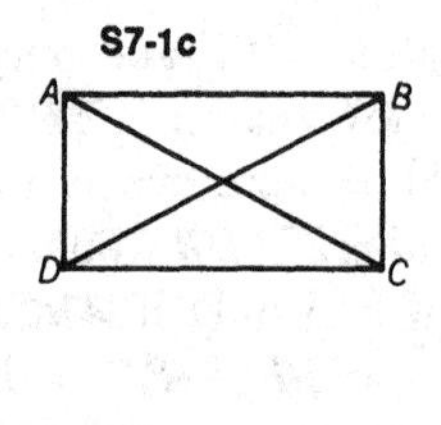

S7-2

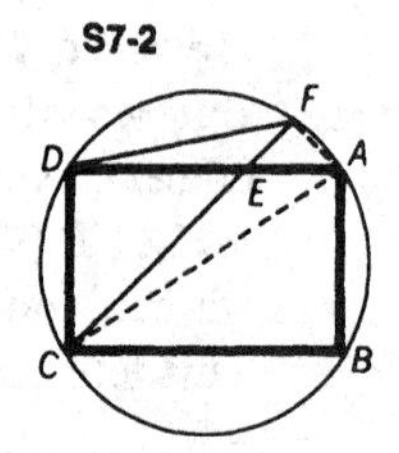

7-2 E *is a point on side* $\overline{AD}$ *of rectangle* ABCD, *so that* DE = 6, *while* DA = 8, *and* DC = 6. *If* $\overline{CE}$ *extended meets the circumcircle of the rectangle at* F, *find the measure of chord* $\overline{DF}$.

Draw $\overline{AF}$ and diagonal $\overline{AC}$. (See Fig. S7-2.) Since $\angle B$ is a right angle, $\overline{AC}$ is a diameter (#36).

Applying the Pythagorean Theorem to right $\triangle ABC$, we obtain $AC = 10$.

Similarly, in isosceles right $\triangle CDE$, $CE = 6\sqrt{2}$ (#55a), and in isosceles right $\triangle EFA$, $EF = FA = \sqrt{2}$ (#55b). Now let us apply Ptolemy's Theorem to quadrilateral $AFDC$.

$$(FC)(DA) = (DF)(AC) + (AF)(DC)$$

Substituting, $(6\sqrt{2} + \sqrt{2})(6 + 2) = DF(10) + (\sqrt{2})(6)$,

$$56\sqrt{2} = 10(DF) + 6\sqrt{2},$$

$$5\sqrt{2} = DF.$$

Challenge *Find the measure of* $\overline{FB}$.

ANSWER: $5\sqrt{2}$

7-3 *On side* $\overline{AB}$ *of square* ABCD, *a right* $\triangle$ABF, *with hypotenuse* $\overline{AB}$, *is drawn externally to the square. If* AF = 6 *and* BF = 8, *find* EF, *where* E *is the point of intersection of the diagonals of the square.*

In right $\triangle AFB$, $AF = 6$, $BF = 8$, and $AB = 10$ (#55). (See Fig. S7-3.)

In isosceles right $\triangle AEB$, $AE = BE = 5\sqrt{2}$ (#55a).

Since $m\angle AFB = m\angle AEB = 90$, quadrilateral $AFBE$ is cyclic (#37).

Therefore, by Ptolemy's Theorem applied to quadrilateral $AFBE$, $(AF)(BE) + (AE)(BF) = (AB)(EF)$.

By substitution, $(6)(5\sqrt{2}) + (5\sqrt{2})(8) = (10)(EF)$ and $EF = 7\sqrt{2}$.

Challenge *Find* EF *when* F *is inside square* ABCD.

ANSWER: $\sqrt{2}$

S7-3

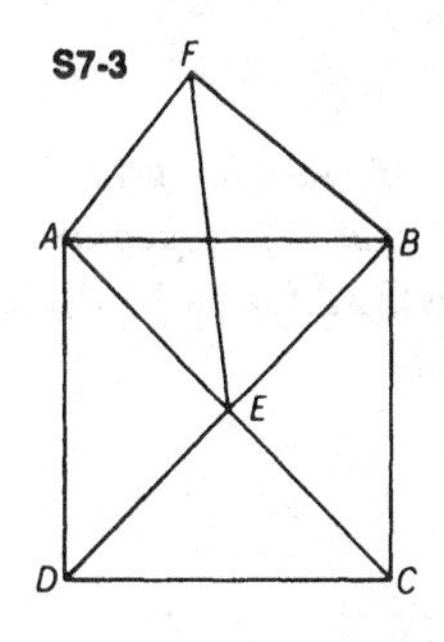

S7-4

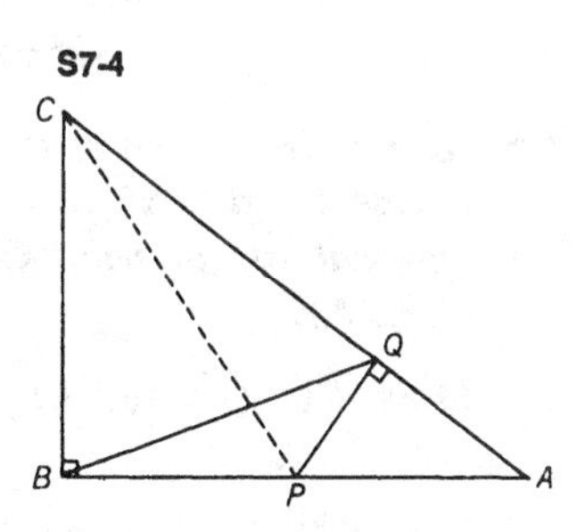

7-4 *Point* P *on side* $\overline{AB}$ *of right* $\triangle$ABC *is placed so that* BP = PA = 2. *Point* Q *is on hypotenuse* $\overline{AC}$ *so that* $\overline{PQ}$ *is perpendicular to* $\overline{AC}$. *If* CB = 3, *find the measure of* $\overline{BQ}$, *using Ptolemy's Theorem.*

Draw $\overline{PC}$. (See Fig. S7-4.)

In right $\triangle PBC$, $PC = \sqrt{13}$, and in right $\triangle ABC$, $AC = 5$ (#55).

Since $\triangle AQP \sim \triangle ABC$ (#48), then $\frac{PQ}{CB} = \frac{PA}{AC}$, and $\frac{PQ}{3} = \frac{2}{5}$, or $PQ = \frac{6}{5}$. Now in right $\triangle PQC$, $(PQ)^2 + (CQ)^2 = (CP)^2$. Therefore $CQ = \frac{17}{5}$.

Since $m\angle CBP \cong m\angle CQP \cong 90$, quadrilateral $BPQC$ is cyclic (#37), and thus we may apply Ptolemy's Theorem to it.

$$(BQ)(CP) = (PQ)(BC) + (BP)(QC)$$

Substituting,

$$(BQ)(\sqrt{13}) = \left(\frac{6}{5}\right)(3) + (2)\left(\frac{17}{5}\right)\cdot$$

Thus, $BQ = \frac{4}{5}\sqrt{13}$.

Challenge 1 *Find the area of quadrilateral* CBPQ.

ANSWER: 5.04

Challenge 2 *As* P *is translated from* B *to* A *along* $\overline{\text{BA}}$, *find the range of values of* BQ *where* $\overline{\text{PQ}}$ *remains perpendicular to* $\overline{\text{CA}}$.

ANSWER: minimum value, 2.4; maximum value, 4

S7-5

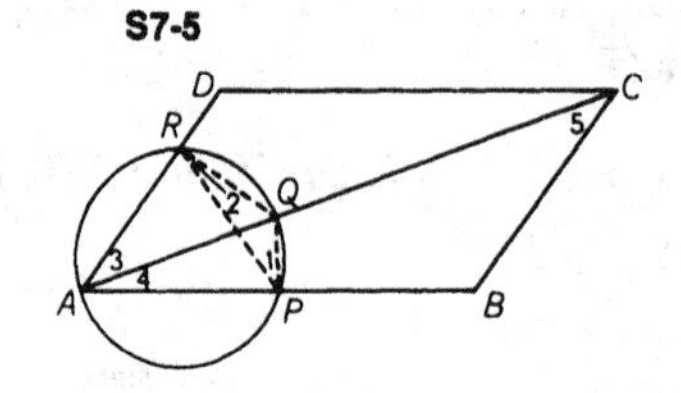

7-5 *If any circle passing through vertex* A *of parallelogram* ABCD *intersects sides* $\overline{\text{AB}}$ *and* $\overline{\text{AD}}$ *at points* P *and* R, *respectively, and diagonal* $\overline{\text{AC}}$ *at point* Q, *prove that* (AQ)(AC) = (AP)(AB) + (AR)(AD).

Draw $\overline{RQ}$, $\overline{QP}$, and $\overline{RP}$, as in Fig. S7-5.

$m\angle 4 = m\angle 2$ (#36).

Similarly, $m\angle 1 = m\angle 3$ (#36).

Since $m\angle 5 = m\angle 3$ (#8), $m\angle 1 = m\angle 5$.

Therefore, $\triangle RQP \sim \triangle ABC$ (#48), and since $\triangle ABC \cong \triangle CDA$, $\triangle RQP \sim \triangle ABC \sim \triangle CDA$.

$$\text{Then } \frac{AC}{RP} = \frac{AB}{RQ} = \frac{AD}{PQ}\cdot \tag{I}$$

Now by Ptolemy's Theorem, in quadrilateral $RQPA$

$$(AQ)(RP) = (RQ)(AP) + (PQ)(AR). \tag{II}$$

By multiplying each of the three equal ratios in (I) by each member of (II),

$$(AQ)(RP)\left(\frac{AC}{RP}\right) = (RQ)(AP)\left(\frac{AB}{RQ}\right) + (PQ)(AR)\left(\frac{AD}{PQ}\right)\cdot$$

Thus, $(AQ)(AC) = (AP)(AB) + (AR)(AD)$.

7-6 *Diagonals* $\overline{AC}$ *and* $\overline{BD}$ *of quadrilateral* ABCD *meet at* E. *If* AE = 2, BE = 5, CE = 10, DE = 4, *and* BC = $\frac{15}{2}$, *find* AB.

In Fig. S7-6, since $\frac{BE}{AE} = \frac{CE}{DE} = \frac{5}{2}$, (I)

$\triangle AED \sim \triangle BEC$ (#50). Therefore, $\frac{BE}{AE} = \frac{BC}{AD}$, or $\frac{5}{2} = \frac{\frac{15}{2}}{AD}$.

Thus, $AD = 3$.

Similarly, from (I), $\triangle AEB \sim \triangle DEC$ (#50). (II)

Therefore, $\frac{AE}{DE} = \frac{AB}{DC}$, or $\frac{1}{2} = \frac{AB}{DC}$. Thus, $DC = 2(AB)$.

Also, from (II), $m\angle BAC = m\angle BDC$. Therefore, quadrilateral $ABCD$ is cyclic (#36a).

Now, applying Ptolemy's Theorem to cyclic quadrilateral $ABCD$,

$$(AB)(DC) + (AD)(BC) = (AC)(BD).$$

Substituting, we find that $AB = \frac{1}{2}\sqrt{171}$.

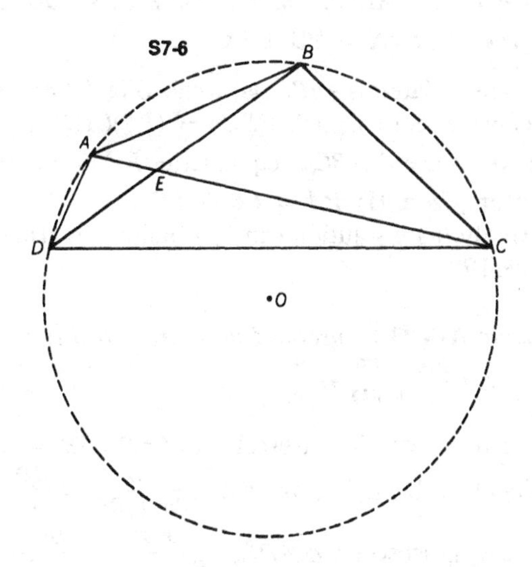

Challenge *Find the radius of the circumcircle of* ABCD *if the measure of the distance from* $\overline{DC}$ *to the center* O *is* $2\frac{1}{2}$.

ANSWER: 7

7-7 *If isosceles* $\triangle ABC$ ($AB = AC$) *is inscribed in a circle, and a point* P *is on* $\widehat{BC}$, *prove that* $\frac{PA}{PB + PC} = \frac{AC}{BC}$, *a constant for the given triangle.*

Applying Ptolemy's Theorem in cyclic quadrilateral $ABPC$ (Fig. S7-7), $(PA)(BC) = (PB)(AC) + (PC)(AB)$.
Since $AB = AC$, $(PA)(BC) = AC(PB + PC)$,
and $\frac{PA}{PB + PC} = \frac{AC}{BC}$.

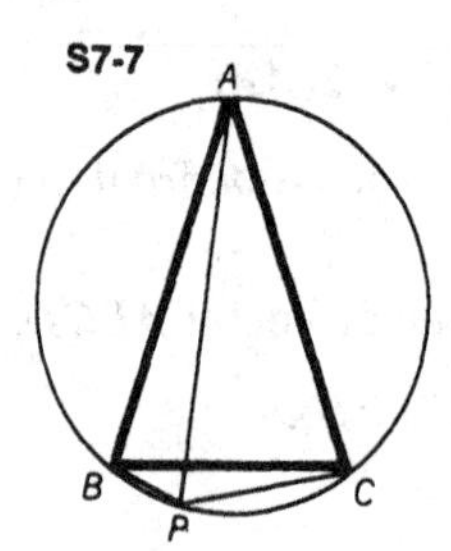

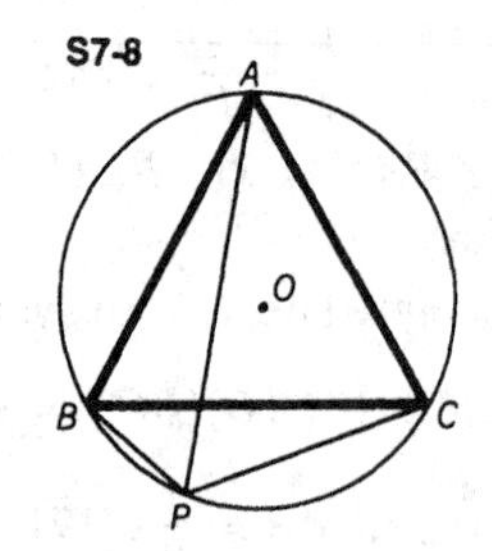

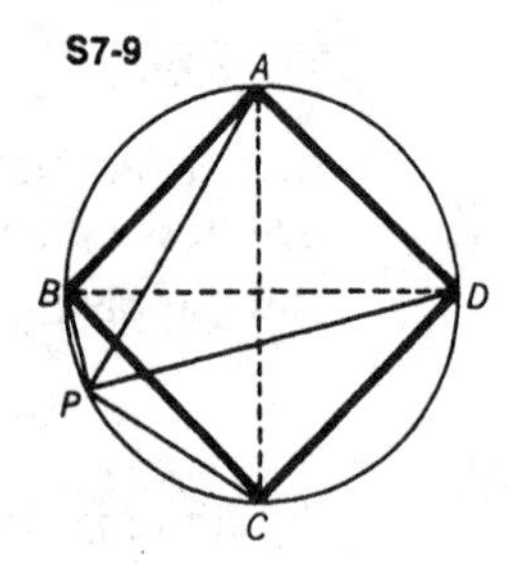

7-8 *If equilateral* $\triangle ABC$ *is inscribed in a circle, and a point* P *is on* $\widehat{BC}$, *prove that* $PA = PB + PC$.

Since quadrilateral $ABPC$ is cyclic (Fig. S7-8), we may apply Ptolemy's Theorem. $(PA)(BC) = (PB)(AC) + (PC)(AB)$ (I)
However, since $\triangle ABC$ is equilateral, $BC = AC = AB$.
Therefore, from (I), $PA = PB + PC$.

An alternate solution can be obtained by using the results of Problem 7-7.

7-9 *If square* ABCD *is inscribed in a circle, and a point* P *is on* $\widehat{BC}$, *prove that* $\frac{PA + PC}{PB + PD} = \frac{PD}{PA}$.

In Fig. S7-9, consider isosceles $\triangle ABD$ ($AB = AD$). Using the results of Problem 7-7, we have $\frac{PA}{PB + PD} = \frac{AD}{DB}$. (I)

Similarly, in isosceles $\triangle ADC$, $\frac{PD}{PA + PC} = \frac{DC}{AC}$. (II)

Since $AD = DC$ and $DB = AC$, $\frac{AD}{DB} = \frac{DC}{AC}$. (III)

From (I), (II) and (III),

$$\frac{PA}{PB + PD} = \frac{PD}{PA + PC}, \text{ or } \frac{PA + PC}{PB + PD} = \frac{PD}{PA}.$$

7-10 *If regular pentagon* ABCDE *is inscribed in a circle, and point* P *is on* $\widehat{BC}$, *prove that* PA + PD = PB + PC + PE.

In quadrilateral $ABPC$, $(PA)(BC) = (BA)(PC) + (PB)(AC)$, (I) by Ptolemy's Theorem. (See Fig. S7-10.)
In quadrilateral $BPCD$, $(PD)(BC) = (CD)(PB) + (PC)(BD)$. (II)
Since $BA = CD$ and $AC = BD$, by adding (I) and (II) we obtain

$$BC(PA + PD) = BA(PB + PC) + AC(PB + PC). \qquad \text{(III)}$$

However, since $\triangle BEC$ is isosceles, based upon Problem 7-7,

$$\frac{CE}{BC} = \frac{PE}{PB + PC}, \text{ or } \frac{(PE)(BC)}{(PB + PC)} = CE = AC. \qquad \text{(IV)}$$

Substituting (IV) into (III),

$$BC(PA + PD) = BA(PB + PC) + \frac{(PE)(BC)}{(PB + PC)}(PB + PC).$$

But $BC = BA$. Therefore $PA + PD = PB + PC + PE$.

S7-10

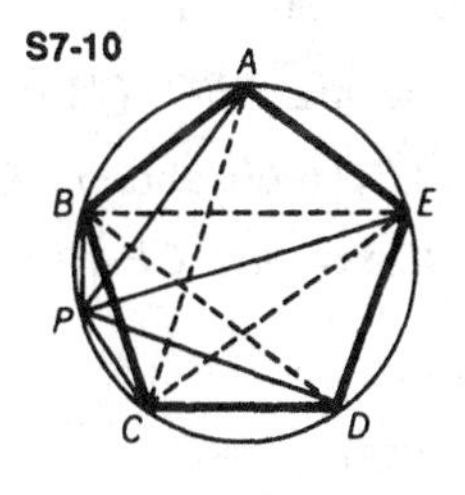

S7-11

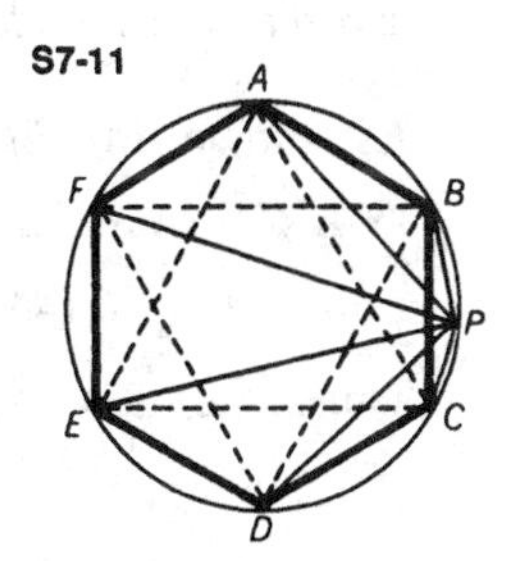

7-11 *If regular hexagon* ABCDEF *is inscribed in a circle, and point* P *is on* $\widehat{BC}$, *prove that* PE + PF = PA + PB + PC + PD.

Lines are drawn between points A, E, and C to make equilateral $\triangle AEC$ (Fig. S7-11). Using the results of Problem 7-8, we have

$$PE = PA + PC. \qquad \text{(I)}$$

In the same way, in equilateral $\triangle BFD$, $PF = PB + PD$. (II)
Adding (I) and (II), $PE + PF = PA + PB + PC + PD$.

7-12 *Equilateral* $\triangle$ADC *is drawn externally on side* $\overline{AC}$ *of* $\triangle$ABC. *Point* P *is taken on* $\overline{BD}$. *Find* m$\angle$APC *such that* BD = PA + PB + PC.

Point P must be the intersection of $\overline{BD}$ with the circumcircle of $\triangle ADC$. Then $m\angle APC = 120$ (#36). (See Fig. S7-12.)
Since $APCD$ is a cyclic quadrilateral, then by Ptolemy's Theorem,

$(PD)(AC) = (PA)(CD) + (PC)(AD)$. (I)

Since $\triangle ADC$ is equilateral, from (I), $PD = PA + PC$. (II)

However, $BD = PB + PD$. (III)

Therefore by substituting (II) into (III), $BD = PA + PB + PC$.

S7-12

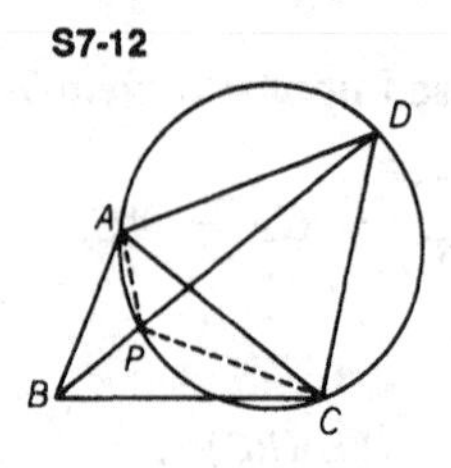

S7-13

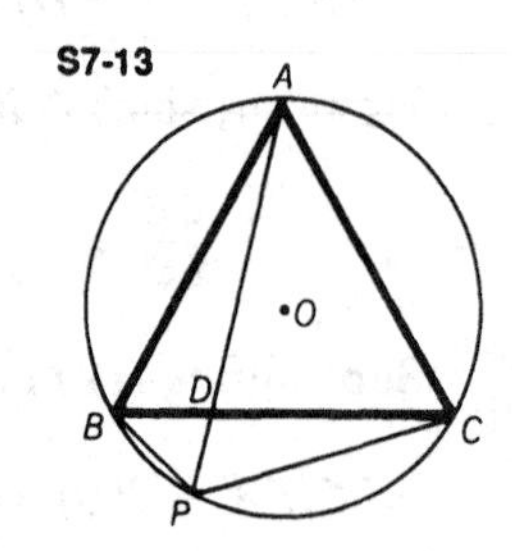

7-13 *A line drawn from vertex A of equilateral $\triangle ABC$, meets $\overline{BC}$ at D and the circumcircle at P. Prove that* $\frac{1}{PD} = \frac{1}{PB} + \frac{1}{PC}$.

As shown in Fig. S7-13, $m\angle PAC = m\angle PBC$ (#36). Since $\triangle ABC$ is equilateral, $m\angle BPA = \frac{1}{2}(m\widehat{AB}) = 60$, and $m\angle CPA = \frac{1}{2}(m\widehat{AC}) = 60$ (#36). Therefore, $m\angle BPA = m\angle CPA$.

Thus, $\triangle APC \sim \triangle BPD$, and $\frac{PA}{PB} = \frac{PC}{PD}$,

or $(PA)(PD) = (PB)(PC)$. (I)

Now, $PA = PB + PC$ (see Solution 7-8). (II)

Substituting (II) into (I),

$(PB)(PC) = PD(PB + PC) = (PD)(PB) + (PD)(PC)$. (III)

Now, dividing each term of (III) by $(PB)(PD)(PC)$, we obtain

$$\frac{1}{PD} = \frac{1}{PC} + \frac{1}{PB}.$$

Challenge 1 *If* BP = 5 *and* PC = 20, *find* AD.

ANSWER: 21

Challenge 2 *If* $m\widehat{BP}:m\widehat{PC} = 1:3$, *find the radius of the circle in challenge I.*

ANSWER: $10\sqrt{2}$

7-14 *Express in terms of the sides of a cylic quadrilateral the ratio of the diagonals.*

On the circumcircle of quadrilateral $ABCD$, choose points P and Q so that $PA = DC$, and $QD = AB$, as in Fig. S7-14.

Applying Ptolemy's Theorem to quadrilateral $ABCP$,

$(AC)(PB) = (AB)(PC) + (BC)(PA)$. (I)

Similarly, by applying Ptolemy's Theorem to quadrilateral $BCDQ$, $(BD)(QC) = (DC)(QB) + (BC)(QD)$. (II)

Since $PA + AB = DC + QD$, $m\widehat{PAB} = m\widehat{QDC}$, and $PB = QC$.

Similarly, since $m\widehat{PBC} = m\widehat{DBA}$, $PC = AD$, and since $m\widehat{QCB} = m\widehat{ACD}$, $QB = AD$.

Finally, dividing (I) by (II), and substituting for all terms containing Q and P, $\dfrac{AC}{BD} = \dfrac{(AB)(AD) + (BC)(DC)}{(DC)(AD) + (BC)(AB)}$.

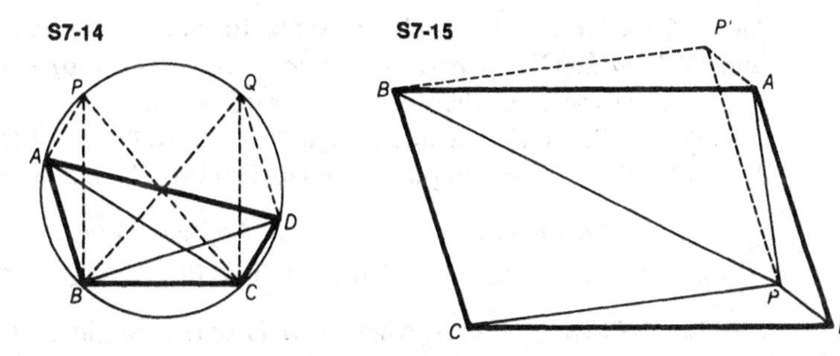

7-15 *A point* P *is chosen inside parallelogram* ABCD *such that* $\angle$APB *is supplementary to* $\angle$CPD.

Prove that (AB)(AD) = (BP)(DP) + (AP)(CP). (Fig. S7-15)

On side $\overline{AB}$ of parallelogram $ABCD$, draw $\triangle AP'B \cong \triangle DPC$, so that $DP = AP'$, $CP = BP'$. (I)

Since $\angle APB$ is supplementary to $\angle CPD$, and $m\angle BP'A = m\angle CPD$, $\angle APB$ is supplementary to $\angle BP'A$. Therefore, quadrilateral $BP'AP$ is cyclic. (#37).

Now, applying Ptolemy's Theorem to cyclic quadrilateral $BP'AP$, $(AB)(P'P) = (BP)(AP') + (AP)(BP')$.

From (I), $(AB)(P'P) = (BP)(DP) + (AP)(CP)$. (II)

Since $m\angle BAP' = m\angle CDP$, and $\overline{CD} \parallel \overline{AB}$, (#21a), $\overline{PD} \parallel \overline{P'A}$.

Therefore $PDAP'$ is a parallelogram (#22), and $P'P = AD$ (#21b).

Thus, from (II), $(AB)(AD) = (BP)(DP) + (AP)(CP)$.

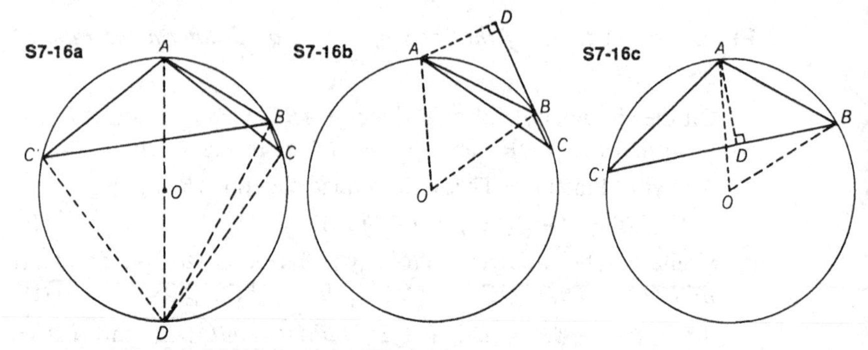

7-16 *A triangle inscribed in a circle of radius 5, has two sides measuring 5 and 6. Find the measure of the third side of the triangle.*

METHOD I: In Fig. S7-16a, we notice that there are two possibilities to consider in this problem. Both $\triangle ABC$, and $\triangle ABC'$ are inscribed in circle O, with $AB = 5$, and $AC = AC' = 6$. We are to find BC and BC'.

Draw diameter $\overline{AOD}$, which measures 10, and draw $\overline{DC}$, $\overline{DB}$, and $\overline{DC'}$. $m\angle AC'D = m\angle ACD = m\angle ABD = 90$ (#36).

Consider the case where $\angle A$ in $\triangle ABC$ is acute.

In right $\triangle ACD$, $DC = 8$, and in right $\triangle ABD$, $BD = 5\sqrt{3}$ (#55).

By Ptolemy's Theorem applied to quadrilateral $ABCD$,

$$(AC)(BD) = (AB)(DC) + (AD)(BC),$$

$$\text{or } (6)(5\sqrt{3}) = (5)(8) + (10)(BC), \text{ and } BC = 3\sqrt{3} - 4.$$

Now consider the case where $\angle A$ is obtuse, as in $\triangle ABC'$.

In right $\triangle AC'D$, $DC' = 8$ (#55).

By Ptolemy's Theorem applied to quadrilateral $ABDC'$,

$$(AC')(BD) + (AB)(DC') = (AD)(BC'),$$

$$(6)(5\sqrt{3}) + (5)(8) = (10)(BC'), \text{ and } BC' = 3\sqrt{3} + 4.$$

METHOD II: In Figs. S7-16b and S7-16c, draw radii $\overline{OA}$ and $\overline{OB}$. Also, draw a line from A perpendicular to $\overline{CB}(\overline{C'B})$ at D.

Since $AB = AO = BO = 5$, $m\angle AOB = 60$ (#6), so $m\widehat{AB} = 60$ (#35). Therefore, $m\angle ACB$ ($\angle AC'B$) $= 30$ (#36).

In right $\triangle ADC$, (right $\triangle ADC'$), since $AC(AC') = 6$, $CD(C'D) = 3\sqrt{3}$, and $AD = 3$ (#55c).

In right $\triangle ADB$, $BD = 4$ (#55).

Since $BC = CD - BD$, then $BC = 3\sqrt{3} - 4$ (in Fig. S7-16b).

In Fig. S7-16c, since $BC' = C'D + BD$, then $BC' = 3\sqrt{3} + 4$.

Challenge *Generalize the result of this problem for any triangle.*

ANSWER: $a = \dfrac{b\sqrt{4R^2 - c^2} \pm c\sqrt{4R^2 - b^2}}{2R}$, where R is the radius of the circumcircle, and the sides b and c are known.

8. Menelaus and Ceva: Collinearity and Concurrency

8-1 *Points P, Q, and R are taken on sides $\overline{AC}$, $\overline{AB}$, and $\overline{BC}$ (extended if necessary) of $\triangle ABC$. Prove that if these points are collinear, then* $\dfrac{AQ}{QB} \cdot \dfrac{BR}{RC} \cdot \dfrac{CP}{PA} = -1.$

This theorem, together with its converse, which is given in the Challenge that follows, constitutes the classic theorem known as Menelaus' Theorem.

METHOD I: In Fig. S8-1a and Fig. S8-1b, points P, Q, and R are collinear. Draw a line through C, parallel to $\overline{AB}$, meeting line segment $\overline{PQR}$ at D.

$\triangle DCR \sim \triangle QBR$ (#49), therefore $\dfrac{DC}{QB} = \dfrac{RC}{BR}$, or

$$DC = \frac{(QB)(RC)}{BR}. \qquad \text{(I)}$$

$\triangle PDC \sim \triangle PQA$ (#49 or #48), therefore $\dfrac{DC}{AQ} = \dfrac{CP}{PA}$, or

$$DC = \frac{(AQ)(CP)}{PA}. \qquad \text{(II)}$$

From (I) and (II), $\dfrac{(QB)(RC)}{BR} = \dfrac{(AQ)(CP)}{PA}$,

and $(QB)(RC)(PA) = (AQ)(CP)(BR)$, or $\left|\dfrac{AQ}{QB} \cdot \dfrac{BR}{RC} \cdot \dfrac{CP}{PA}\right| = 1.$

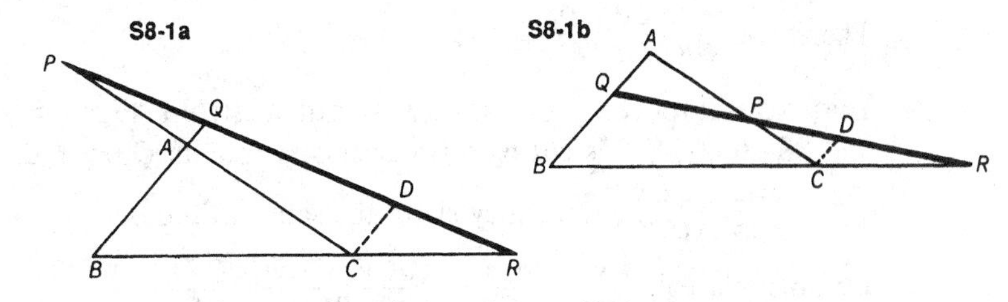

Taking direction into account in Fig. S8-1a, $\frac{AQ}{QB}$, $\frac{BR}{RC}$, and $\frac{CP}{PA}$ are each negative ratios, and in Fig. S8-1b $\frac{BR}{RC}$ is a negative ratio, while $\frac{AQ}{QB}$ and $\frac{CP}{PA}$ are positive ratios.

Therefore, $\frac{AQ}{QB} \cdot \frac{BR}{RC} \cdot \frac{CP}{PA} = -1$, since in each case there is an odd number of negative ratios.

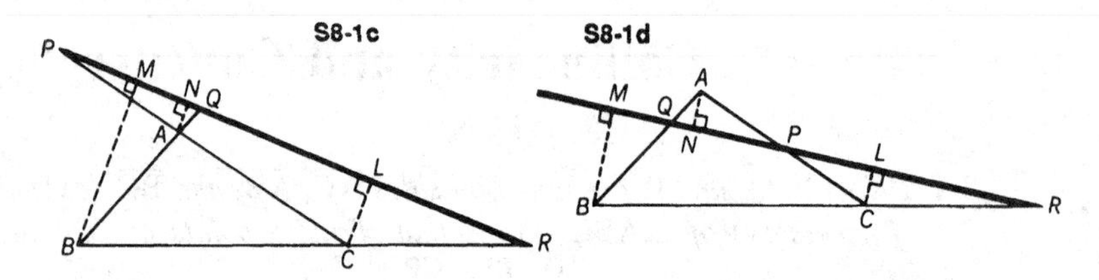

METHOD II: In Fig. S8-1c and Fig. S8-1d, $\overline{PQR}$ is a straight line. Draw $\overline{BM} \perp \overleftrightarrow{PR}$, $\overline{AN} \perp \overleftrightarrow{PR}$, and $\overline{CL} \perp \overleftrightarrow{PR}$.

Since $\triangle BMQ \sim \triangle ANQ$ (#48), $\frac{AQ}{QB} = \frac{AN}{BM}$. (I)

Also $\triangle LCP \sim \triangle NAP$ (#48), and $\frac{CP}{PA} = \frac{LC}{AN}$. (II)

$\triangle MRB \sim \triangle LRC$ (#49), and $\frac{BR}{RC} = \frac{BM}{LC}$. (III)

By multiplying (I), (II), and (III), we get, numerically,

$$\frac{AQ}{QB} \cdot \frac{CP}{PA} \cdot \frac{BR}{RC} = \frac{AN}{BM} \cdot \frac{LC}{AN} \cdot \frac{BM}{LC} = 1.$$

In Fig. S8-1c, $\frac{AQ}{QB}$ is negative, $\frac{CP}{PA}$ is negative, and $\frac{BR}{RC}$ is negative.

Therefore, $\frac{AQ}{QB} \cdot \frac{CP}{PA} \cdot \frac{BR}{RC} = -1$.

In Fig. S8-1d, $\frac{AQ}{QB}$ is positive, $\frac{CP}{PA}$ is positive, and $\frac{BR}{RC}$ is negative.

Therefore, $\frac{AQ}{QB} \cdot \frac{CP}{PA} \cdot \frac{BR}{RC} = -1$.

TRIGONOMETRIC FORM OF MENELAUS' THEOREM: In Figs. S8-1a and S8-1b, $\triangle ABC$ is cut by a transversal at points Q, P, and R.

$\frac{AQ}{BQ} = \frac{\text{area } \triangle QCA}{\text{area } \triangle QCB}$, since they share the same altitude.

By Formula #5b, $\frac{\text{area } \triangle QCA}{\text{area } \triangle QCB} = \frac{(QC)(AC) \sin \angle QCA}{(QC)(BC) \sin \angle QCB}$.

$$\text{Therefore, } \frac{AQ}{BQ} = \frac{AC \sin \angle QCA}{BC \sin \angle QCB}. \quad \text{(I)}$$

$$\text{Similarly, } \frac{BR}{CR} = \frac{AB \sin \angle BAR}{AC \sin \angle CAR}, \quad \text{(II)}$$

$$\text{and } \frac{PC}{PA} = \frac{BC \sin \angle PBC}{AB \sin \angle PBA}. \quad \text{(III)}$$

Multiplying (I), (II), and (III),

$$\frac{AQ}{BQ} \cdot \frac{BR}{CR} \cdot \frac{PC}{PA} = \frac{(AC)(AB)(BC)(\sin \angle QCA)(\sin \angle BAR)(\sin \angle PBC)}{(BC)(AC)(AB)(\sin \angle QCB)(\sin \angle CAR)(\sin \angle PBA)}.$$

However, $\frac{AQ}{BQ} \cdot \frac{BR}{CR} \cdot \frac{PC}{PA} = -1$ (Menelaus' Theorem).

Thus, $\frac{(\sin \angle QCA)(\sin \angle BAR)(\sin \angle PBC)}{(\sin \angle QCB)(\sin \angle CAR)(\sin \angle PBA)} = -1.$

Challenge *In* $\triangle ABC$ *points* P, Q, *and* R *are situated respectively on sides* $\overline{AC}$, $\overline{AB}$, *and* $\overline{BC}$ *(extended when necessary). Prove that if*

$$\frac{AQ}{QB} \cdot \frac{BR}{RC} \cdot \frac{CP}{PA} = -1,$$

then P, Q, *and* R *are collinear. This is part of Menelaus' Theorem.*

In Fig. S8-1a and Fig. S8-1b, let the line through R and Q meet $\overline{AC}$ at P'.

Then, by the theorem just proved in Problem 8-1,

$$\frac{AQ}{QB} \cdot \frac{BR}{RC} \cdot \frac{CP'}{P'A} = -1.$$

However, from our hypothesis,

$$\frac{AQ}{QB} \cdot \frac{BR}{RC} \cdot \frac{CP}{PA} = -1.$$

Therefore, $\frac{CP'}{P'A} = \frac{CP}{PA}$, and P and P' must coincide.

8-2 *Prove that three lines drawn from the vertices* A, B, *and* C *of* $\triangle ABC$ *meeting the opposite sides in points* L, M, *and* N, *respectively, are concurrent if and only if*

$$\frac{AN}{NB} \cdot \frac{BL}{LC} \cdot \frac{CM}{MA} = 1.$$

This is known as Ceva's Theorem.

METHOD I: In Fig. S8-2a and Fig. S8-2b, $\overline{AL}$, $\overline{BM}$, and $\overline{CN}$ meet in point P.

$$\frac{BL}{LC} = \frac{\text{area } \triangle ABL}{\text{area } \triangle ACL}, \text{ (share same altitude)} \tag{I}$$

Similarly, $$\frac{BL}{LC} = \frac{\text{area } \triangle PBL}{\text{area } \triangle PCL}. \tag{II}$$

Therefore from (I) and (II), $\dfrac{\text{area } \triangle ABL}{\text{area } \triangle ACL} = \dfrac{\text{area } \triangle PBL}{\text{area } \triangle PCL}$.

Thus, $$\frac{BL}{LC} = \frac{\text{area } \triangle ABL - \text{area } \triangle PBL}{\text{area } \triangle ACL - \text{area } \triangle PCL} = \frac{\text{area } \triangle ABP}{\text{area } \triangle ACP}. \tag{III}$$

Similarly, $\dfrac{CM}{MA} = \dfrac{\text{area } \triangle BMC}{\text{area } \triangle BMA} = \dfrac{\text{area } \triangle PMC}{\text{area } \triangle PMA}$.

Therefore, $$\frac{CM}{MA} = \frac{\text{area } \triangle BMC - \text{area } \triangle PMC}{\text{area } \triangle BMA - \text{area } \triangle PMA} = \frac{\text{area } \triangle BCP}{\text{area } \triangle ABP}. \tag{IV}$$

Also, $\dfrac{AN}{NB} = \dfrac{\text{area } \triangle ACN}{\text{area } \triangle BCN} = \dfrac{\text{area } \triangle APN}{\text{area } \triangle BPN}$.

Therefore, $$\frac{AN}{NB} = \frac{\text{area } \triangle ACN - \text{area } \triangle APN}{\text{area } \triangle BCN - \text{area } \triangle BPN} = \frac{\text{area } \triangle ACP}{\text{area } \triangle BCP}. \tag{V}$$

By multiplying (III), (IV), and (V) we get

$$\frac{BL}{LC} \cdot \frac{CM}{MA} \cdot \frac{AN}{NB} = 1. \tag{VI}$$

Since in Fig. S8-2a, all the ratios are positive, (VI) is positive. In Fig. S8-2b, $\dfrac{BL}{LC}$ and $\dfrac{AN}{NB}$ are negative, while $\dfrac{CM}{MA}$ is positive. Therefore, again, (VI) is positive.

Since Ceva's Theorem is an equivalence, it is necessary to prove the converse of the implication we have just proved. Let $\overline{BM}$ and $\overline{AL}$ meet at P. Join $\overline{PC}$ and extend it to meet $\overline{AB}$ at N'. Since $\overline{AL}$, $\overline{BM}$, and $\overline{CN'}$ are concurrent by the part of Ceva's Theorem we have already proved,

$$\frac{BL}{LC} \cdot \frac{CM}{MA} \cdot \frac{AN'}{N'B} = 1.$$

However, our hypothesis is $\dfrac{BL}{LC} \cdot \dfrac{CM}{MA} \cdot \dfrac{AN}{NB} = 1$.

Therefore, $\dfrac{AN'}{N'B} = \dfrac{AN}{NB}$, so that N and N' must coincide.

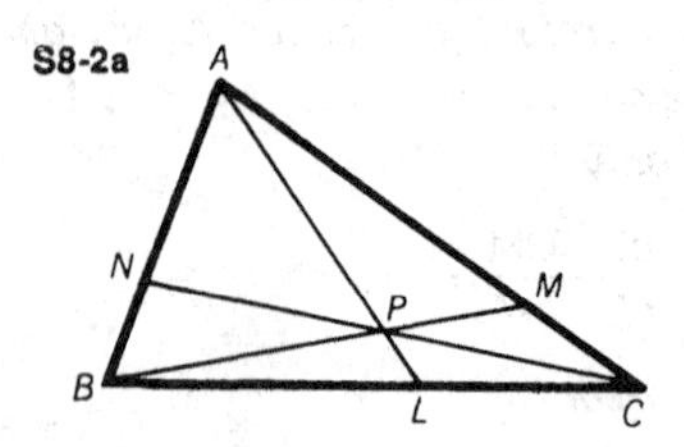

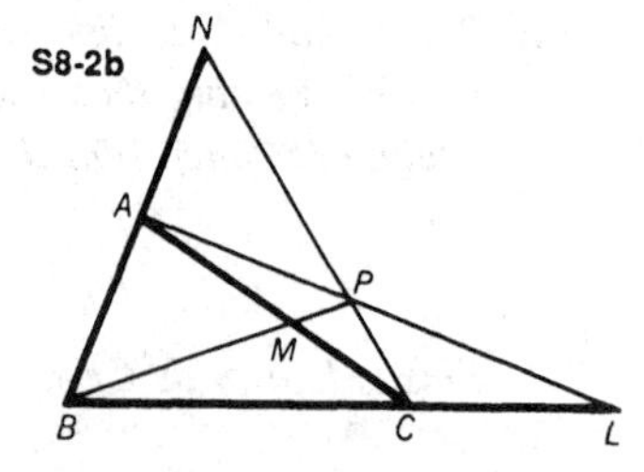

METHOD II: In Fig. S8-2c and Fig. S8-2d, draw a line through A, parallel to $\overline{BC}$ meeting $\overleftrightarrow{CP}$ at S and $\overleftrightarrow{BP}$ at R.

$$\triangle AMR \sim \triangle CMB \text{ (\#48), therefore } \frac{AM}{MC} = \frac{AR}{CB}. \quad \text{(I)}$$

$$\triangle BNC \sim \triangle ANS \text{ (\#48), therefore } \frac{BN}{NA} = \frac{CB}{SA}. \quad \text{(II)}$$

$$\triangle CLP \sim \triangle SAP \text{ (\#48), therefore } \frac{CL}{SA} = \frac{LP}{AP}. \quad \text{(III)}$$

$$\triangle BLP \sim \triangle RAP \text{ (\#48), therefore } \frac{BL}{RA} = \frac{LP}{AP}. \quad \text{(IV)}$$

From (III) and (IV), $\frac{CL}{SA} = \frac{BL}{RA}$, or $\frac{CL}{BL} = \frac{SA}{RA}$. (V)

By multiplying (I), (II), and (V),

$$\frac{AM}{MC} \cdot \frac{BN}{NA} \cdot \frac{CL}{BL} = \frac{AR}{CB} \cdot \frac{CB}{SA} \cdot \frac{SA}{RA} = 1, \quad \text{or } \frac{AN}{NB} \cdot \frac{BL}{LC} \cdot \frac{CM}{MA} = 1.$$

For a discussion about the sign of the resulting product, see Method I. The converse is proved as in Method I.

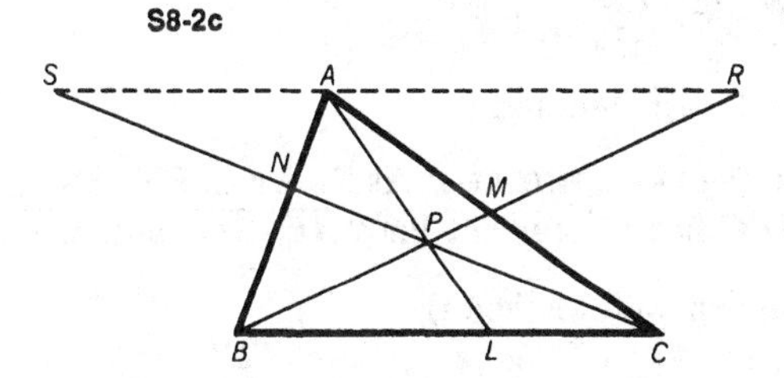

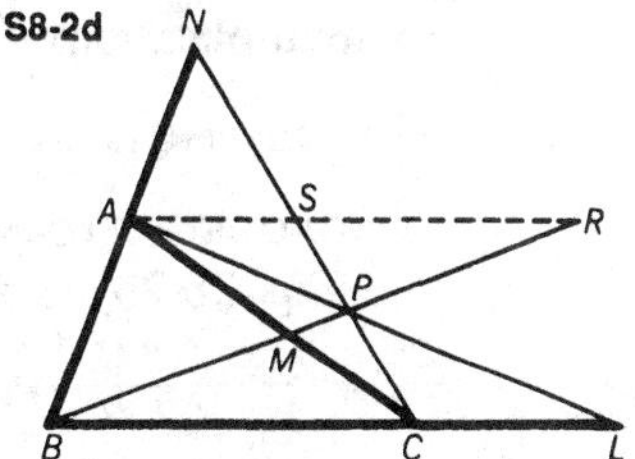

METHOD III: In Fig. S8-2e and Fig. S8-2f, draw a line through A and a line through C parallel to $\overline{BP}$ meeting $\overleftrightarrow{CP}$ and $\overleftrightarrow{AP}$ at S and R, respectively.

$$\triangle ASN \sim \triangle BPN \text{ (\#48 or \#49), and } \frac{AN}{NB} = \frac{AS}{BP}. \quad \text{(I)}$$

$$\triangle BPL \sim \triangle CRL \text{ (\#48 or 49), and } \frac{BL}{LC} = \frac{BP}{CR}. \quad \text{(II)}$$

$$\triangle PAM \sim \triangle RAC, \frac{CA}{MA} = \frac{RC}{PM} \text{ (\#49), and } CA = \frac{(RC)(MA)}{PM}. \quad \text{(III)}$$

$$\triangle PCM \sim \triangle SCA, \frac{CM}{CA} = \frac{PM}{AS} \text{ (\#49), and } CA = \frac{(AS)(CM)}{PM}. \quad \text{(IV)}$$

From (III) and (IV), $\frac{(RC)(MA)}{PM} = \frac{(AS)(CM)}{(PM)}$, or $\frac{CM}{MA} = \frac{RC}{AS}$. (V)

By multiplying (I), (II), and (V),

$$\frac{AN}{NB} \cdot \frac{BL}{LC} \cdot \frac{CM}{MA} = \frac{AS}{BP} \cdot \frac{BP}{CR} \cdot \frac{RC}{AS} = 1.$$

This proves that if the lines are concurrent, the ratio holds. The converse is proved as in Method I.

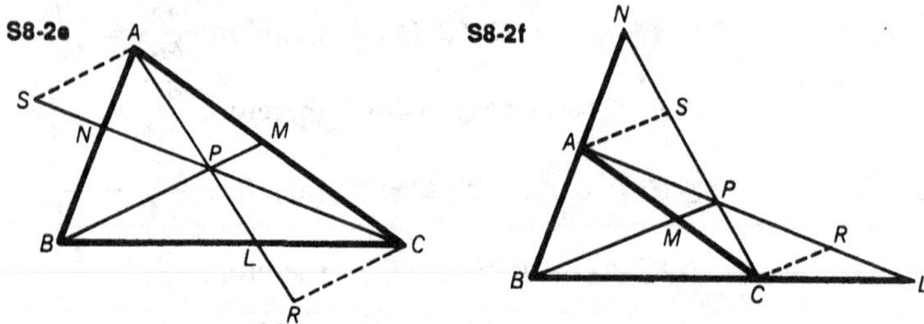

METHOD IV: In Figs. S8-2a and S8-2b, $\overline{BPM}$ is a transversal of $\triangle ACL$.

Applying Menelaus' Theorem, $\frac{AP}{PL} \cdot \frac{LB}{BC} \cdot \frac{CM}{MA} = -1$.

Similarly in $\triangle ALB$, $\overline{CPN}$ may be considered a transversal.

Thus, $\frac{AN}{NB} \cdot \frac{BC}{CL} \cdot \frac{LP}{PA} = -1$.

By multiplication, $\frac{AN}{NB} \cdot \frac{BL}{CL} \cdot \frac{CM}{MA} = 1$.

The converse is proved as in Method I.

TRIGONOMETRIC FORM OF CEVA'S THEOREM: As shown in Fig. S8-2a and Fig. S8-2b, $\triangle ABC$ has concurrent lines $\overline{AL}$, $\overline{BM}$, and $\overline{CN}$.

$\frac{BL}{LC} = \frac{\text{area } \triangle BAL}{\text{area } \triangle LAC}$ (Problem 8-2, Method I)

$$\frac{\frac{1}{2}(AL)(AB) \sin \angle BAL}{\frac{1}{2}(AL)(AC) \sin \angle LAC} = \frac{AB \sin \angle BAL}{AC \sin \angle LAC} \text{ (Formula \#5b)}$$

Similarly, $\frac{CM}{MA} = \frac{CB \sin \angle CBM}{AB \sin \angle ABM}$ and $\frac{AN}{NB} = \frac{AC \sin \angle ACN}{BC \sin \angle BCN}$.

By multiplying, $\frac{BL}{LC} \cdot \frac{CM}{MA} \cdot \frac{AN}{NB} =$

$$\frac{(AB)(BC)(AC)(\sin \angle BAL)(\sin \angle CBM)(\sin \angle ACN)}{(AC)(AB)(BC)(\sin \angle LAC)(\sin \angle ABM)(\sin \angle BCN)}.$$

However, since by Ceva's Theorem $\frac{BL}{LC} \cdot \frac{CM}{MA} \cdot \frac{AN}{NB} = 1$,

$$\frac{(\sin \angle BAL)(\sin \angle CBM)(\sin \angle ACN)}{(\sin \angle LAC)(\sin \angle ABM)(\sin \angle BCN)} = 1.$$

The converse is also true, that if

$\frac{(\sin \angle BAL)(\sin \angle CBM)(\sin \angle ACN)}{(\sin \angle LAC)(\sin \angle ABM)(\sin \angle BCN)} = 1$, then lines $\overline{AL}$, $\overline{BM}$, and $\overline{CN}$ are concurrent.

8-3 *Prove that the medians of any triangle are concurrent.*

In $\triangle ABC$, $\overline{AL}$, $\overline{BM}$, and $\overline{CN}$ are medians, as in Fig. S8-3. Therefore, $AN = NB$, $BL = LC$, and $CM = MA$.

So $(AN)(BL)(MC) = (NB)(LC)(MA)$,

$$\text{or } \frac{(AN)(BL)(CM)}{(NB)(LC)(MA)} = 1.$$

Thus, by Ceva's Theorem, $\overline{AL}$, $\overline{BM}$, and $\overline{CN}$ are concurrent.

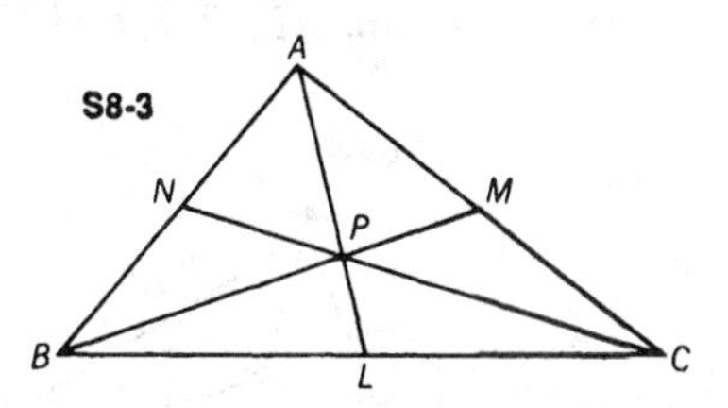

8-4 *Prove that the altitudes of any triangle are concurrent.*

In $\triangle ABC$, $\overline{AL}$, $\overline{BM}$, and $\overline{CN}$ are altitudes. (See Fig. S8-4a and Fig. S8-4b.)

$$\triangle ANC \sim \triangle AMB \text{ (\#48), and } \frac{AN}{MA} = \frac{AC}{AB}. \qquad \text{(I)}$$

$$\triangle BLA \sim \triangle BNC \text{ (\#48), and } \frac{BL}{NB} = \frac{AB}{BC}. \qquad \text{(II)}$$

$$\triangle CMB \sim \triangle CLA \text{ (\#48), and } \frac{CM}{LC} = \frac{BC}{AC}. \qquad \text{(III)}$$

By multiplying (I), (II), and (III),

$$\frac{AN}{MA} \cdot \frac{BL}{NB} \cdot \frac{CM}{LC} = \frac{AC}{AB} \cdot \frac{AB}{BC} \cdot \frac{BC}{AC} = 1.$$

Thus, by Ceva's Theorem, altitudes $\overline{AL}$, $\overline{BM}$, and $\overline{CN}$ are concurrent.

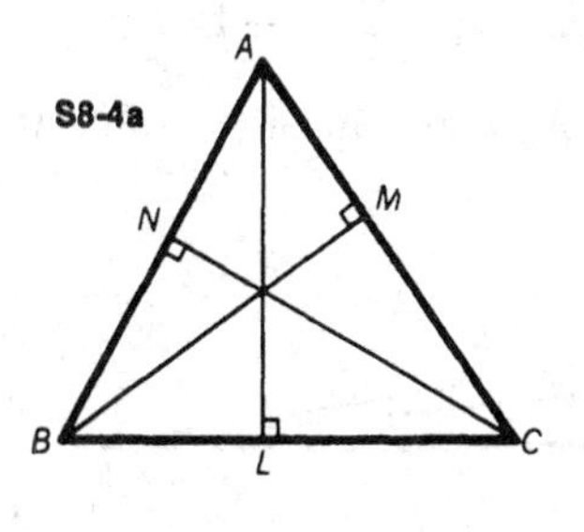

S8-4b

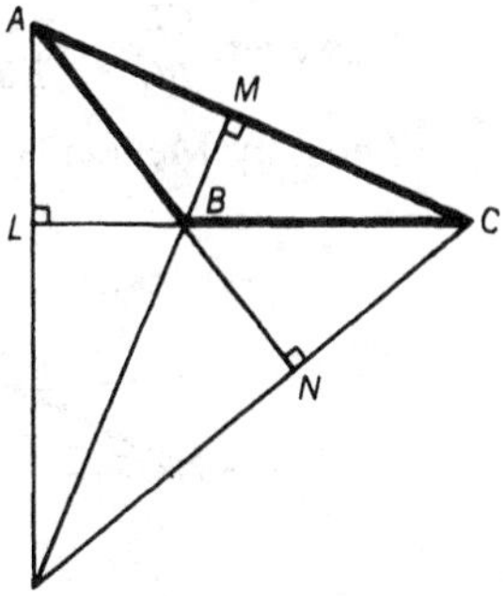

8-5 *Prove that the interior angle bisectors of a triangle are concurrent.*

In $\triangle ABC$, $\overline{AL}$, $\overline{BM}$, and $\overline{CN}$ are interior angle bisectors, as in Fig. S8-5.

Therefore, $\frac{AN}{NB} = \frac{AC}{BC}$ (#47), $\frac{BL}{LC} = \frac{AB}{AC}$ (#47), and $\frac{CM}{MA} = \frac{BC}{AB}$ (#47).

Thus, by multiplying,

$$\frac{AN}{NB} \cdot \frac{BL}{LC} \cdot \frac{CM}{MA} = \frac{AC}{BC} \cdot \frac{AB}{AC} \cdot \frac{BC}{AB} = 1.$$

Then, by Ceva's Theorem, $\overline{AL}$, $\overline{BM}$, and $\overline{CN}$ are concurrent.

S8-5

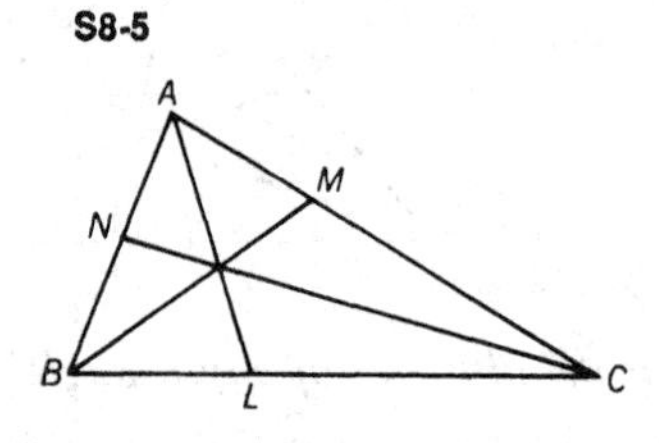

8-6 *Prove that the interior angle bisectors of two angles of a non-isosceles triangle and the exterior angle bisector of the third angle meet the opposite sides in three collinear points.*

In $\triangle ABC$, $\overline{BM}$ and $\overline{CN}$ are the interior angle bisectors, while $\overline{AL}$ bisects the exterior angle at A. (see Fig. S8-6.)

$\frac{AM}{MC} = \frac{AB}{BC}$ (#47), $\frac{BN}{NA} = \frac{BC}{AC}$ (#47), and $\frac{CL}{BL} = \frac{AC}{AB}$ (#47).

Therefore, by multiplication,

$$\frac{AM}{MC} \cdot \frac{BN}{NA} \cdot \frac{CL}{BL} = \frac{AB}{BC} \cdot \frac{BC}{AC} \cdot \frac{AC}{AB} = 1.$$

However, $\frac{CL}{BL} = \frac{-CL}{LB}$ therefore $\frac{AM}{MC} \cdot \frac{BN}{NA} \cdot \frac{CL}{LB} = -1$.

Thus, by Menelaus' Theorem, N, M, and L must be collinear.

S8-6

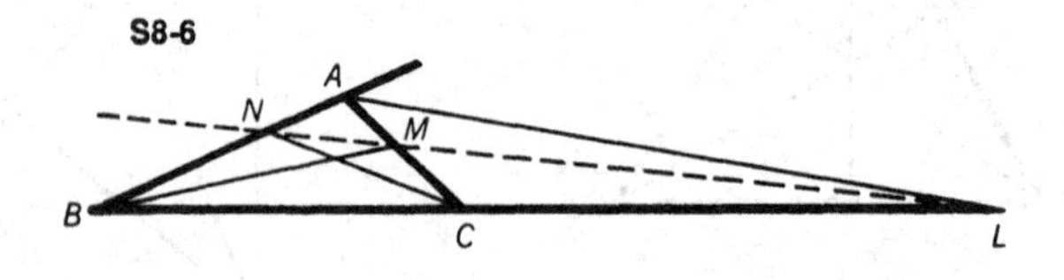

8-7 *Prove that the exterior angle bisectors of any non-isosceles triangle meet the opposite sides in three collinear points.*

In $\triangle ABC$, the bisectors of the exterior angles at A, B, and C meet the opposite sides (extended) at points N, L, and M respectively (Fig. S8-7).

$\frac{CL}{AL} = \frac{BC}{AB}$ (#47), $\frac{AM}{BM} = \frac{AC}{BC}$ (#47), and $\frac{BN}{CN} = \frac{AB}{AC}$ (#47).

Therefore, $\frac{CL}{AL} \cdot \frac{AM}{BM} \cdot \frac{BN}{CN} = \frac{BC}{AB} \cdot \frac{AC}{BC} \cdot \frac{AB}{AC} = -1$, since all three ratios are negative.

Thus, by Menelaus' Theorem, L, M, and N are collinear.

S8-7

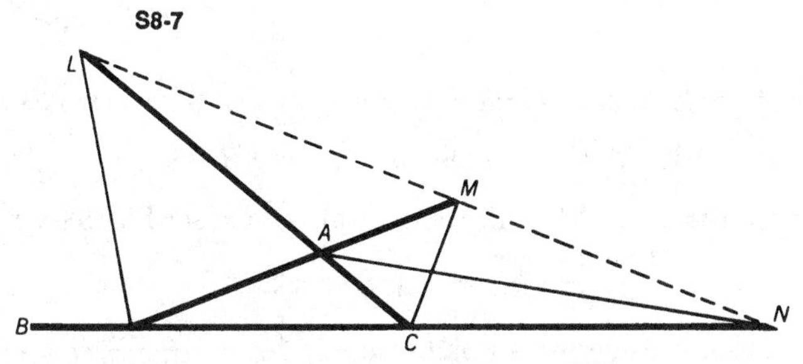

8-8 *In right* $\triangle$ABC, P *and* Q *are on* $\overline{BC}$ *and* $\overline{AC}$, *respectively, such that* CP = CQ = 2. *Through the point of intersection,* R, *of* $\overline{AP}$ *and* $\overline{BQ}$, *a line is drawn also passing through* C *and meeting* $\overline{AB}$ *at* S. $\overline{PQ}$ *extended meets* $\overleftrightarrow{AB}$ *at* T. *If hypotenuse* AB = 10 *and* AC = 8, *find* TS. *(See Fig. S8-8.)*

In right $\triangle ABC$, hypotenuse $AB = 10$, and $AC = 8$, so $BC = 6$ (#55).

In $\triangle ABC$, since $\overline{AP}$, $\overline{BQ}$, and $\overline{CS}$ are concurrent,

$$\frac{AQ}{QC} \cdot \frac{CP}{PB} \cdot \frac{BS}{SA} = 1, \text{ by Ceva's Theorem.}$$

Substituting, $\frac{6}{2} \cdot \frac{2}{4} \cdot \frac{BS}{10 - BS} = 1$, and $BS = 4$.

Now consider $\triangle ABC$ with transversal $\overline{QPT}$.

$$\frac{AQ}{QC} \cdot \frac{CP}{PB} \cdot \frac{BT}{TA} = -1 \text{ (Menelaus' Theorem).}$$

Since we are not dealing with directed line segments, this may be restated as $(AQ)(CP)(BT) = (QC)(PB)(AT)$.

Substituting, $(6)(2)(BT) = (2)(4)(BT + 10)$.

Then $BT = 20$, and $TS = 24$.

Challenge 1 *By how much is* TS *decreased if* P *is taken at the midpoint of* $\overline{BC}$?

ANSWER: $24 - 7\frac{1}{2} = 16\frac{1}{2}$

Challenge 2 *What is the minimum value of* TS?

ANSWER: $TS = 0$

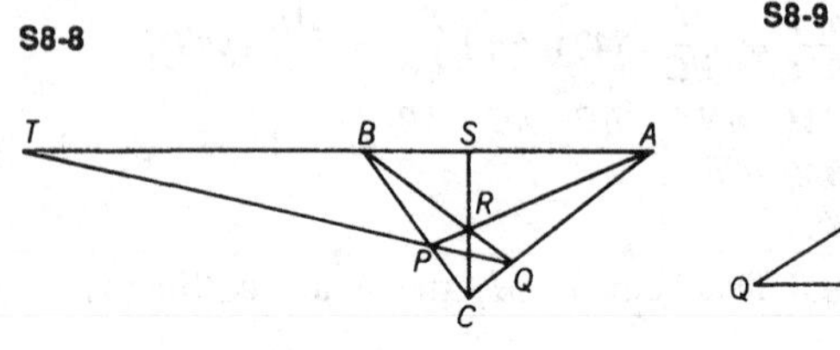

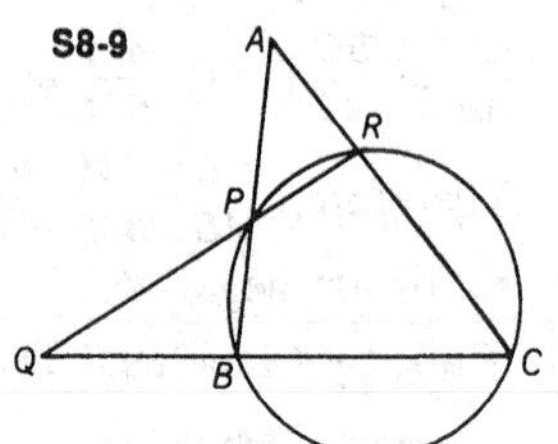

8-9 *A circle through vertices* B *and* C *of* $\triangle$ABC *meets* $\overline{AB}$ *at* P *and* $\overline{AC}$ *at* R. *If* $\overleftrightarrow{PR}$ *meets* $\overleftrightarrow{BC}$ *at* Q, *prove that* $\frac{QC}{QB} = \frac{(RC)(AC)}{(PB)(AB)}$.

Consider $\triangle ABC$ with transversal $\overline{QPR}$. (See Fig. S8-9.)

$$\frac{RC}{AR} \cdot \frac{AP}{PB} \cdot \frac{QB}{CQ} = -1 \text{ (Menelaus' Theorem)}$$

Then, considering absolute values, $\frac{QC}{QB} = \frac{RC}{AR} \cdot \frac{AP}{PB}$. (I)

However, $(AP)(AB) = (AR)(AC)$ (#54), or $\frac{AP}{AR} = \frac{AC}{AB}$. (II)

By substituting (II) in (I), we get $\frac{QC}{QB} = \frac{(RC)(AC)}{(PB)(AB)}$.

8-10 *In quadrilateral* ABCD, $\overleftrightarrow{AB}$ *and* $\overleftrightarrow{CD}$ *meet at* P; *while* $\overleftrightarrow{AD}$ *and* $\overleftrightarrow{BC}$ *meet at* Q. *Diagonals* $\overleftrightarrow{AC}$ *and* $\overleftrightarrow{BD}$ *meet* $\overleftrightarrow{PQ}$ *at* X *and* Y, *respectively. Prove that* $\frac{PX}{XQ} = -\frac{PY}{YQ}$. (See Fig. S8-10.)

Consider $\triangle PQC$ with $\overline{PB}$, $\overline{QD}$, and $\overline{CX}$ concurrent. By Ceva's Theorem,

$$\frac{PX}{XQ} \cdot \frac{QB}{BC} \cdot \frac{CD}{DP} = 1. \qquad \text{(I)}$$

Now consider $\triangle PQC$ with $\overline{DBY}$ as a transversal. By Menelaus' Theorem,

$$\frac{PY}{YQ} \cdot \frac{QB}{BC} \cdot \frac{CD}{DP} = -1. \qquad \text{(II)}$$

Therefore, from (I) and (II), $\frac{PX}{XQ} = -\frac{PY}{YQ}$.

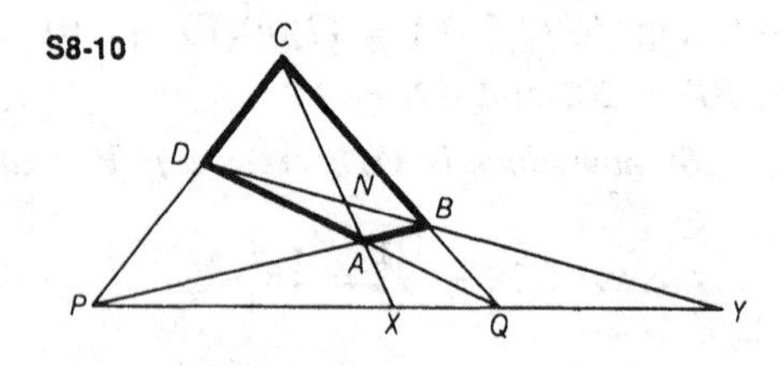

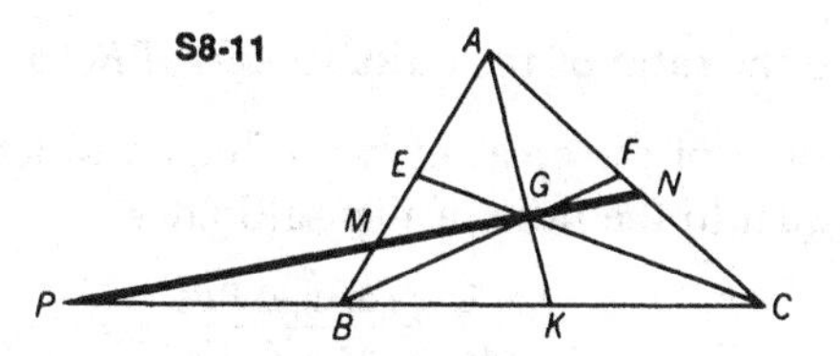

8-11 *Prove that a line drawn through the centroid,* G, *of* △ABC, *cuts sides* $\overline{AB}$ *and* $\overline{AC}$ *at points* M *and* N, *respectively, so that* (AM)(NC) + (AN)(MB) = (AM)(AN).

In Fig. S8-11, line $\overleftrightarrow{MGN}$ cuts $\overleftrightarrow{BC}$ at P. G is the centroid of $\triangle ABC$. Consider $\overline{NGP}$ as a transversal of $\triangle AKC$.

$$\frac{NC}{AN}\cdot\frac{AG}{GK}\cdot\frac{PK}{CP} = -1, \text{ by Menelaus' Theorem.}$$

Since $\frac{AG}{GK} = \frac{2}{1}$ (#29), $\frac{NC}{AN}\cdot\frac{2PK}{CP} = 1$, or $\frac{NC}{AN} = \frac{PC}{2PK}\cdot$ (I)

Now taking $\overline{GMP}$ as a transversal of $\triangle AKB$,

$$\frac{MB}{AM}\cdot\frac{AG}{GK}\cdot\frac{PK}{BP} = -1 \text{ (Menelaus' Theorem).}$$

Since $\frac{AG}{GK} = \frac{2}{1}$ (#29), $\frac{MB}{AM}\cdot\frac{2PK}{PB} = 1$ or $\frac{MB}{AM} = \frac{PB}{2PK}\cdot$ (II)

By adding (I) and (II), $\frac{NC}{AN} + \frac{MB}{AM} = \frac{PC + PB}{2PK}\cdot$

Since $PC = PB + 2BK$, then $PC + PB = 2(PB + BK) = 2PK$.

Thus, $\frac{(AM)(NC) + (AN)(MB)}{(AM)(AN)} = 1,$

and $(AM)(NC) + (AN)(MB) = (AM)(AN)$.

8-12 *In* △ABC, *points* L, M, *and* N *lie on* $\overline{BC}$, $\overline{AC}$, *and* $\overline{AB}$, *respectively, and* $\overline{AL}$, $\overline{BM}$, *and* $\overline{CN}$ *are concurrent. (See Fig. S8-12.)*

(a) *Find the numerical value of* $\frac{PL}{AL} + \frac{PM}{BM} + \frac{PN}{CN}\cdot$

(b) *Find the numerical value of* $\frac{AP}{AL} + \frac{BP}{BM} + \frac{CP}{CN}\cdot$

(a) Consider $\triangle PBC$ and $\triangle ABC$. Draw altitudes $\overline{PE}$ and $\overline{AD}$ of $\triangle PBC$ and $\triangle ABC$, respectively. Since $\overline{PE} \parallel \overline{AD}$ (#9), $\triangle PEL \sim \triangle ADL$ (#49), and $\frac{PE}{AD} = \frac{PL}{AL}\cdot$

Therefore the ratio of the altitudes of $\triangle PBC$ and $\triangle ABC$ is $\frac{PL}{AL}$.

The ratio of the areas of two triangles which share the same base is equal to the ratio of their altitudes.

$$\frac{PL}{AL} = \frac{\text{area } \triangle PBC}{\text{area } \triangle ABC}. \quad \text{(I)}$$

Similarly,

$$\frac{PM}{BM} = \frac{\text{area } \triangle CPA}{\text{area } \triangle ABC}, \quad \text{(II)}$$

and

$$\frac{PN}{CN} = \frac{\text{area } \triangle APB}{\text{area } \triangle ABC}. \quad \text{(III)}$$

By adding (I), (II), and (III), $\frac{PL}{AL} + \frac{PM}{BM} + \frac{PN}{CN}$

$$= \frac{\text{area } \triangle PBC}{\text{area } \triangle ABC} + \frac{\text{area } \triangle CPA}{\text{area } \triangle ABC} + \frac{\text{area } \triangle APB}{\text{area } \triangle ABC} = 1. \quad \text{(IV)}$$

(b)

$$\frac{AP}{AL} = \frac{AL - PL}{AL} = 1 - \frac{PL}{AL} \quad \text{(V)}$$

$$\frac{BP}{BM} = \frac{BM - BP}{BM} = 1 - \frac{BP}{BM} \quad \text{(VI)}$$

$$\frac{CP}{CN} = \frac{CN - PN}{CN} = 1 - \frac{PN}{CN} \quad \text{(VII)}$$

By adding (V), (VI), and (VII),

$$\frac{AP}{AL} + \frac{BP}{BM} + \frac{CP}{CN} = 3 - \left[\frac{PL}{AL} + \frac{BP}{BM} + \frac{PN}{CN}\right]. \quad \text{(VIII)}$$

However, from (IV), $\frac{PL}{AL} + \frac{BP}{BM} + \frac{PN}{CN} = 1.$

Substituting into (VIII), $\frac{AP}{AL} + \frac{BP}{BM} + \frac{CP}{CN} = 2.$

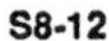

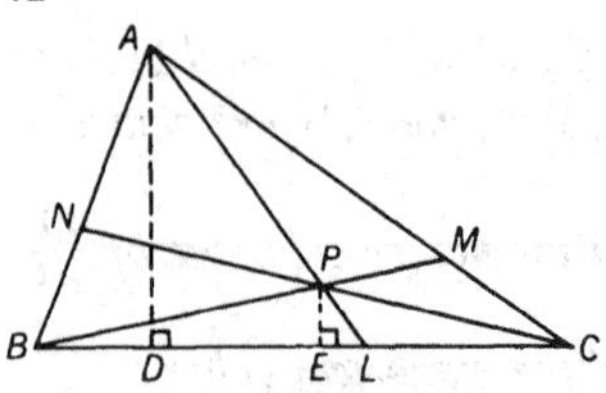

8-13 *Congruent line segments* $\overline{AE}$ *and* $\overline{AF}$ *are taken on sides* $\overline{AB}$ *and* $\overline{AC}$, *respectively, of* $\triangle ABC$. *The median* $\overline{AM}$ *intersects* $\overline{EF}$ *at point* Q. *Prove that* $\frac{QE}{QF} = \frac{AC}{AB}$.

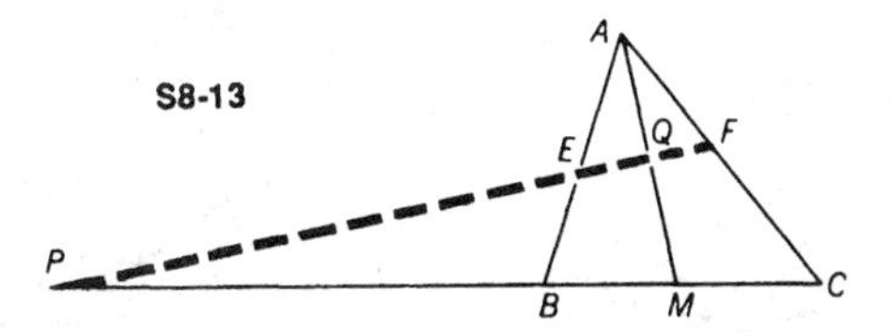

For $AB = AC$, the proof is trivial. Consider $AB \neq AC$.

Extend $\overline{FE}$ to meet $\overline{BC}$ (extended) at P. $\overline{FE}$ meets median $\overline{AM}$ at Q, as in Fig. S8-13.

Consider $\overline{AM}$ as a transversal of $\triangle PFC$.

$$\frac{PQ}{QF} \cdot \frac{FA}{AC} \cdot \frac{CM}{MP} = -1, \text{ by Menelaus' Theorem.} \tag{I}$$

Taking $\overline{AM}$ as a transversal of $\triangle PEB$, we have

$$\frac{QE}{PQ} \cdot \frac{AB}{EA} \cdot \frac{MP}{BM} = -1. \tag{II}$$

By multiplying (I) and (II), we obtain $\dfrac{QE}{QF} \cdot \dfrac{FA}{AC} \cdot \dfrac{AB}{EA} \cdot \dfrac{CM}{BM} = 1.$

However, since $FA = EA$ and $BM = CM$, $\dfrac{QE}{QF} = \dfrac{AC}{AB}$.

8-14 *In $\triangle ABC$, $\overleftrightarrow{AL}$, $\overleftrightarrow{BM}$, and $\overleftrightarrow{CN}$ are concurrent at P. Express the ratio $\frac{AP}{PL}$ in terms of segments made by the concurrent lines on the sides of $\triangle ABC$.*

In the proof of Ceva's Theorem (Problem 8-2, Method I), it was established that

$$\frac{BL}{LC} = \frac{\text{area } \triangle ABP}{\text{area } \triangle ACP}, \tag{III}$$

$$\frac{CM}{MA} = \frac{\text{area } \triangle BCP}{\text{area } \triangle ABP}, \tag{IV}$$

$$\text{and } \frac{AN}{NB} = \frac{\text{area } \triangle ACP}{\text{area } \triangle BCP}. \tag{V}$$

$$\frac{AP}{PL} = \frac{\text{area } \triangle ABP}{\text{area } \triangle LBP}, \tag{VI}$$

$$\text{and } \frac{AP}{PL} = \frac{\text{area } \triangle ACP}{\text{area } \triangle LCP}. \tag{VII}$$

Therefore from (VI) and (VII),

$$\frac{AP}{PL} = \frac{\text{area } \triangle ABP}{\text{area } \triangle LBP} = \frac{\text{area } \triangle ACP}{\text{area } \triangle LCP}$$

$$= \frac{\text{area } \triangle ABP + \text{area } \triangle ACP}{\text{area } \triangle BCP} = \frac{\text{area } \triangle ABP}{\text{area } \triangle BCP} + \frac{\text{area } \triangle ACP}{\text{area } \triangle BCP}.$$

From (IV) and (V), $\dfrac{AP}{PL} = \dfrac{MA}{CM} + \dfrac{AN}{NB}$, for Fig. S8-14a;

and $\dfrac{AP}{PL} = \dfrac{MA}{CM} - \dfrac{AN}{NB}$, for Fig. S8-14b.

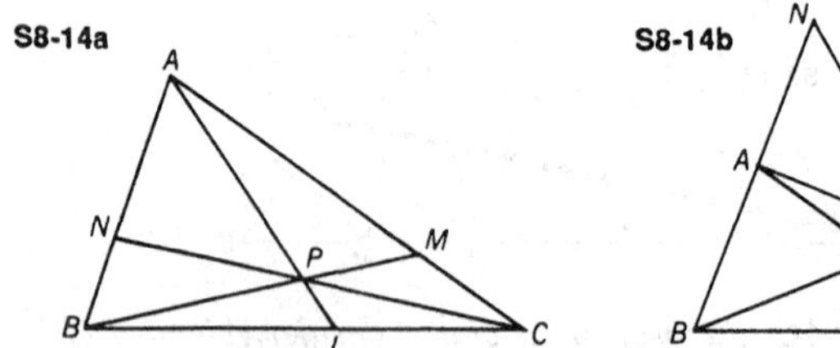

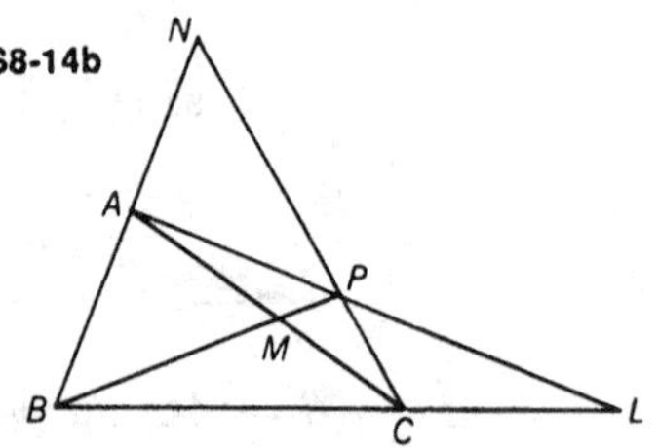

Thus, we have established the ratio into which the point of concurrency divides any cevian (i.e. the line segment from any vertex to the opposite side).

8-15 *Side* $\overline{AB}$ *of square* ABCD *is extended to* P *so that* BP = 2(AB). *With* M *the midpoint of* $\overline{DC}$, $\overline{BM}$ *is drawn meeting* $\overline{AC}$ *at* Q. $\overline{PQ}$ *meets* $\overline{BC}$ *at* R. *Using Menelaus' Theorem, find the ratio* $\frac{CR}{RB}$.

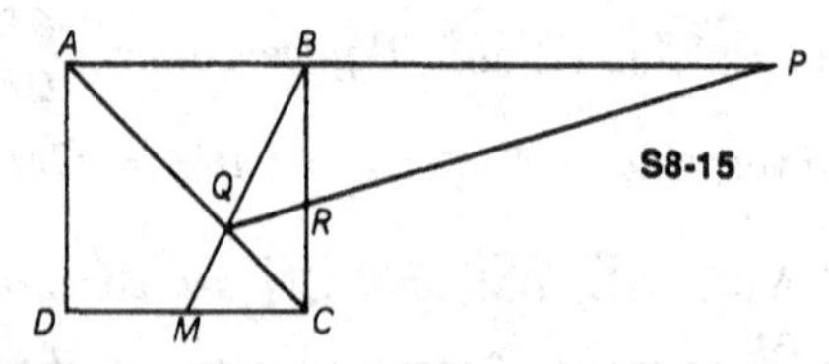

Applying Menelaus' Theorem to $\triangle ABC$ (Fig. S8-15) with transversal $\overline{PRQ}$, $\frac{CR}{RB} \cdot \frac{AQ}{QC} \cdot \frac{BP}{PA} = -1.$ (I)

Since $m\angle BAC = m\angle MCA$ (#8), and $m\angle MQC = m\angle AQB$ (#1), $\triangle MQC \sim \triangle BQA$ (#48), and $\frac{AQ}{QC} = \frac{AB}{MC}$. (II)

But $2(MC) = DC = AB$, or $\frac{AB}{MC} = \frac{2}{1}$. (III)

From (II) and (III), $\frac{AQ}{QC} = \frac{2}{1}$. (IV)

Since $BP = 2(AB)$, $\frac{BP}{AP} = \frac{2}{3}$ or $\frac{PB}{PA} = \frac{-2}{3}$. (V)

Substituting (IV) and (V) into (I), $\frac{CR}{RB} \cdot \frac{2}{1} \cdot \frac{-2}{3} = -1$, or $\frac{CR}{RB} = \frac{3}{4}$.

Challenge 1 *Find* $\frac{CR}{RB}$ *when* BP = AB.

ANSWER: 1

Challenge 2 *Find* $\frac{CR}{RB}$ *when* BP = k(AB).

ANSWER: $\frac{k+1}{2k}$

8-16 *Sides* $\overleftrightarrow{AB}$, $\overleftrightarrow{BC}$, $\overleftrightarrow{CD}$, *and* $\overleftrightarrow{DA}$ *of quadrilateral* ABCD *are cut by a straight line at points* K, L, M, *and* N *respectively. Prove that* $\frac{BL}{LC} \cdot \frac{AK}{KB} \cdot \frac{DN}{NA} \cdot \frac{CM}{MD} = 1$.

Draw diagonal $\overline{AC}$ meeting $\overline{KLNM}$ at P. (See Fig. S8-16.)

Consider $\overline{KLP}$ as a transversal of $\triangle ABC$.

$$\frac{BL}{LC} \cdot \frac{AK}{KB} \cdot \frac{CP}{PA} = -1 \text{ (Menelaus' Theorem)} \qquad \text{(I)}$$

Now consider $\overline{MNP}$ as a transversal of $\triangle ADC$.

$$\frac{DN}{NA} \cdot \frac{CM}{MD} \cdot \frac{PA}{CP} = -1 \text{ Then, } \frac{DN}{NA} \cdot \frac{CM}{MD} = -\frac{CP}{PA}. \qquad \text{(II)}$$

Substituting (II) into (I), we get $\frac{BL}{LC} \cdot \frac{AK}{KB} \cdot \frac{DN}{NA} \cdot \frac{CM}{MD} = 1.$

S8-16

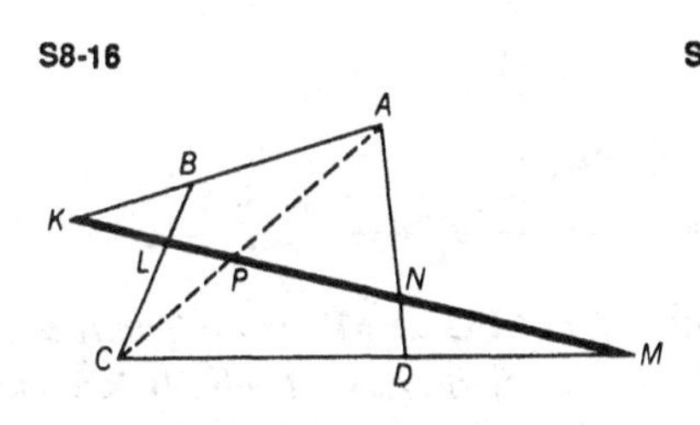

S8-17

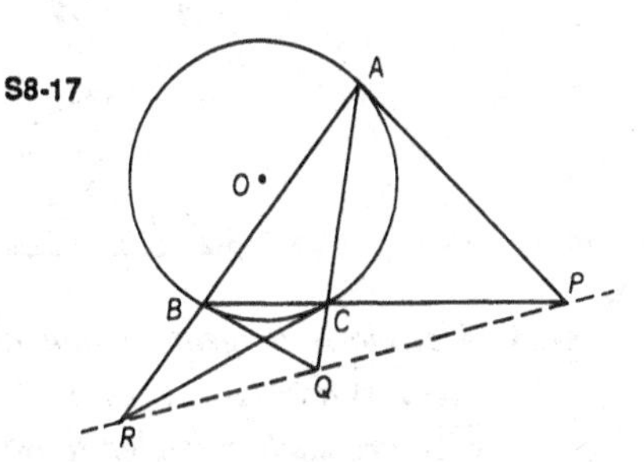

8-17 *Tangents to the circumcircle of* $\triangle$ABC, *at points* A, B, *and* C, *meet sides* $\overleftrightarrow{BC}$, $\overleftrightarrow{AC}$, *and* $\overleftrightarrow{AB}$ *at points* P, Q, *and* R *respectively. Prove that points* P, Q, *and* R *are collinear.*

In Fig. S8-17, since both $\angle BAC$ and $\angle QBC$ are equal in measure to one-half $m\widehat{BC}$ (#36, #38), $m\angle BAC = m\angle QBC$. Therefore, $\triangle ABQ \sim \triangle BCQ$ (#48), and $\frac{AQ}{BQ} = \frac{BA}{BC}$, or $\frac{(AQ)^2}{(BQ)^2} = \frac{(BA)^2}{(BC)^2}$. (I)

However, $(BQ)^2 = (AQ)(CQ)$ (#53). (II)

By substituting (II) into (I), we get $\frac{AQ}{CQ} = \frac{(BA)^2}{(BC)^2}$. (III)

Similarly, since $\angle BCR$ and $\angle BAC$ are equal in measure to one-half $m\widehat{BC}$ (#36, #38), $m\angle BCR = m\angle BAC$. Therefore, $\triangle CRB \sim \triangle ARC$ (#48), and $\frac{CR}{AR} = \frac{BC}{AC}$, or $\frac{(CR)^2}{(AR)^2} = \frac{(BC)^2}{(AC)^2}$. (IV)

However, $(CR)^2 = (AR)(RB)$ (#53). (V)

By substituting (V) into (IV), $\frac{RB}{AR} = \frac{(BC)^2}{(AC)^2}$. (VI)

Also, since $\angle CAP$ and $\angle ABC$ are equal in measure to one-half $m\widehat{AC}$ (#36, #38), $m\angle CAP = m\angle ABC$. Therefore,

$\triangle CAP \sim \triangle ABP$ and $\frac{AP}{BP} = \frac{AC}{BA}$, or $\frac{(AP)^2}{(BP)^2} = \frac{(AC)^2}{(BA)^2}$. (VII)

However, $(AP)^2 = (BP)(PC)$ (#53). (VIII)

By substituting (VIII) into (VII), $\frac{PC}{BP} = \frac{(AC)^2}{(BA)^2}$. (IX)

Now multiplying (III), (VI), and (IX),

$$\left|\frac{AQ}{CQ}\right| \cdot \left|\frac{RB}{AR}\right| \cdot \left|\frac{PC}{BP}\right| = \frac{(BA)^2}{(BC)^2} \cdot \frac{(BC)^2}{(AC)^2} \cdot \frac{(AC)^2}{(BA)^2} = |-1|.$$

Therefore, $\frac{AQ}{CQ} \cdot \frac{RB}{AR} \cdot \frac{PC}{BP} = -1$, since all the ratios on the left side are negative. Thus, by Menelaus' Theorem, P, Q, and R are collinear.

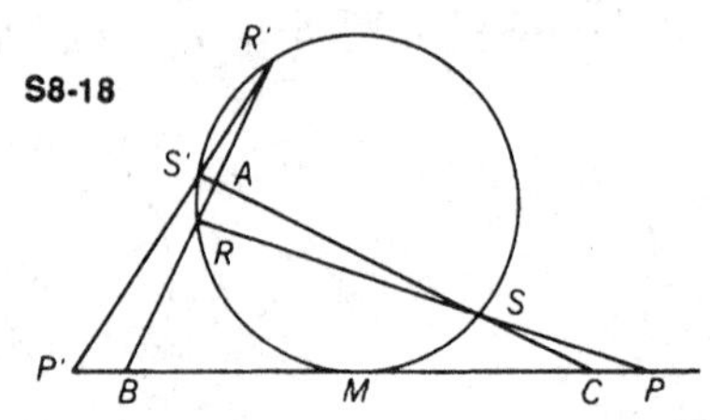

8-18 *A circle is tangent to side* $\overline{BC}$ *of* $\triangle ABC$ *at* M, *its midpoint, and cuts* $\overleftrightarrow{AB}$ *and* $\overleftrightarrow{AC}$ *at points* R, R′, *and* S, S′, *respectively. If* $\overline{RS}$ *and* $\overline{R'S'}$ *are each extended to meet* $\overleftrightarrow{BC}$ *at points* P *and* P′ *respectively, prove that* (BP)(BP′) = (CP)(CP′).

Consider $\overline{RSP}$ as a transversal of $\triangle ABC$ (Fig. S8-18).

$$\frac{BP}{CP} \cdot \frac{AR}{BR} \cdot \frac{CS}{AS} = -1, \text{ (Menelaus' Theorem)}$$

$$\text{or } \frac{BP}{CP} = -\frac{BR}{AR} \cdot \frac{AS}{CS}. \qquad \text{(I)}$$

Now consider $\overline{R'S'P'}$ as a transversal of $\triangle ABC$.

$$\frac{CP'}{BP'} \cdot \frac{BR'}{AR'} \cdot \frac{AS'}{CS'} = -1, \text{ or } \frac{CP'}{BP'} = \frac{-AR'}{BR'} \cdot \frac{CS'}{AS'}. \qquad \text{(II)}$$

However, $(AS')(AS) = (AR')(AR)$ (#52), or $\frac{AR'}{AS'} = \frac{AS}{AR}$. (III)

Also, $(BM)^2 = (BR)(BR')$ and $(MC)^2 = (CS)(CS')$ (#53).

But $BM = MC$; therefore $(BR)(BR') = (CS)(CS')$

$$\text{or } \frac{CS'}{BR'} = \frac{BR}{CS}. \qquad \text{(IV)}$$

By substituting (III) and (IV) into (II), we get from (I),

$$\frac{CP'}{BP'} = -\frac{BR}{CS} \cdot \frac{AS}{AR} = \frac{BP}{CP}.$$

Therefore, $(BP)(BP') = (CP)(CP')$.

8-19 *In* $\triangle$ABC, P, Q, *and* R *are the midpoints of the sides* $\overline{AB}$, $\overline{BC}$, *and* $\overline{AC}$. *Lines* $\overleftrightarrow{AN}$, $\overleftrightarrow{BL}$, *and* $\overleftrightarrow{CM}$ *are concurrent, meeting the opposite sides in* N, L, *and* M, *respectively. If* $\overleftrightarrow{PL}$ *meets* $\overleftrightarrow{BC}$ *at* J, $\overleftrightarrow{MQ}$ *meets* $\overleftrightarrow{AC}$ *at* I, *and* $\overleftrightarrow{RN}$ *meets* AB *at* H, *prove that* H, I, *and* J *are collinear.*

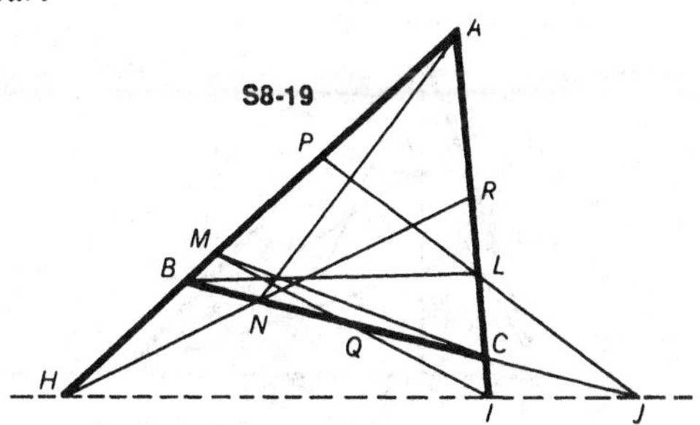

Since $\overline{RNH}$ is a transversal of $\triangle ABC$, as shown in Fig. S8-19, $\frac{AH}{HB} \cdot \frac{CR}{RA} \cdot \frac{BN}{NC} = -1$, by Menelaus' Theorem.

However, $RA = CR$.

Therefore, $\frac{AH}{HB} = -\frac{NC}{BN}$. (I)

Consider $\overline{PLJ}$ as a transversal of $\triangle ABC$.

$\frac{CL}{LA} \cdot \frac{AP}{PB} \cdot \frac{BJ}{JC} = -1$ (Menelaus' Theorem)

However $AP = PB$, therefore $\frac{BJ}{JC} = -\frac{LA}{CL}$. (II)

Now consider $\overline{MQI}$ as a transversal of $\triangle ABC$

$\frac{CI}{IA} \cdot \frac{BQ}{QC} \cdot \frac{AM}{MB} = -1$ (by Menelaus' Theorem)

Since $BQ = QC$, $\frac{CI}{IA} = -\frac{MB}{AM}$. (III)

By multiplying (I), (II), and (III), we get

$$\frac{AH}{HB} \cdot \frac{BJ}{JC} \cdot \frac{CI}{IA} = -\frac{NC}{BN} \cdot \frac{LA}{CL} \cdot \frac{MB}{AM}.$$

However, since $\overline{AN}$, $\overline{BL}$, and $\overline{CM}$ are concurrent,

$$\frac{NC}{BN} \cdot \frac{LA}{CL} \cdot \frac{MB}{AM} = 1 \text{ (Ceva's Theorem).}$$

Therefore, $\frac{AH}{HB} \cdot \frac{BJ}{JC} \cdot \frac{CI}{IA} = -1$, and by Menelaus' Theorem, H, I, and J are collinear.

8-20 $\triangle ABC$ *cuts a circle at points* E, E′, D, D′, F, F′, *as in Fig.* S8-20. *Prove that if* $\overline{AD}$, $\overline{BF}$, *and* $\overline{CE}$ *are concurrent, then* $\overline{AD'}$, $\overline{BF'}$, *and* $\overline{CE'}$ *are also concurrent.*

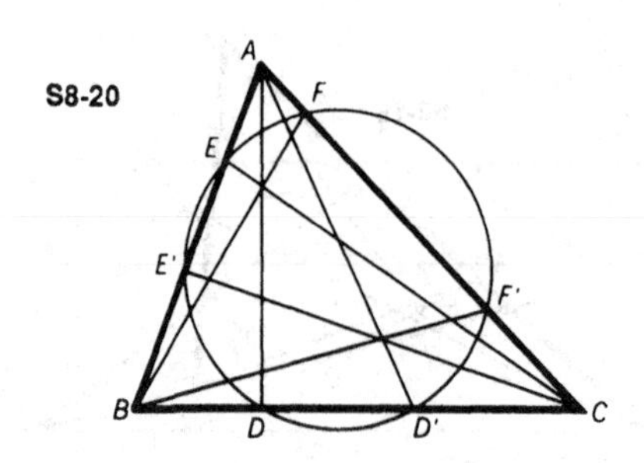

Since $\overline{AD}$, $\overline{BF}$, and $\overline{CE}$ are concurrent, then

$$\frac{AE}{EB}\cdot\frac{BD}{DC}\cdot\frac{CF}{FA} = 1 \text{ (Ceva's Theorem).} \qquad \text{(I)}$$

$$(AE)(AE') = (AF)(AF') \text{ (\#54), or } \frac{AE}{AF} = \frac{AF'}{AE'}. \qquad \text{(II)}$$

$$(BE')(BE) = (BD)(BD') \text{ (\#54), or } \frac{BD}{BE} = \frac{BE'}{BD'}. \qquad \text{(III)}$$

$$(CD')(CD) = (CF')(CF) \text{ (\#54), or } \frac{CF}{CD} = \frac{CD'}{CF'}. \qquad \text{(IV)}$$

By multiplying (II), (III), and (IV), we get

$$\frac{AE}{AF}\cdot\frac{BD}{BE}\cdot\frac{CF}{CD} = \frac{AF'}{AE'}\cdot\frac{BE'}{BD'}\cdot\frac{CD'}{CF'}.$$

But from (I) we know that $\frac{AE}{AF}\cdot\frac{BD}{BE}\cdot\frac{CF}{CD} = 1.$

Therefore, $\frac{AF'}{AE'}\cdot\frac{BE'}{BD'}\cdot\frac{CD'}{CF'} = 1$, and by Ceva's Theorem, $\overline{AD'}$, $\overline{BF'}$, and $\overline{CE'}$ are concurrent.

8-21 *Prove that the three pairs of common external tangents to three circles, taken two at a time, meet in three collinear points.*

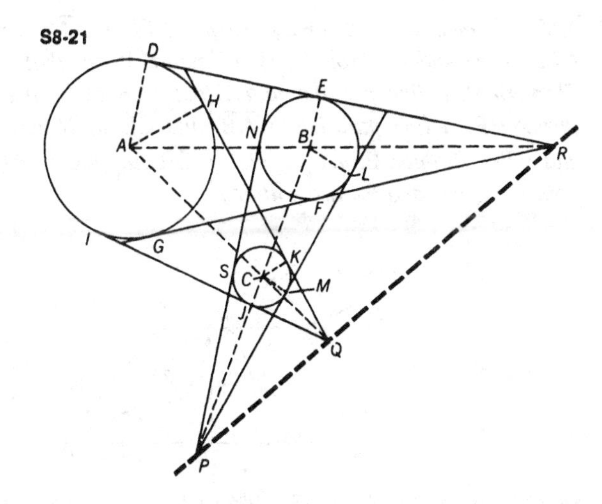

In Fig. S8-21, common external tangents to circles A and B meet at R, and intersect the circles at points D, E, F, and G.

Common external tangents to circles A and C meet at Q, and intersect the circles at points H, I, J, and K.

Common external tangents to circles B and C meet at P, and intersect the circles at points L, M, N, and S.

Draw $\overline{AD}$, $\overline{AH}$, $\overline{BE}$, $\overline{BL}$, $\overline{CK}$, and $\overline{CM}$. $\overline{AD} \perp \overline{DR}$, $\overline{BE} \perp \overline{DR}$ (#32a), so $\overline{AD} \parallel \overline{BE}$ (#9), $\triangle RAD \sim \triangle RBE$ (#49),

$$\text{and } \frac{AR}{RB} = \frac{AD}{BE}. \tag{I}$$

Similarly, $\overline{BL} \perp \overline{PL}$, $\overline{CM} \perp \overline{PL}$ and $\overline{BL} \parallel \overline{CM}$, so that

$$\triangle PBL \sim \triangle PCM \text{ (\#49), and } \frac{BP}{PC} = \frac{BL}{CM}. \tag{II}$$

Also $\overline{AH} \perp \overline{QH}$, and $\overline{CK} \perp \overline{QH}$, and $\overline{AH} \parallel \overline{CK}$, so that

$$\triangle QAH \sim \triangle QCK, \text{ (\#49), and } \frac{QC}{AQ} = \frac{CK}{AH}. \tag{III}$$

By multiplying (I), (II), and (III), we get

$$\frac{AR}{RB} \cdot \frac{BP}{PC} \cdot \frac{QC}{AQ} = \frac{AD}{BE} \cdot \frac{BL}{CM} \cdot \frac{CK}{AH}. \tag{IV}$$

Since $AH = AD$, $CK = CM$, and $BL = BE$,

$\frac{AR}{RB} \cdot \frac{BP}{PC} \cdot \frac{QC}{AQ} = -1$. (Note that they are all negative ratios.)

Thus, by Menelaus' Theorem, P, Q, and R are collinear.

8-22 $\overline{AM}$ *is a median of* $\triangle ABC$, *and point* G *on* $\overline{AM}$ *is the centroid.* $\overline{AM}$ *is extended through* M *to point* P *so that* GM = MP. *Through* P, *a line parallel to* $\overline{AC}$ *cuts* $\overline{AB}$ *at* Q, *and* $\overline{BC}$ *at* P_1; *through* P, *a line parallel to* $\overline{AB}$ *cuts* $\overline{CB}$ *at* N *and* $\overline{AC}$ *at* P_2; *and a line through* P *and parallel to* $\overline{CB}$ *cuts* $\overleftrightarrow{AB}$ *at* P_3. *Prove that points* P_1, P_2, *and* P_3 *are collinear.*

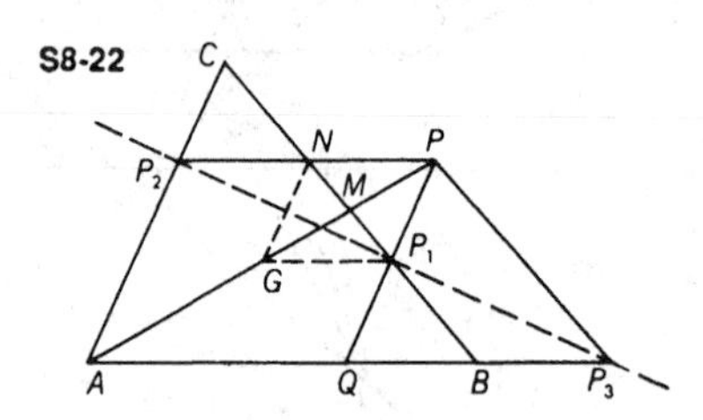

In Fig. S8-22, since $\overline{PP_1Q} \parallel \overline{AC}$, $\triangle CMA \sim \triangle P_1MP$ (#48), and

$$\frac{CM}{MP_1} = \frac{AM}{MP} = \frac{3}{1} \text{ (\#29).} \qquad \text{(I)}$$

Similarly, $\triangle AMB \sim \triangle PMN$, and

$$\frac{MB}{MN} = \frac{AM}{MP} = \frac{3}{1}. \qquad \text{(II)}$$

From (I) and (II), $\dfrac{CM}{MP_1} = \dfrac{MB}{MN}$. (III)

However, since $CM = MB$, from (III), $MP_1 = MN$,

and $CN = P_1B$. (IV)

Thus, $PNGP_1$ is a parallelogram (#21f).

Since $\overline{NG} \parallel \overline{AC}$, in $\triangle CMA$, $\dfrac{CN}{NM} = \dfrac{AG}{GM} = \dfrac{2}{1}$ (#46).

Therefore, $\dfrac{CN}{NB} = \dfrac{2}{1}$.

In $\triangle ABC$, where $\overline{P_2N} \parallel \overline{AB}$, $\dfrac{CP_2}{P_2A} = \dfrac{CN}{NB} = \dfrac{1}{2}$ (#46). (V)

Similarly, $\dfrac{BP_1}{P_1C} = \dfrac{CN}{NB} = \dfrac{1}{2}$. (VI)

Also in $\triangle APP_3$, since $\overline{MB} \parallel PP_3$, $\dfrac{AP_3}{P_3B} = \dfrac{AP}{MP} = \dfrac{4}{1}$ (#46). (VII)

Multiplying (V), (VI), and (VII), we get

$$\frac{CP_2}{P_2A} \cdot \frac{BP_1}{P_1C} \cdot \frac{AP_3}{P_3B} = \left(\frac{1}{2}\right)\left(\frac{1}{2}\right)(4) = -1,$$

taking direction into account. Thus, by Menelaus' Theorem, points P_1, P_2, and P_3 are collinear.

8-23 *If $\triangle A_1B_1C_1$ and $\triangle A_2B_2C_2$ are situated so that the lines joining the corresponding vertices, $\overleftrightarrow{A_1A_2}$, $\overleftrightarrow{B_1B_2}$, and $\overleftrightarrow{C_1C_2}$, are concurrent, then the pairs of corresponding sides intersect in three collinear points. (Desargues' Theorem)*

In Fig. S8-23, lines $\overleftrightarrow{A_1A_2}$, $\overleftrightarrow{B_1B_2}$, $\overleftrightarrow{C_1C_2}$ all meet at P, by the hypothesis.

Lines $\overleftrightarrow{B_2C_2}$ and $\overleftrightarrow{B_1C_1}$ meet at A'; lines $\overleftrightarrow{A_2C_2}$ and $\overleftrightarrow{A_1C_1}$ meet at B'; and lines $\overleftrightarrow{B_2A_2}$ and $\overleftrightarrow{B_1A_1}$ meet at C'.

Consider $\overline{A'C_1B_1}$ to be a transversal of $\triangle PB_2C_2$. Therefore,

$$\frac{PB_1}{B_1B_2} \cdot \frac{B_2A'}{A'C_2} \cdot \frac{C_2C_1}{C_1P} = -1 \text{ (Menelaus' Theorem).} \quad \text{(I)}$$

Similarly, considering $\overline{C'B_1A_1}$ as a transversal of $\triangle PB_2A_2$,

$$\frac{PA_1}{A_1A_2} \cdot \frac{A_2C'}{C'B_2} \cdot \frac{B_2B_1}{B_1P} = -1. \text{ (Menelaus' Theorem)} \quad \text{(II)}$$

And taking $\overline{B'A_1C_1}$ as a transversal of $\triangle PA_2C_2$,

$$\frac{PC_1}{C_1C_2} \cdot \frac{C_2B'}{B'A_2} \cdot \frac{A_2A_1}{A_1P} = -1. \text{ (Menelaus' Theorem)} \quad \text{(III)}$$

By multiplying (I), (II), and (III), we get

$$\frac{B_2A'}{A'C_2} \cdot \frac{A_2C'}{C'B_2} \cdot \frac{C_2B'}{B'A_2} = -1.$$

Thus, by Menelaus' Theorem, applied to $\triangle A_2B_2C_2$, we have points A', B', and C' collinear.

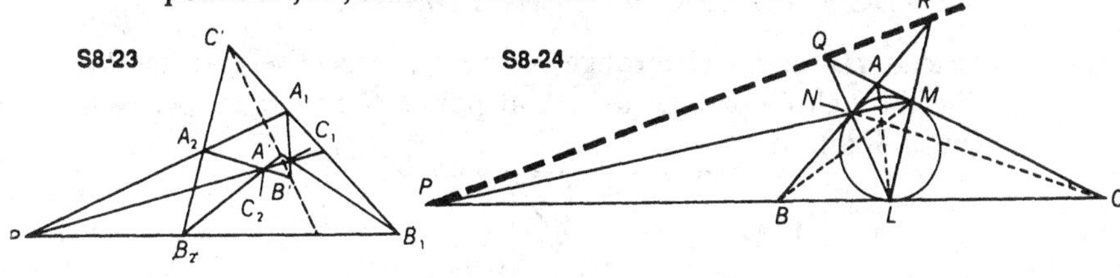

8-24 *A circle inscribed in $\triangle ABC$ is tangent to sides $\overline{BC}$, $\overline{CA}$, and $\overline{AB}$ at points L, M, and N, respectively. If $\overline{MN}$ extended meets $\overleftrightarrow{BC}$ at P,*

(a) *prove that* $\frac{BL}{LC} = -\frac{BP}{PC}$,

(b) *prove that if $\overleftrightarrow{NL}$ meets $\overleftrightarrow{AC}$ at Q and $\overleftrightarrow{ML}$ meets $\overleftrightarrow{AB}$ at R, then P, Q, and R are collinear.*

(a) By Menelaus' Theorem applied to $\triangle ABC$ with transversal $\overline{PNM}$, $\frac{AN}{NB} \cdot \frac{BP}{PC} \cdot \frac{MC}{AM} = -1$ (Fig. S8-24).

However, $AN = AM$, $NB = BL$, and $MC = LC$ (#34). (I)

By substitution, $\frac{AN}{BL} \cdot \frac{BP}{PC} \cdot \frac{LC}{AN} = -1$, so $\frac{BL}{LC} = -\frac{BP}{PC}$. (II)

(b) Similarly, $\frac{AN}{NB} = -\frac{AR}{RB}$, and $\frac{MC}{AM} = -\frac{QC}{AQ}$. (III), (IV)

By multiplication of (II), (III) and (IV), we get

$$\frac{BL}{LC} \cdot \frac{AN}{NB} \cdot \frac{MC}{AM} = \frac{-BP}{PC} \cdot \frac{-AR}{RB} \cdot \frac{-QC}{AQ}.$$

However from (I), $\frac{BL}{LC} \cdot \frac{AN}{NB} \cdot \frac{MC}{AM} = 1$.

Therefore, $\frac{BP}{PC} \cdot \frac{AR}{RB} \cdot \frac{QC}{AQ} = -1$, and points P, Q, and R are collinear, by Menelaus' Theorem.

Another method of proof following equation (II) reasons in this fashion. From (I), $\frac{AN}{NB} \cdot \frac{BL}{LC} \cdot \frac{MC}{AM} = 1$. Therefore, by Ceva's Theorem, $\overleftrightarrow{AL}$, $\overleftrightarrow{BM}$, and $\overleftrightarrow{CN}$ are concurrent. Since these are the lines joining the corresponding vertices of $\triangle ABC$ and $\triangle LMN$, by Desargues' Theorem (Problem 8-23), the intersections of the corresponding sides are collinear; therefore P, Q, and R are collinear.

8-25 *In* $\triangle$ABC, *where* $\overline{\text{CD}}$ *is the altitude to* $\overline{\text{AB}}$ *and* P *is any point on* $\overline{\text{DC}}$, $\overline{\text{AP}}$ *meets* $\overline{\text{CB}}$ *at* Q, *and* $\overline{\text{BP}}$ *meets* $\overline{\text{CA}}$ *at* R. *Prove that* m$\angle$RDC = m$\angle$QDC, *using Ceva's Theorem.*

Extend $\overline{DR}$ and $\overline{DQ}$ through R and Q, respectively, to meet a line through C, parallel to $\overline{AB}$, at points G and H, respectively (Fig. S8-25).

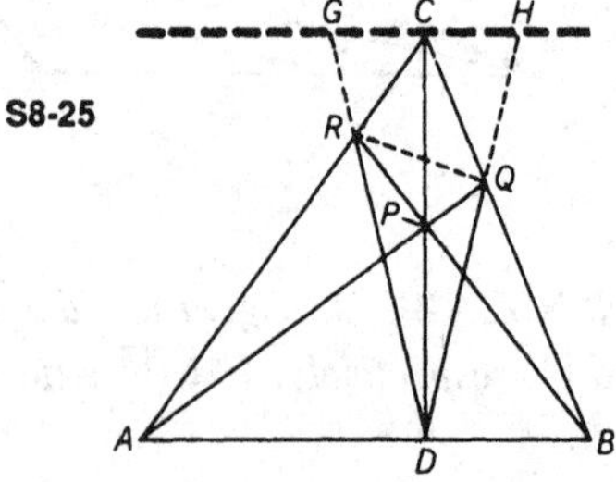

S8-25

$\triangle CGR \sim \triangle ADR$ (#48), and $\frac{CR}{RA} = \frac{GC}{AD}$. (I)

Similarly, $\triangle BDQ \sim \triangle CHQ$, and $\frac{BQ}{QC} = \frac{DB}{CH}$. (II)

However, in $\triangle ABC$ $\frac{CR}{RA} \cdot \frac{AD}{DB} \cdot \frac{BQ}{QC} = 1$ (Ceva's Theorem). (III)

By substituting (I) and (II) into (III), we get
$\frac{GC}{AD} \cdot \frac{AD}{DB} \cdot \frac{DB}{CH} = 1$, or $\frac{GC}{CH} = 1$. Thus, $GC = CH$.
Since $\overline{CD}$ is the perpendicular bisector of $\overline{GH}$ (#10),

$$\triangle GCD \cong \triangle HCD, \text{ and } m\angle GDC = m\angle HDC,$$
$$\text{or } m\angle RDC = m\angle QDC.$$

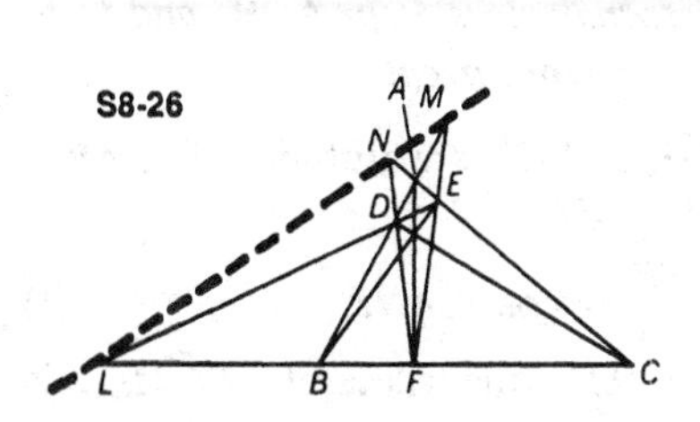

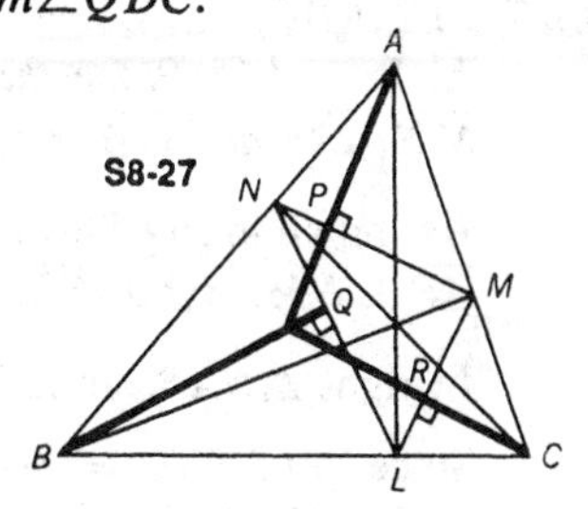

8-26 *In* $\triangle$ABC, *points* F, E, *and* D *are the feet of the altitudes drawn from the vertices* A, B, *and* C, *respectively. The sides of the pedal* $\triangle$FED, $\overline{\text{EF}}$, $\overline{\text{DF}}$, *and* $\overline{\text{DE}}$, *when extended, meet the sides of* $\triangle$ABC, $\overleftrightarrow{\text{AB}}$, $\overleftrightarrow{\text{AC}}$, *and* $\overleftrightarrow{\text{BC}}$ *at points* M, N, *and* L, *respectively. Prove that* M, N, *and* L *are collinear.* (*See Fig. S*8-26.)

METHOD I: In Problem 8-25, it was proved that the altitude of a triangle bisects the corresponding angle of the pedal triangle. Therefore, $\overline{BE}$ bisects $\angle DEF$, and $m\angle DEB = m\angle BEF$. (I)

$\angle DEB$ is complementary to $\angle NED$. (II)

Therefore since $\overleftrightarrow{MEF}$ is a straight line,

$\angle NEM$ is complementary to $\angle BEF$. (III)

Therefore from (I), (II), and (III), $m\angle NED = m\angle NEM$, or $\overline{NE}$ is an exterior angle bisector of $\triangle FED$. It then follows that

$$\frac{NF}{ND} = \frac{EF}{DE} \text{ (\#47)}. \qquad \text{(IV)}$$

Similarly, $\overline{FL}$ is an exterior angle bisector of $\triangle FED$ and

$$\frac{LD}{LE} = \frac{DF}{EF}. \qquad \text{(V)}$$

Also, $\overline{DM}$ is an exterior angle bisector of $\triangle FED$ and so

$$\frac{ME}{MF} = \frac{DE}{DF} \text{ (\#47)}. \qquad \text{(VI)}$$

By multiplying (IV), (V), and (VI), we get

$$\frac{NF}{ND} \cdot \frac{LD}{LE} \cdot \frac{ME}{MF} = \frac{EF}{DE} \cdot \frac{DF}{EF} \cdot \frac{DE}{DF} = -1,$$

taking direction into account.

Thus, by Menelaus' Theorem, M, N, and L are collinear.

METHOD II: Let D, E, F and C, B, A be corresponding vertices of $\triangle DEF$ and $\triangle CBA$, respectively. Since $\overline{AF}$, $\overline{CD}$, and $\overline{BE}$ are concurrent (Problem 8-4), the intersections of the corresponding sides $\overline{DE}$ and $\overline{BC}$, $\overline{FE}$ and $\overline{BA}$, and $\overline{FD}$ and $\overline{CA}$, are collinear by Desargues' Theorem (Problem 8-23).

8-27 *In $\triangle ABC$, L, M, and N are the feet of the altitudes from vertices A, B, and C. Prove that the perpendiculars from A, B, and C to $\overline{MN}$, $\overline{LN}$, and $\overline{LM}$, respectively, are concurrent.*

As shown in Fig. S8-27, $\overline{AL}$, $\overline{BM}$, and $\overline{CN}$ are altitudes of $\triangle ABC$. $\overline{AP} \perp \overline{NM}$, $\overline{BQ} \perp \overline{NL}$, and $\overline{CR} \perp \overline{ML}$.

In right $\triangle NAP$, $\sin \angle NAP = \dfrac{NP}{NA} = \cos \angle ANP$. (I)

Since $m\angle BNC = m\angle BMC = 90$, quadrilateral $BNMC$ is cyclic (#36a).

Therefore, $\angle MCB$ is supplementary to $\angle BNM$.

But $\angle ANP$ is also supplementary to $\angle BNM$. Thus, $m\angle MCB = m\angle ANP$, and $\cos \angle MCB = \cos \angle ANP$. (II)

From (I) and (II), by transitivity,

$$\sin \angle NAP = \cos \angle MCB. \quad \text{(III)}$$

Now, in right $\triangle AMP$, $\sin \angle MAP = \dfrac{MP}{MA} = \cos \angle AMP$. (IV)

Since quadrilateral $BNMC$ is cyclic, $\angle NBC$ is supplementary to $\angle NMC$, while $\angle AMP$ is supplementary to $\angle NMC$. Therefore, $m\angle NBC = m\angle AMP$ and $\cos \angle NBC = \cos \angle AMP$. (V)

From (IV) and (V), it follows that $\sin \angle MAP = \cos \angle NBC$. (VI)

From (III) and (VI), $\dfrac{\sin \angle NAP}{\sin \angle MAP} = \dfrac{\cos \angle MCB}{\cos \angle NBC}$. (VII)

In a similar fashion we are able to get the following proportions:

$$\frac{\sin \angle CBQ}{\sin \angle ABQ} = \frac{\cos \angle BAC}{\cos \angle ACB}, \quad \text{(VIII)}$$

$$\text{and } \frac{\sin \angle ACR}{\sin \angle BCR} = \frac{\cos \angle ABC}{\cos \angle BAC}. \quad \text{(IX)}$$

By multiplying (VII), (VIII), and (IX), we get

$$\frac{\sin \angle NAP}{\sin \angle MAP} \cdot \frac{\sin \angle CBQ}{\sin \angle ABQ} \cdot \frac{\sin \angle ACR}{\sin \angle BCR} =$$

$$\frac{\cos \angle ACB}{\cos \angle ABC} \cdot \frac{\cos \angle BAC}{\cos \angle ACB} \cdot \frac{\cos \angle ABC}{\cos \angle BAC} = 1.$$

Thus, by Ceva's Theorem (trigonometric form) $\overline{AP}$, $\overline{BQ}$, and $\overline{CR}$ are concurrent.

8-28 *Prove that the perpendicular bisectors of the interior angle bisectors of any triangle meet the sides opposite the angles being bisected in three collinear points.*

Let $\overline{AA'}$, $\overline{BB'}$, and $\overline{CC'}$ be the bisectors of angles A, B, and C, respectively, terminating at the opposite side. The perpendicular bisector of $\overline{AA'}$ meets $\overleftrightarrow{AC}$, $\overleftrightarrow{AB}$, and $\overleftrightarrow{CB}$ at points M, M', and P_1, respectively; the perpendicular bisector of $\overline{BB'}$ meets $\overleftrightarrow{CB}$, $\overleftrightarrow{AB}$, and $\overleftrightarrow{AC}$ at points L, L', and P_2, respectively; and the perpendicular bisector of $\overline{CC'}$ meets $\overleftrightarrow{AC}$, $\overleftrightarrow{CB}$, and $\overleftrightarrow{AB}$ at points N, N', and P_3, respectively. (See Fig. S8-28.)

Draw $\overline{B'L}$. Since $B'L = LB$ (#18), $m\angle LB'B = m\angle LBB'$ (#5). However, $m\angle ABB' = m\angle LBB'$; therefore $m\angle LB'B = m\angle ABB'$, and $\overline{B'L} \parallel \overline{AB}$ (#8).

Then $\dfrac{CB'}{AB'} = \dfrac{a}{c} = \dfrac{CL}{LB}$ (#46). (I)

However, $m\angle B'P_2L' = m\angle BP_2L'$, and $\dfrac{CP_2}{BP_2} = \dfrac{CL}{LB}$ (#47). (II)

Therefore, $\dfrac{CB'}{AB'} = \dfrac{a}{c} = \dfrac{CP_2}{BP_2}$. (III)

Similarly, since $\overline{B'L'} \parallel \overline{CB}$, $\dfrac{CB'}{AB'} = \dfrac{a}{c} = \dfrac{L'B}{AL'} = \dfrac{BP_2}{AP_2}$. (IV)

Thus, multiplying (III) and (IV), we get (V)

$$\frac{CP_2}{AP_2} = \frac{a^2}{c^2}.$$

Since $\overline{A'M'} \parallel \overline{AC}$, $\dfrac{CA'}{BA'} = \dfrac{b}{c} = \dfrac{AM'}{M'B} = \dfrac{AP_1}{BP_1}$. (VI)

And since $\overline{A'M} \parallel \overline{AB}$, $\dfrac{CA'}{BA'} = \dfrac{b}{c} = \dfrac{CM}{MA} = \dfrac{CP_1}{AP_1}$. (VII)

Now, multiplying (VI) and (VII), we get

$$\frac{CP_1}{BP_1} = \frac{c^2}{b^2}. \quad \text{(VIII)}$$

In a similar fashion we obtain $\dfrac{AP_3}{BP_3} = \dfrac{b^2}{a^2}$. (IX)

By multiplying (V), (VIII), and (IX), we get

$$\frac{CP_2}{AP_2} \cdot \frac{BP_1}{CP_1} \cdot \frac{AP_3}{BP_3} = \frac{a^2}{c^2} \cdot \frac{c^2}{b^2} \cdot \frac{b^2}{a^2} = -1,$$

taking direction into account. Therefore, by Menelaus' Theorem, P_1, P_2, and P_3 are concurrent.

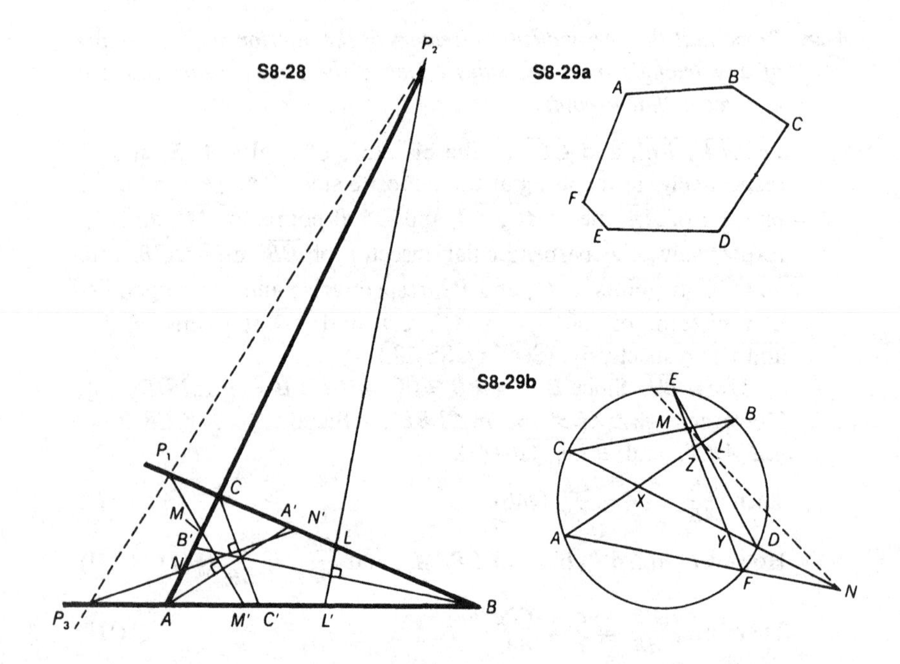

8-29 *Figure 8-29a shows a hexagon* ABCDEF *whose pairs of opposite sides are:* [$\overline{\text{AB}}$, $\overline{\text{DE}}$], [$\overline{\text{CB}}$, $\overline{\text{EF}}$], *and* [$\overline{\text{CD}}$, $\overline{\text{AF}}$]. *If we place points* A, B, C, D, E, *and* F *in any order on a circle, the above pairs of opposite sides intersect at points* L, M, *and* N. *Prove that* L, M, *and* N *are collinear.*

Pairs of opposite sides (see Fig. S8-29a) $\overline{AB}$ and $\overline{DE}$ meet at L, $\overline{CB}$ and $\overline{EF}$ meet at M, and $\overline{CD}$ and $\overline{AF}$ meet at N. (See Fig. S8-29b.) Also $\overline{AB}$ meets $\overline{CN}$ at X, $\overline{EF}$ meets $\overline{CN}$ at Y, and $\overline{EF}$ meets $\overline{AB}$ at Z. Consider $\overline{BC}$ to be a transversal of $\triangle XYZ$. Then

$$\frac{ZB}{BX}\cdot\frac{XC}{CY}\cdot\frac{YM}{MZ} = -1, \text{ by Menelaus' Theorem.} \qquad \text{(I)}$$

Now taking $\overline{AF}$ to be a transversal of $\triangle XYZ$,

$$\frac{ZA}{AX}\cdot\frac{YF}{FZ}\cdot\frac{XN}{NY} = -1. \qquad \text{(II)}$$

Also since $\overline{DE}$ is a transversal of $\triangle XYZ$,

$$\frac{XD}{DY}\cdot\frac{YE}{EZ}\cdot\frac{ZL}{LX} = -1. \qquad \text{(III)}$$

By multiplying (I), (II), and (III), we get

$$\frac{YM}{MZ}\cdot\frac{XN}{NY}\cdot\frac{ZL}{LX}\cdot\frac{(ZB)(ZA)}{(EZ)(FZ)}\cdot\frac{(XD)(XC)}{(AX)(BX)}\cdot\frac{(YE)(YF)}{(DY)(CY)} = -1. \qquad \text{(IV)}$$

$$\text{However, } \frac{(ZB)(ZA)}{(EZ)(FZ)} = 1, \quad \text{(V)}$$

$$\frac{(XD)(XC)}{(AX)(BX)} = 1, \quad \text{(VI)}$$

$$\text{and } \frac{(YE)(YF)}{(DY)(CY)} = 1 \text{ (\#52).} \quad \text{(VII)}$$

By substituting (V), (VI), and (VII) into (IV), we get

$$\frac{YM}{MZ} \cdot \frac{XN}{NY} \cdot \frac{ZL}{LX} = -1.$$

Thus, by Menelaus' Theorem, points M, N, and L must be collinear.

S8-30

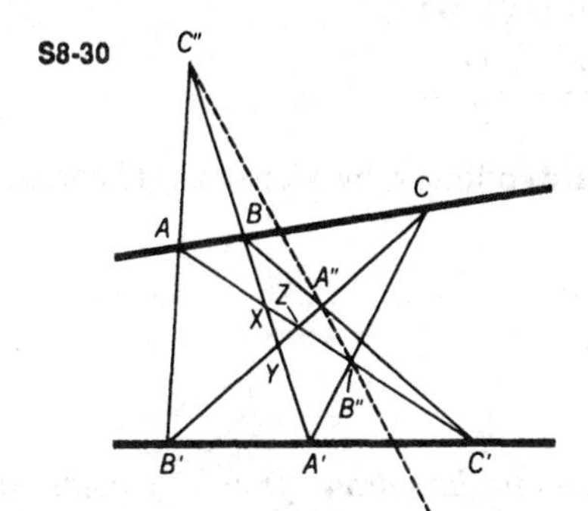

S9-1a

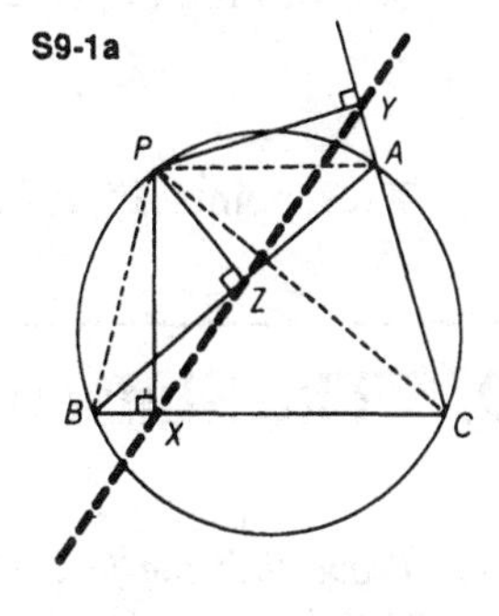

8-30 *Points A, B, and C are on one line and points A′, B′, and C′ are on another line (in any order). If $\overline{AB'}$ and $\overline{A'B}$ meet at C″, while $\overline{AC'}$ and $\overline{A'C}$ meet at B″, and $\overline{BC'}$ and $\overline{B'C}$ meet at A″, prove that points A″, B″, and C″ are collinear.*

(This theorem was first published by Pappus of Alexandria about 300 A.D.)

In Fig. S8-30, $\overline{B'C}$ meets $\overline{A'B}$ at Y, $\overline{AC'}$ meets $\overline{A'B}$ at X, and $\overline{B'C}$ meets $\overline{AC'}$ at Z.

Consider $\overline{C''AB'}$ as a transversal of $\triangle XYZ$.

$$\frac{ZB'}{YB'} \cdot \frac{XA}{ZA} \cdot \frac{YC''}{XC''} = -1 \text{ (Menelaus' Theorem)} \quad \text{(I)}$$

Now taking $\overline{A'B''C}$ as a transversal of $\triangle XYZ$,

$$\frac{YA'}{XA'} \cdot \frac{XB''}{ZB''} \cdot \frac{ZC}{YC} = -1. \quad \text{(II)}$$

$\overline{BA''C'}$ is also a transversal of $\triangle XYZ$, so that

$$\frac{YB}{XB} \cdot \frac{ZA''}{YA''} \cdot \frac{XC'}{ZC'} = -1. \quad \text{(III)}$$

Multiplying (I), (II), and (III) gives us equation (IV),

$$\frac{YC''}{XC''}\cdot\frac{XB''}{ZB''}\cdot\frac{ZA''}{YA''}\cdot\frac{ZB'}{YB'}\cdot\frac{YA'}{XA'}\cdot\frac{XC'}{ZC'}\cdot\frac{XA}{ZA}\cdot\frac{ZC}{YC}\cdot\frac{YB}{XB} = -1.$$

Since points A, B, C and A', B', C' are collinear, we obtain the following two relationships by Menelaus' Theorem when we consider each line a transversal of $\triangle XYZ$.

$$\frac{ZB'}{YB'}\cdot\frac{YA'}{XA'}\cdot\frac{XC'}{ZC'} = -1 \qquad \text{(V)}$$

$$\frac{XA}{ZA}\cdot\frac{ZC}{YC}\cdot\frac{YB}{XB} = -1 \qquad \text{(VI)}$$

Substituting (V) and (VI) into (IV), we get

$$\frac{YC''}{XC''}\cdot\frac{XB''}{ZB''}\cdot\frac{ZA''}{YA''} = -1.$$

Thus, points A'', B'', and C'' are collinear, by Menelaus' Theorem.

9. The Simson Line

9-1 *Prove that the feet of the perpendiculars drawn from any point on the circumcircle of a given triangle to the sides of the triangle are collinear (Simson's Theorem).*

METHOD I: From any point P, on the circumcircle of $\triangle ABC$ perpendiculars $\overline{PX}$, $\overline{PY}$, and $\overline{PZ}$ are drawn to sides $\overline{BC}$, $\overline{AC}$, and $\overline{AB}$, respectively (Fig. S9-1a). Since $\angle PYA$ is supplementary to $\angle PZA$, quadrilateral $PZAY$ is cyclic (#37). Draw $\overline{PA}$, $\overline{PB}$, and $\overline{PC}$.

Therefore, $m\angle PYZ = m\angle PAZ$ (#36). (I)

Similarly, since $\angle PYC$ is supplementary to $\angle PXC$, quadrilateral $PXCY$ is cyclic, and $m\angle PYX = m\angle PCB$. (II)

However, quadrilateral $PACB$ is also cyclic, since it is inscribed in the given circumcircle, and therefore

$m\angle PAZ(m\angle PAB) = m\angle PCB$ (#36). (III)

From (I), (II), and (III), $m\angle PYZ = m\angle PYX$, and thus points X, Y, and Z are collinear. The line through X, Y, and Z is called the Simson Line of $\triangle ABC$ with respect to point P.

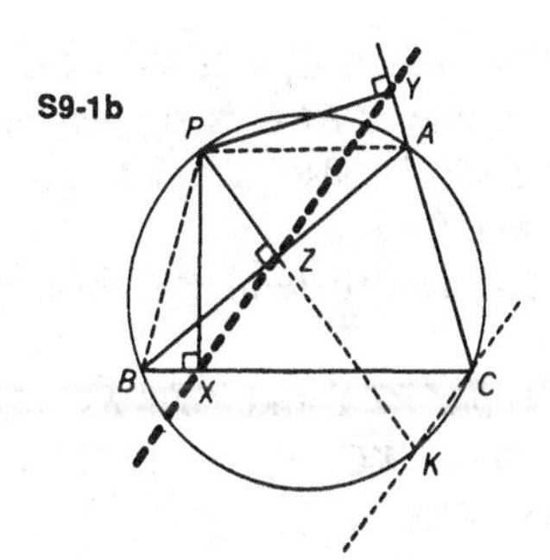

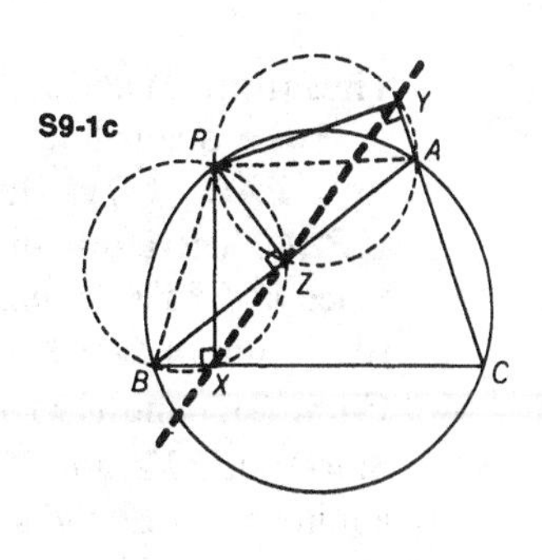

METHOD II: From any point P on the circumcircle of $\triangle ABC$ (inside $\angle ACB$), perpendiculars $\overline{PX}$, $\overline{PY}$, and $\overline{PZ}$ are drawn to sides $\overline{BC}$, $\overline{AC}$, and $\overline{AB}$, respectively. (See Fig. S9-1b.) Draw circles with $\overline{PA}$ and $\overline{PB}$ as diameters. Since $m\angle PYA = m\angle PXB = m\angle PZA = 90$, points Y and Z lie on the circle with $\overline{PA}$ as diameter (#37). Also points X and Z lie on the circle with $\overline{PB}$ as diameter (#36a).

Since $m\angle PXC = m\angle PYC = 90$, in quadrilateral $XPYC$, $\angle C$ is supplementary to $\angle XPY$ (#15).

However $\angle C$ is also supplementary to $\angle APB$ (#37).

Therefore, $m\angle XPY = m\angle APB$. (I)

By subtracting each member of (I) from $m\angle BPY$,

we get $m\angle BPX = m\angle APY$. (II)

Now $m\angle BPX = m\angle BZX$ (#36),

and $m\angle APY = m\angle AZY$ (#36). (III)

Substituting (III) into (II), $m\angle BZX = m\angle AZY$.

Since $\overleftrightarrow{AZB}$ is a straight line, points X, Y, and Z must be collinear, making $\angle BZX$ and $\angle AZY$ vertical angles.

METHOD III: From any point, P, on the circumcircle of $\triangle ABC$, $\overline{PX}$, $\overline{PY}$, and $\overline{PZ}$ are drawn to the sides $\overline{BC}$, $\overline{AC}$, and $\overline{AB}$, respectively. $\overline{PZ}$ extended meets the circle at K. Draw $\overleftrightarrow{CK}$, as shown in Fig. S9-1c.

Since $m\angle PZB \cong m\angle PXB \cong 90$, quadrilateral $PZXB$ is cyclic (#36a), and so $\angle PBC$ is supplementary to $\angle PZX$ (#37).

However $\angle KZX$ is supplementary to $\angle PZX$;

therefore, $m\angle PBC = m\angle KZX$. (I)

But $m\angle PBC = m\angle PKC$ (#36). (II)

Thus from (I) and (II) $m\angle KZX = m\angle PKC$, and $\overline{XZ} \parallel \overline{KC}$ (#8).

Since quadrilateral $PACK$ is cyclic, $\angle PKC$ is supplementary to $\angle PAC$ (#37). However, $\angle PAY$ is also supplementary to $\angle PAC$. Therefore, $m\angle PKC = m\angle PAY$. (III)

Since $m\angle PYA \cong m\angle PZA \cong 90$, quadrilateral $PYAZ$ is cyclic (#37), and $m\angle PZY = m\angle PAY$. (IV)

From (III) and (IV), $m\angle PKC = m\angle PZY$ and $\overleftrightarrow{ZY} \parallel \overleftrightarrow{KC}$ (#7).

Since both $\overleftrightarrow{XZ}$ and $\overleftrightarrow{ZY}$ are parallel to $\overleftrightarrow{KC}$, X, Y, and Z must be collinear, by Euclid's parallel postulate.

Challenge 1 *State and prove the converse of Simson's Theorem.*

If the feet of the perpendiculars from a point to the sides of a given triangle are collinear, then the point must lie on the circumcircle of the triangle.

Collinear points X, Y, and Z, are the feet of perpendiculars $\overline{PX}$, $\overline{PY}$, and $\overline{PZ}$ to sides $\overline{BC}$, $\overline{AC}$, and $\overline{AB}$, respectively, of $\triangle ABC$ (Fig. S9-1d). Draw $\overline{PA}$, $\overline{PB}$, and $\overline{PC}$.

S9-1d

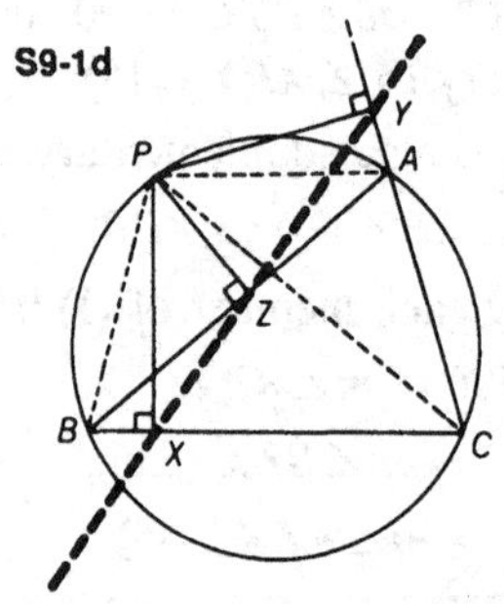

Since $m\angle PZB \cong m\angle PXB \cong 90$, quadrilateral $PZXB$ is cyclic (#36a), and $\angle PBX$ is supplementary to $\angle PZX$ (#37). However, $\angle PZX$ is supplementary to $\angle PZY$, since X, Z, and Y are collinear.

Therefore, $m\angle PBX = m\angle PZY$. (I)

Since $\angle PZA$ is supplementary to $\angle PYA$, quadrilateral $PZAY$ is also cyclic (#37), and

$m\angle PAY = m\angle PZY$ (#36). (II)

From (I) and (II), $m\angle PBX = m\angle PAY$ or $m\angle PBC = m\angle PAY$.

Therefore $\angle PBC$ is supplementary to $\angle PAC$ and quadrilateral $PACB$ is cyclic (#37); in other words point P lies on the circumcircle of $\triangle ABC$.

Another proof of the converse of Simson's Theorem can be obtained by simply reversing the steps shown in the proof of the theorem itself, Method II.

Challenge 2 *Which points on the circumcircle of a given triangle lie on their own Simson Lines with respect to the given triangle?*

ANSWER: The three vertices of the triangle are the only points which lie on their own Simson Lines.

9-2 *Altitude* $\overline{AD}$ *of* $\triangle ABC$ *meets the circumcircle at* P. *Prove that the Simson Line of* P *with respect to* $\triangle ABC$ *is parallel to the line tangent to the circle at* A.

Since $\overline{PX}$, and $\overline{PZ}$ are perpendicular respectively to sides $\overleftrightarrow{AC}$, and $\overleftrightarrow{AB}$ of $\triangle ABC$, points X, D, and Z determine the Simson Line of P with respect to $\triangle ABC$.
Draw $\overline{PB}$ (Fig. S9-2).
Consider quadrilateral $PDBZ$, where $m\angle PDB \cong m\angle PZB \cong 90$, thus making $PDBZ$ a cyclic quadrilateral (#37).

In $PDBZ$, $m\angle DZB = m\angle DPB$ (#36). (I)
However, in the circumcircle of $\triangle ABC$, $m\angle GAB = \frac{1}{2}(m\widehat{AB})$ (#38), and $m\angle DPB$ $(\angle APB) = \frac{1}{2}(m\widehat{AB})$(#36).
Therefore, $m\angle GAB = m\angle DPB$. (II)
From (I) and (II), by transitivity, $m\angle DZB = m\angle GAB$, and thus Simson Line $\overleftrightarrow{XDZ} \parallel$ tangent $\overleftrightarrow{GA}$ (#8).

S9-2

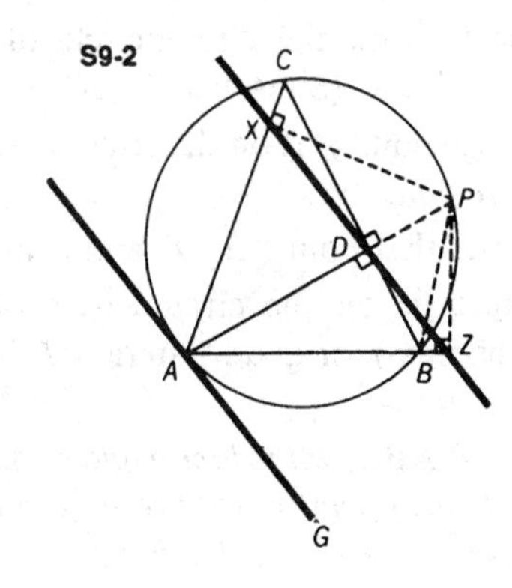

9-3 *From point* P *on the circumcircle of* $\triangle$ABC, *perpendiculars* $\overline{PX}$, $\overline{PY}$, *and* $\overline{PZ}$ *are drawn to sides* $\overleftrightarrow{AC}$, $\overleftrightarrow{AB}$, *and* $\overleftrightarrow{BC}$, *respectively. Prove that* (PA)(PZ) = (PB)(PX). (*See Fig. S9-3.*)

Since $m\angle PYB \cong m\angle PZB \cong 90$, quadrilateral $PYZB$ is cyclic (#36a), and $m\angle PBY = m\angle PZY$ (#36). (I)

Since $m\angle PXA \cong m\angle PYA \cong 90$, quadrilateral $PXAY$ is cyclic (#37), and $m\angle PXY = m\angle PAY$. (II)

Since X, Y, and Z are collinear (the Simson Line), $\triangle PAB \sim \triangle PXZ$ (#48), and $\frac{PA}{PX} = \frac{PB}{PZ}$, or $(PA)(PZ) = (PB)(PX)$.

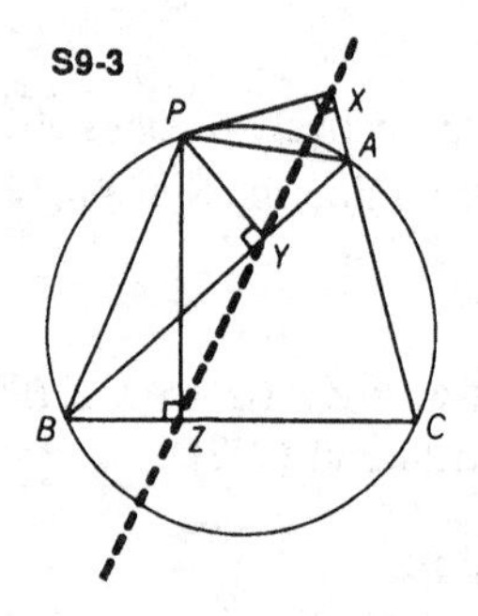

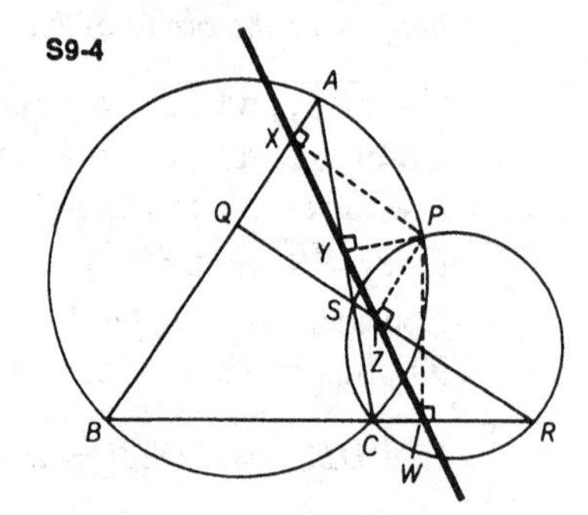

9-4 *Sides* $\overleftrightarrow{AB}$, $\overleftrightarrow{BC}$, *and* $\overleftrightarrow{CA}$ *of* $\triangle$ABC *are cut by a transversal at points* Q, R, *and* S, *respectively. The circumcircles of* $\triangle$ABC *and* $\triangle$SCR *intersect at* P. *Prove that quadrilateral* APSQ *is cyclic.*

Draw perpendiculars $\overline{PX}$, $\overline{PY}$, $\overline{PZ}$, and $\overline{PW}$ to $\overleftrightarrow{AB}$, $\overleftrightarrow{AC}$, $\overleftrightarrow{QR}$, and $\overleftrightarrow{BC}$, respectively, as in Fig. S9-4.

Since point P is on the circumcircle of $\triangle ABC$, points X, Y, and W are collinear (Simson's Theorem).

Similarly, since point P is on the circumcircle of $\triangle SCR$, points Y, Z, and W are collinear.

It then follows that points X, Y, and Z are collinear.

Thus, P must lie on the circumcircle of $\triangle AQS$ (converse of Simson's Theorem), or quadrilateral $APSQ$ is cyclic.

9-5 *In Fig. S9-5,* $\triangle$ABC, *with right angle at* A, *is inscribed in circle* O. *The Simson Line of point* P, *with respect to* $\triangle$ABC *meets* $\overline{PA}$ *at* M. *Prove that* $\overline{MO}$ *is perpendicular to* $\overline{PA}$.

In Fig. S9-5, $\overline{PZ}$, $\overline{PY}$, and $\overline{PX}$ are perpendicular to lines $\overleftrightarrow{AB}$, $\overleftrightarrow{AC}$, and $\overleftrightarrow{BC}$, respectively. $\overleftrightarrow{XYZ}$ is the Simson Line of $\triangle ABC$ and point P, and meets $\overline{PA}$ at M. Since $\angle BAC$ is a right angle, $AZPY$ is a rectangle (it has three right angles). Therefore, M is the midpoint of $\overline{PA}$ (#21f). It then follows that $\overline{MO}$ is perpendicular to $\overline{PA}$ (#31).

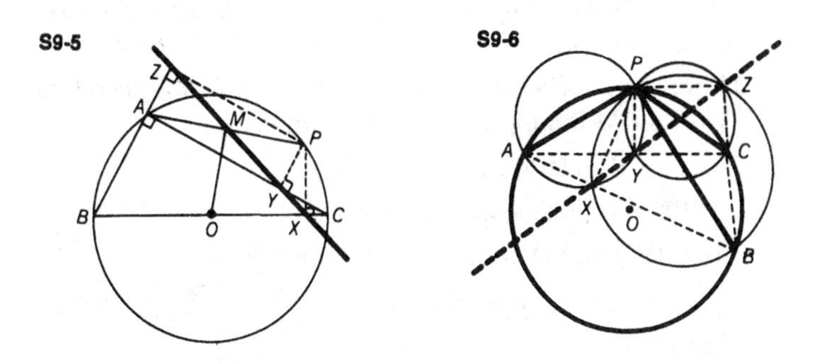

9-6 *From a point* P *on the circumference of circle* O, *three chords are drawn meeting the circle in points* A, B, *and* C. *Prove that the three points of intersection of the three circles with* $\overline{\text{PA}}$, $\overline{\text{PB}}$, *and* $\overline{\text{PC}}$ *as diameters, are collinear.*

In Fig. S9-6, the circle on $\overline{PA}$ meets the circle on $\overline{PB}$ at X, and the circle on $\overline{PC}$ at Y, while the circle on $\overline{PB}$ meets the circle on $\overline{PC}$ at Z.

Draw $\overline{AB}$, $\overline{BC}$, and $\overline{AC}$, also $\overline{PX}$, $\overline{PY}$, and $\overline{PZ}$. In the circle on $\overline{PA}$, $\angle PXA$ is a right angle (#36). Similarly, $\angle PYC$ and $\angle PZC$ are right angles. Since $\overline{PX}$, $\overline{PY}$, and $\overline{PZ}$ are drawn from a point on the circumcircle of $\triangle ABC$ perpendicular to the sides of $\triangle ABC$, X, Y, and Z determine a Simson Line and are therefore collinear.

9-7 P *is any point on the circumcircle of cyclic quadrilateral* ABCD. *If* $\overline{\text{PK}}$, $\overline{\text{PL}}$, $\overline{\text{PM}}$, *and* $\overline{\text{PN}}$ *are the perpendiculars from* P *to sides* $\overleftrightarrow{\text{AB}}$, $\overleftrightarrow{\text{BC}}$, $\overleftrightarrow{\text{CD}}$, *and* $\overleftrightarrow{\text{DA}}$, *respectively, prove that* (PK)(PM) = (PL)(PN).

Draw $\overline{DB}$, $\overline{AP}$, and $\overline{CP}$, as shown in Fig. S9-7. Draw $\overline{PS} \perp \overline{BD}$.

Since $m\angle ANP \cong m\angle AKP \cong 90$, quadrilateral $AKPN$ is cyclic (#37), and $m\angle NAP = m\angle NKP$ (#36). (I)

$\overleftrightarrow{NSK}$ is the Simson Line of $\triangle ABD$ with respect to point P.
Also $m\angle NAP$ ($\angle DAP$) $= m\angle PCM$ ($\angle PCD$) (#36). (II)
Since $m\angle PLC \cong m\angle PMC \cong 90$, quadrilateral $PLCM$ is cyclic (#37), $m\angle PCM = m\angle PLM$ (#36), (III)
and $\overleftrightarrow{LMS}$ is the Simson Line of $\triangle DBC$ with respect to point P.
From (I), (II), and (III), $m\angle PLM = m\angle NKP$. (IV)
Since $\angle LCM$ is supplementary to $\angle BCD$, and $\angle BAD$ is supplementary to $\angle BCD$ (#37), $m\angle LCM = m\angle BAD$. (V)
However, $\angle LPM$ is supplementary to $\angle LCM$, therefore, from (V), $\angle LPM$ is supplementary to $\angle BAD$. (VI)
Since quadrilateral $AKPN$ is cyclic,
$\angle NPK$ is supplementary to $\angle BAD$. (VII)
From (VI) and (VII), $m\angle LPM = m\angle NPK$.
Thus, $\triangle LPM \sim \triangle KPN$ (#48), and $\frac{PL}{PK} = \frac{PM}{PN}$, or $(PK)(PM) = (PL)(PN)$.

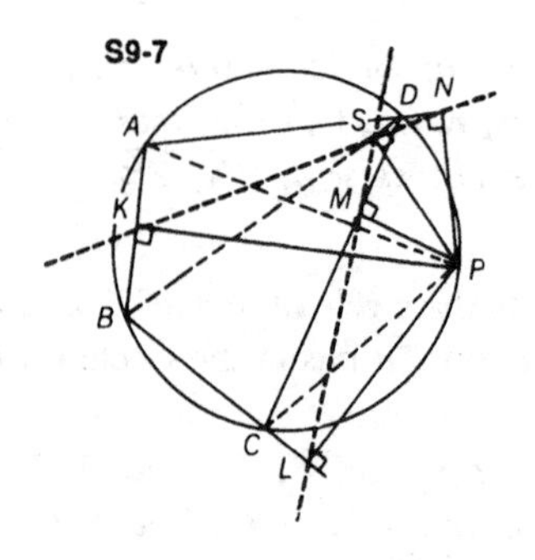

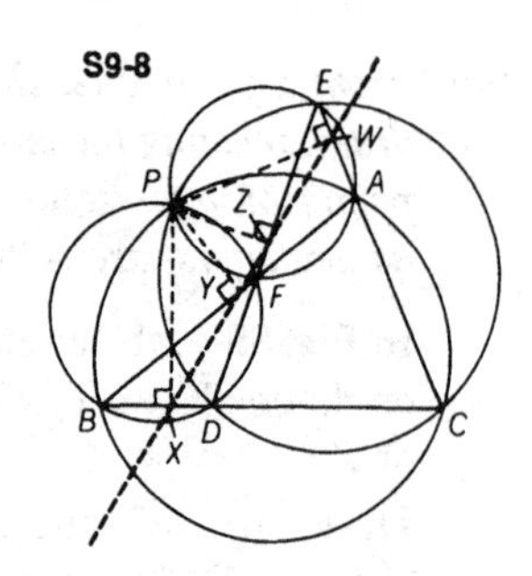

9-8 *Line segments* $\overline{AB}$, $\overline{BC}$, $\overline{EC}$, *and* $\overline{ED}$ *form triangles* ABC, FBD, EFA, *and* EDC. *Prove that the four circumcircles of these triangles meet at a common point.*

Consider the circumcircles of $\triangle ABC$ and $\triangle FBD$, which meet at B and P.

From point P draw perpendiculars $\overline{PX}$, $\overline{PY}$, $\overline{PZ}$, and $\overline{PW}$ to $\overline{BC}$, $\overline{AB}$, $\overline{ED}$, and $\overline{EC}$, respectively (Fig. S9-8). Since P is on the circumcircle of $\triangle FBD$, X, Y, and Z are collinear (Simson Line). Similarly, since P is on the circumcircle of $\triangle ABC$, X, Y, and W are collinear. Therefore X, Y, Z, and W are collinear.

Since Y, Z, and W are collinear, P must lie on the circumcircle of $\triangle EFA$ (converse of Simson's Theorem). By the same reasoning, since X, Z, and W are collinear, P lies on the circumcircle of $\triangle EDC$. Thus all four circles pass through point P.

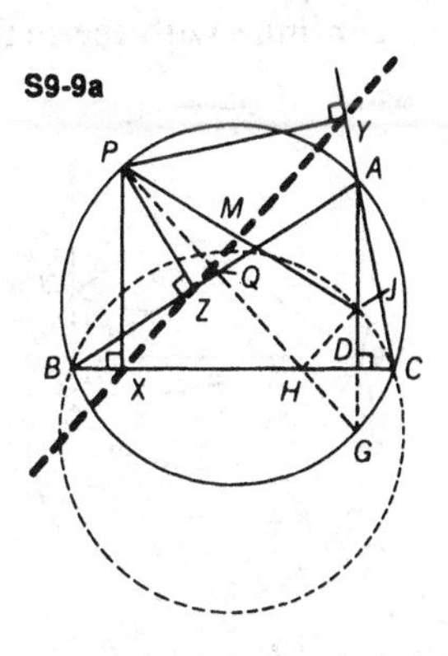

9-9 *The line joining the orthocenter of a given triangle with a point on the circumcircle of the triangle is bisected by the Simson Line, (with respect to that point).*

METHOD I: As in Fig. S9-9a, point P is on the circumcircle of $\triangle ABC$. $\overline{PX}$, $\overline{PY}$, and $\overline{PZ}$ are perpendicular to $\overleftrightarrow{BC}$, $\overleftrightarrow{AC}$, and $\overleftrightarrow{AB}$, respectively. Points X, Y, and Z are therefore collinear and define the Simson Line. Let J be the orthocenter of $\triangle ABC$. $\overline{PG}$ meets the Simson Line at Q and $\overline{BC}$ at H. $\overline{PJ}$ meets the Simson Line at M. Draw $\overline{HJ}$.

Since $m\angle PZB \cong m\angle PXB \cong 90$, quadrilateral $PZXB$ is cyclic (#36a),

and $m\angle PXQ$ ($\angle PXZ$) $= m\angle PBZ$ (#36). (I)

In the circumcircle, $m\angle PBZ = m\angle PGA$ (#36). (II)

Since $\overline{PX} \parallel \overline{AG}$ (#9), $m\angle PGA = m\angle QPX$ (#8). (III)

From (I), (II), and (III),

$$m\angle PXQ = m\angle QPX. \qquad \text{(IV)}$$

Therefore, $PQ = XQ$ (#5). Since $\angle QXH$ is complementary to $\angle PXQ$, and $\angle QHX$ is complementary to $\angle QPX$ (#14), $m\angle QXH = m\angle QHX$, and $XQ = HQ$ (#5). Thus Q is the midpoint of hypotenuse $\overline{PH}$ of right $\triangle PXH$.

Consider a circle passing through points B, J, and C. $\overline{BC}$, the common chord of the new circle and the original circle, is the perpendicular bisector of line segment $\overline{JG}$. To prove this last statement, it is necessary to set up an auxiliary proof (called a Lemma), before we continue with the main proof.

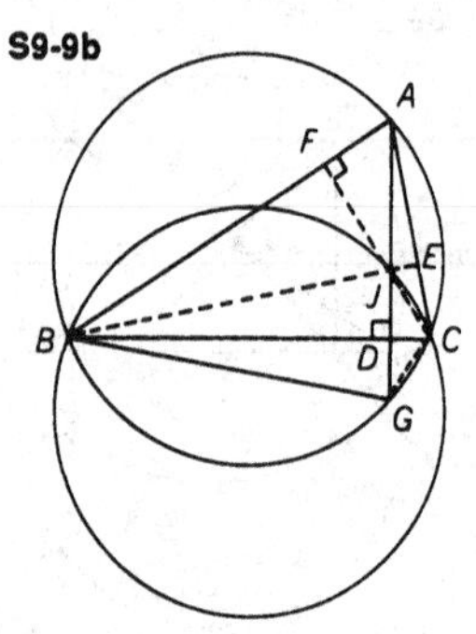

LEMMA: Draw altitudes $\overline{BE}$, and $\overline{CF}$; also draw $\overline{BG}$, $\overline{CJ}$, and $\overline{CG}$. (See Fig. S9-9b.)

$\overline{JD} \perp \overline{BC}$, therefore $m\angle JDB = m\angle GDB = 90$ (V)

$\angle JBC$ ($\angle EBC$) is complementary to $\angle C$ (#14). (VI)

$m\angle GBC = m\angle GAC$ ($\angle DAC$) (#36). Therefore, since $\angle GAC$ ($\angle DAC$) is complementary to $\angle C$ (#14),

$\angle GBC$ is complementary to $\angle C$. (VII)

Thus, from (VI), and (VII), $m\angle JBC = m\angle GBC$. Hence, $\triangle BJD \cong \triangle BGD$; therefore $JD = GD$, and $\overline{BC}$ is the perpendicular bisector of $\overline{JG}$.

Continuing with the main proof, we can now say that $HJ = HG$ (#18), and $m\angle HJG = m\angle HGJ$ (#5). (VIII)

$\angle JHD$ is complementary to $\angle HJD$.

But $m\angle HJD = m\angle HGD$ (IX), and $m\angle HGD = m\angle QPX$ (III), and $m\angle QPX = m\angle PXQ$ (IV).

Therefore, $\angle JHD$ is complementary to $\angle QXP$.

However, $\angle QXH$ is complementary to $\angle QXP$; therefore

$$m\angle JHD = m\angle QXH.$$

Thus $\overline{JH}$ is parallel to the Simson Line $\overleftrightarrow{XYZ}$ (#7).

Therefore, in $\triangle PJH$, since Q is the midpoint of $\overline{PH}$, and $\overline{QM}$ is parallel to $\overline{JH}$, M is the midpoint of $\overline{PJ}$, (#46).

Thus the Simson Line bisects $\overline{PJ}$ at M.

METHOD II: In Fig. S9-9c, point P is on the circumcircle of $\triangle ABC$. $\overline{PX}$, $\overline{PY}$, and $\overline{PZ}$ are perpendicular to sides $\overleftrightarrow{BC}$, $\overleftrightarrow{AC}$, and $\overleftrightarrow{AB}$, respectively. Therefore points X, Y, and Z are collinear and define the Simson Line. $\overline{PY}$ extended meets the circle at K. Let J be the orthocenter of $\triangle ABC$. The altitude from B meets $\overline{AC}$ at E and the circle at N. $\overline{PJ}$ meets the Simson line at M. Draw a line parallel to $\overline{KB}$, and through the orthocenter, J, meeting $\overline{PYK}$ at L.

Since $\overline{PK} \parallel \overline{NB}$ (#9), $KBJL$ is a parallelogram, and $LJ = KB$ (#21b). Also $\overset{\frown}{PN} \cong \overset{\frown}{KB}$ (#33), and $PN = KB$. Therefore $LJ = PN$ and trapezoid $PNJL$ is isosceles.

Consider a circle passing through points A, J, and C. The common chord $\overline{AC}$ is then the perpendicular bisector of $\overline{JN}$. (See Method I Lemma.) Thus E is the midpoint of $\overline{JN}$. Since $\overline{AC}$ is perpendicular to both bases of isosceles trapezoid $PNJL$, it may easily be shown that Y is the midpoint of $\overline{PL}$.

Since quadrilateral $AYPZ$ is cyclic (#37), $m\angle KBA = m\angle KPA = m\angle YPA = m\angle YZA$ (#36), and $\overline{KB}$ is parallel to Simson Line $\overleftrightarrow{XYZ}$ (#8). Now, in $\triangle PLJ$, M, the point of intersection of $\overline{PJ}$ with the Simson Line, is the midpoint of $\overline{PJ}$ (#25).

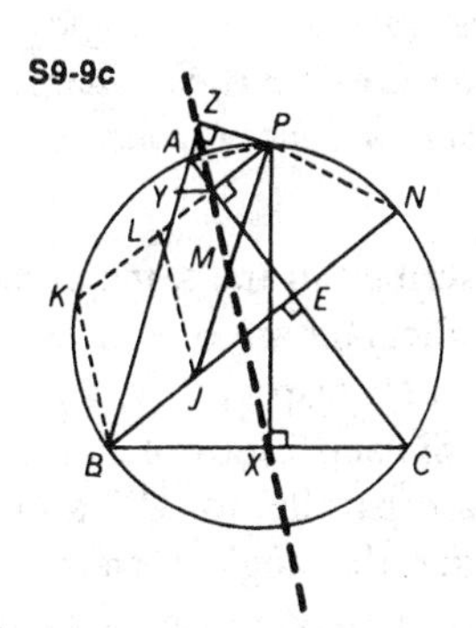

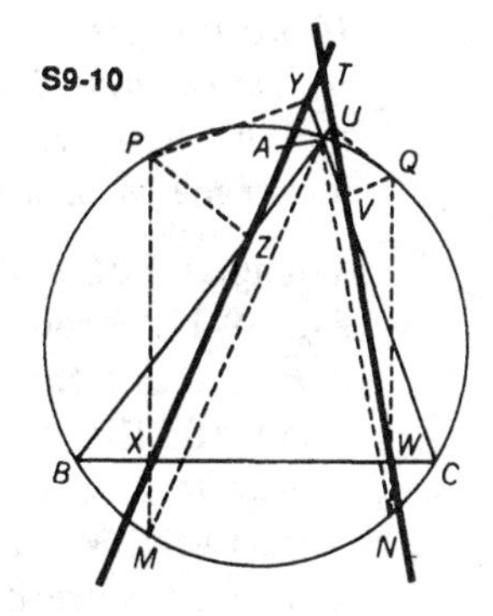

9-10 *The measure of the angle determined by the Simson Lines of two given points on the circumcircle of a given triangle is equal to one-half the measure of the arc determined by the two points.*

In Fig. S9-10, $\overleftrightarrow{XYZ}$ is the Simson Line for point P, and $\overleftrightarrow{UVW}$ is the Simson Line for point Q. Extend $\overline{PX}$ and $\overline{QW}$ to meet the circle at M and N, respectively. Then draw $\overline{AM}$ and $\overline{AN}$.

Since $m\angle PZB \cong m\angle PXB \cong 90$, quadrilateral $PZXB$ is cyclic (#36a), and $m\angle ZXP = m\angle ZBP$ (#36). (I)
Also $m\angle ABP = m\angle AMP$ (#36), or $m\angle ZBP = m\angle AMP$. (II)
From (I) and (II), $m\angle ZXP = m\angle AMP$, and $\overleftrightarrow{XYZ} \parallel \overline{AM}$. (III)
In a similar fashion it may be shown that $\overleftrightarrow{UVW} \parallel \overline{AN}$.
Hence, if T is the point of intersection of the two Simson Lines, then $m\angle XTW = m\angle MAN$, since their corresponding sides are parallel. Now, $m\angle MAN = \frac{1}{2}(m\widehat{MN})$, but since $\overline{PM} \parallel \overline{QN}$ (#9), $m\widehat{MN} = m\widehat{PQ}$ (#33), and therefore $m\angle MAN = \frac{1}{2}(m\widehat{PQ})$. Thus, $m\angle XTW = \frac{1}{2}(m\widehat{PQ})$.

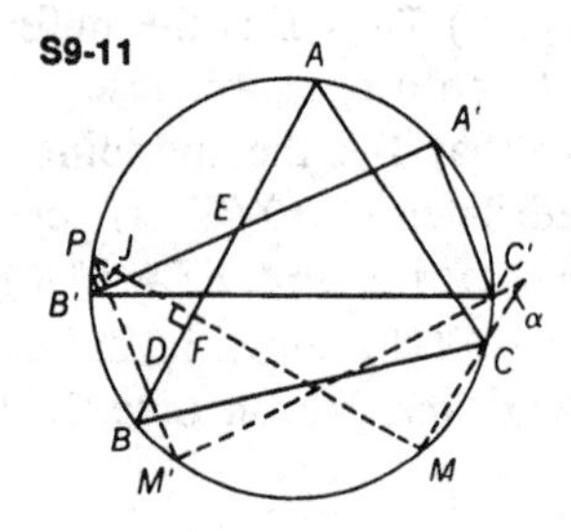

9-11 *If two triangles are inscribed in the same circle, a single point on the circumcircle determines a Simson Line for each triangle. Prove that the angle formed by these two Simson Lines is constant, regardless of the position of the point.*

Triangles ABC and $A'B'C'$ are inscribed in the same circle. (See Fig. S9-11.) From point P, perpendiculars are drawn to $\overline{AB}$ and $\overline{A'B'}$, meeting the circle at M and M', respectively. From Solution 9-10 (III), we know that the Simson Lines of point P with respect to $\triangle ABC$ and $\triangle A'B'C'$ are parallel to $\overline{MC}$ and $\overline{M'C'}$, respectively. We may now consider the angle formed by $\overline{MC}$ and $\overline{M'C'}$, since it is congruent to the angle formed by the two Simson Lines. The angle α formed by $\overline{MC}$ and $\overline{M'C'} = \frac{1}{2}(m\widehat{MM'} - m\widehat{CC'})$ (#40). In Fig. S9-11, $\triangle PFD \sim \triangle EJD$ (#48), and $m\angle M'PM = m\angle B'EB$. Now, $m\angle M'PM = \frac{1}{2}(m\widehat{MM'})$ (#36), while $m\angle B'EB = \frac{1}{2}(m\widehat{BB'} + m\widehat{AA'})$ (#39). Therefore, $m\widehat{MM'} = m\widehat{BB'} + m\widehat{AA'}$. Thus, $m\angle\alpha = \frac{1}{2}(m\widehat{BB'} + m\widehat{AA'} - m\widehat{CC'})$. Since $\widehat{CC'}$, $\widehat{BB'}$, and $\widehat{AA'}$ are independent of the position of point P, the theorem is proved.

9-12 *In the circumcircle of* $\triangle ABC$, *chord* $\overline{PQ}$ *is drawn parallel to side* $\overline{BC}$. *Prove that the Simson Lines of* $\triangle ABC$, *with respect to points* P *and* Q, *are concurrent with the altitude* $\overline{AD}$ *of* $\triangle ABC$.

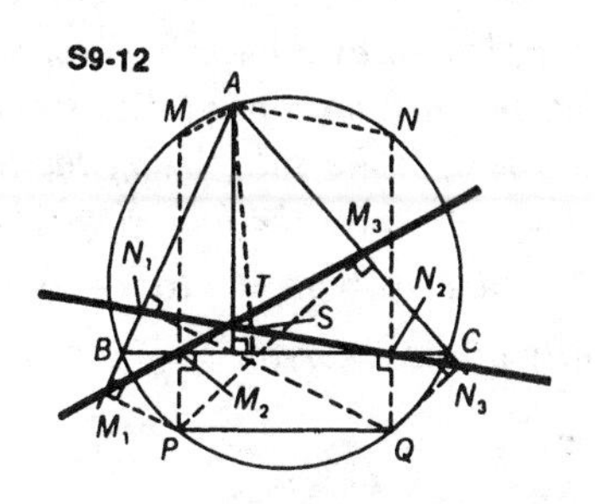

As illustrated in Fig. S9-12, $\overleftrightarrow{M_1M_2M_3}$ is the Simson Line of point P, and $\overleftrightarrow{N_1N_2N_3}$ is the Simson Line of point Q.

Extend $\overline{PM_2}$ and $\overline{QN_2}$ to meet the circle at points M and N, respectively. In Solution 9-10 (III), it was proved that $\overline{AM} \parallel$ Simson Line $\overleftrightarrow{M_1M_2M_3}$ and $\overline{AN} \parallel$ Simson Line $\overleftrightarrow{N_1N_2N_3}$.

Draw altitude $\overline{AD}$, cutting $\overleftrightarrow{M_1M_2M_3}$ and $\overleftrightarrow{N_1N_2N_3}$, at points T and S.

Since $\overline{MM_2} \parallel \overline{AD} \parallel \overline{NN_2}$ (#9), quadrilaterals ATM_2M, and ASN_2N are parallelograms, (#21a). Therefore, $MM_2 = AT$

$$\text{and } NN_2 = AS \text{ (\#21b).} \tag{I}$$

However, since $\overline{PM} \parallel \overline{QN}$, $m\widehat{MN} = m\widehat{PQ}$, and $MN = PQ$. As $\overline{MP} \perp \overline{PQ}$ (#10), then quadrilaterals $MNQP$ and M_2N_2QP are rectangles,

$$\text{and } MM_2 = NN_2. \tag{II}$$

From (I) and (II), $AT = AS$.

Therefore, altitude $\overline{AD}$ crosses Simson Lines $\overleftrightarrow{M_1M_2M_3}$ and $\overleftrightarrow{N_1N_2N_3}$ at the same point. Thus, the Simson Lines are concurrent with the altitude $\overline{AD}$.

10. The Theorem of Stewart

10-1 *A classic theorem, known as Stewart's Theorem, is very useful as a means of finding the measure of any line segment from the vertex of a triangle to the opposite side. Using the letter designations in Fig. S10-1, the theorem states the following relationship:*

$$a^2n + b^2m = c(d^2 + mn).$$

Prove the validity of the theorem.

S10-1

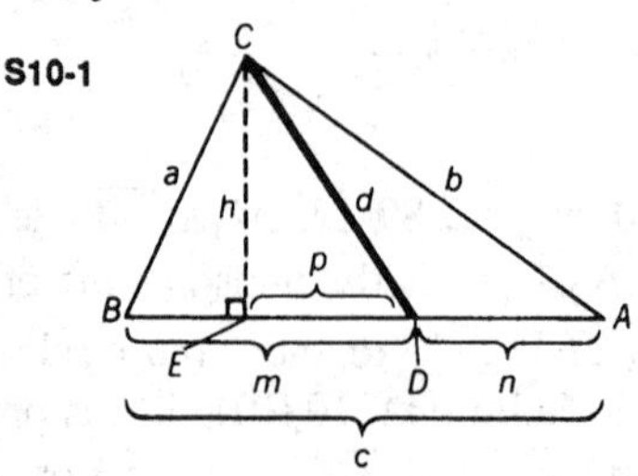

In $\triangle ABC$, let $BC = a$, $AC = b$, $AB = c$, $CD = d$. Point D divides $\overline{AB}$ into two segments; $BD = m$ and $DA = n$. Draw altitude $CE = h$ and let $ED = p$.

In order to proceed with the proof of Stewart's Theorem we first derive two necessary formulas. The first one is applicable to $\triangle CBD$. We apply the Pythagorean Theorem to $\triangle CEB$ to obtain

$$(CB)^2 = (CE)^2 + (BE)^2.$$

Since $BE = m - p$, $a^2 = h^2 + (m - p)^2$. (I)

However, by applying the Pythagorean Theorem to $\triangle CED$, we have $(CD)^2 = (CE)^2 + (ED)^2$, or $h^2 = d^2 - p^2$.

Replacing h^2 in equation (I), we obtain

$$a^2 = d^2 - p^2 + (m - p)^2,$$

$$a^2 = d^2 - p^2 + m^2 - 2mp + p^2.$$

$$\text{Thus, } a^2 = d^2 + m^2 - 2mp. \qquad \text{(II)}$$

A similar argument is applicable to $\triangle CDA$.

Applying the Pythagorean Theorem to $\triangle CEA$, we find that

$$(CA)^2 = (CE)^2 + (EA)^2.$$

Since $EA = (n + p)$, $b^2 = h^2 + (n + p)^2$. (III)

However, $h^2 = d^2 - p^2$, substitute for h^2 in (III) as follows:

$$b^2 = d^2 - p^2 + (n + p)^2,$$

$$b^2 = d^2 - p^2 + n^2 + 2np + p^2.$$

$$\text{Thus, } b^2 = d^2 + n^2 + 2np. \tag{IV}$$

Equations (II) and (IV) give us the formulas we need.
Now multiply equation (II) by n to get

$$a^2n = d^2n + m^2n - 2mnp, \tag{V}$$

and multiply equation (IV) by m to get

$$b^2m = d^2m + n^2m + 2mnp. \tag{VI}$$

Adding (V) and (VI), we have

$$a^2n + b^2m = d^2n + d^2m + m^2n + n^2m + 2mnp - 2mnp.$$

Therefore, $a^2n + b^2m = d^2(n + m) + mn(m + n)$.

Since $m + n = c$, we have $a^2n + b^2m = d^2c + mnc$, or $a^2n + b^2m = c(d^2 + mn)$.

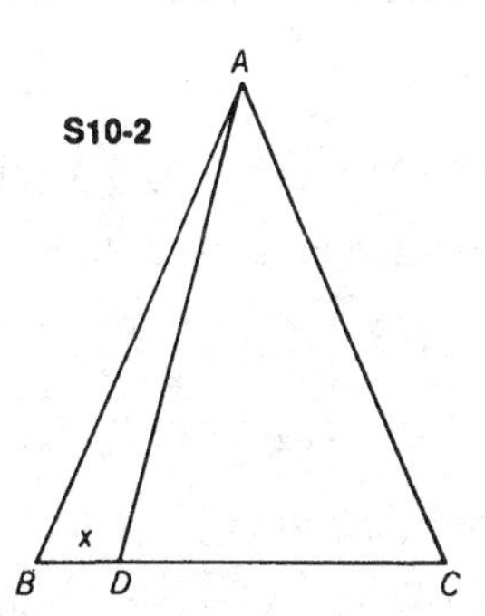

10-2 *In an isosceles triangle with two sides of measure* 17, *a line measuring* 16 *is drawn from the vertex to the base. If one segment of the base, as cut by this line, exceeds the other by* 8, *find the measures of the two segments.*

In Fig. S10-2, $AB = AC = 17$, and $AD = 16$. Let $BD = x$ so that $DC = x + 8$.

By Stewart's Theorem,

$$(AB)^2(DC) + (AC)^2(BD) = BC[(AD)^2 + (BD)(DC)].$$

Therefore,

$$(17)^2(x + 8) + (17)^2(x) = (2x + 8)[(16)^2 + x(x + 8)],$$

and $x = 3$. Therefore, $BD = 3$ and $DC = 11$.

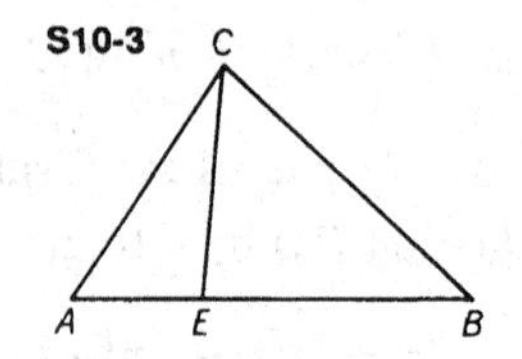

10-3 *In* $\triangle$ABC, *point* E *is on* $\overline{\text{AB}}$, *so that* AE $= \frac{1}{2}$ EB. *Find* CE *if* AC $= 4$, CB $= 5$, *and* AB $= 6$.

METHOD I: By applying Stewart's Theorem to $\triangle ABC$ (Fig. S10-3), we get

$$(AC)^2(EB) + (CB)^2(AE) = AB[(CE)^2 + (AE)(EB)].$$

Therefore, $(4)^2(4) + (5)^2(2) = 6[(CE)^2 + (2)(4)]$,
$114 = 6(CE)^2 + 48$, and $CE = \sqrt{11}$.

METHOD II: Since $\triangle ACE$ and $\triangle ACB$ share the same altitude, and $AE = \frac{1}{3} AB$, the area of $\triangle ACE = \frac{1}{3}$ the area of $\triangle ACB$.

By Heron's Formula,

$$\frac{1}{3} \text{ the area } \triangle ACB = \frac{1}{3}\sqrt{\frac{15}{2}\left(\frac{7}{2}\right)\left(\frac{5}{2}\right)\left(\frac{3}{2}\right)} = \frac{5}{4}\sqrt{7}. \qquad \text{(I)}$$

Let $CE = x$. Then the area of $\triangle ACE$

$$= \sqrt{\left(\frac{6+x}{2}\right)\left(\frac{6-x}{2}\right)\left(\frac{x+2}{2}\right)\left(\frac{x-2}{2}\right)}$$

$$= \frac{1}{4}\sqrt{-(x^2 - 36)(x^2 - 4)}. \qquad \text{(II)}$$

Let $y = x^2$. From (I) and (II),

$$\frac{5}{4}\sqrt{7} = \frac{1}{4}\sqrt{-(y^2 - 40y + 144)}.$$

Therefore, $y^2 - 40y + 319 = 0$, and $y = 11$ or, $y = 29$ (reject). Therefore, $CE = \sqrt{11}$.

COMMENT: Compare the efficiency of Method II with that of Method I.

Challenge *Find the measure of the segment from* E *to the midpoint of* $\overline{\text{CB}}$.

ANSWER: $\frac{1}{2}\sqrt{29}$

10-4 *Prove that the sum of the squares of the distances from the vertex of the right angle, in a right triangle, to the trisection points along the hypotenuse is equal to* $\frac{5}{9}$ *the square of the measure of the hypotenuse.*

Applying Stewart's Theorem to Fig. S10-4,
using p as the internal line segment,

$$2a^2n + b^2n = c(p^2 + 2n^2); \qquad \text{(I)}$$

using q as the internal line segment,

$$a^2n + 2b^2n = c(q^2 + 2n^2). \qquad \text{(II)}$$

By adding (I) and (II), we get

$$3a^2n + 3b^2n = c(4n^2 + p^2 + q^2).$$

Since $a^2 + b^2 = c^2$, $3n(c^2) = c(4n^2 + p^2 + q^2)$.
Since $3n = c$, $c^2 = (2n)^2 + p^2 + q^2$.
But $2n = \frac{2}{3}c$; therefore, $p^2 + q^2 = c^2 - \left(\frac{2}{3}c\right)^2 = \frac{5}{9}c^2$.

S10-4

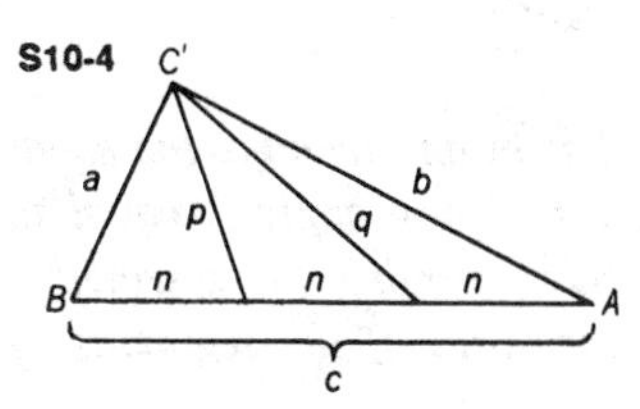

S10-5

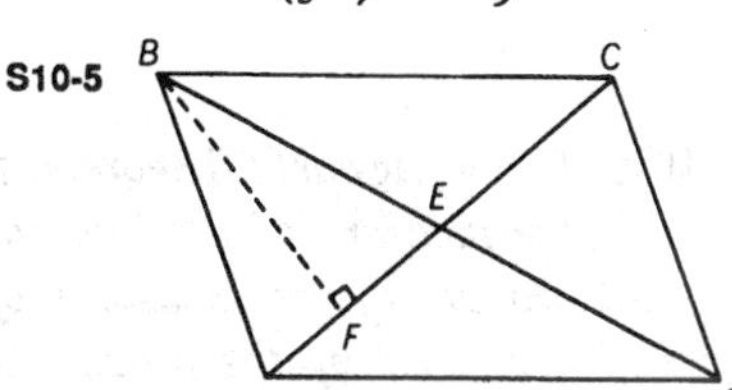

10-5 *Prove that the sum of the squares of the measures of the sides of a parallelogram equals the sum of the squares of the measures of the diagonals.*

In Fig. S10-5, consider $\triangle ABE$.
Draw altitude $\overline{BF}$.

$$(AB)^2 = (BE)^2 + (AE)^2 - 2(AE)(FE), \qquad \text{(I)}$$

and

$$(BC)^2 = (BE)^2 + (EC)^2 + 2(EC)(FE). \qquad \text{(II)}$$

[See the proof of Stewart's Theorem, Solution 10-1, equations (II) and (IV).]
Since the diagonals of $ABCD$ bisect each other, $AE = EC$.
Therefore, by adding equations (I) and (II), we get

$$(AB)^2 + (BC)^2 = 2(BE)^2 + 2(AE)^2. \qquad \text{(III)}$$

Similarly, in $\triangle CAD$,

$$(CD)^2 + (DA)^2 = 2(DE)^2 + 2(CE)^2. \qquad \text{(IV)}$$

By adding lines (III) and (IV), we get

$$(AB)^2 + (BC)^2 + (CD)^2 + (DA)^2 = 2(BE)^2 + 2(AE)^2 + 2(DE)^2 + 2(CE)^2.$$

Since $AE = EC$ and $BE = ED$,

$$(AB)^2 + (BC)^2 + (CD)^2 + (DA)^2 = 4(BE)^2 + 4(AE)^2,$$
$$(AB)^2 + (BC)^2 + (CD)^2 + (DA)^2 = (2BE)^2 + (2AE)^2,$$
$$(AB)^2 + (BC)^2 + (CD)^2 + (DA)^2 = (BD)^2 + (AC)^2.$$

Challenge *A given parallelogram has sides measuring 7 and 9, and a shorter diagonal measuring 8. Find the measure of the longer diagonal.*

ANSWER: 14

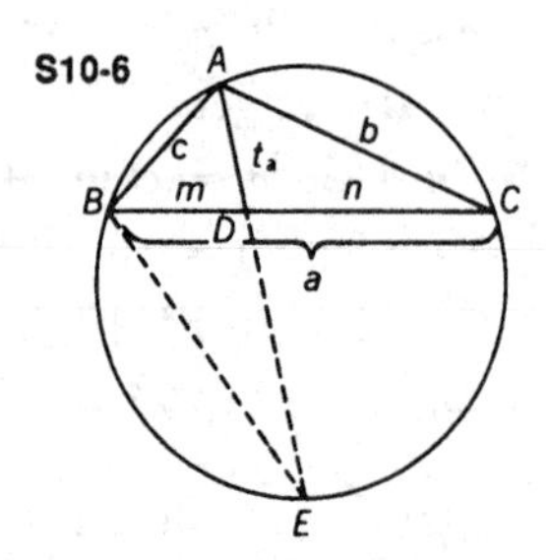

10-6 *Using Stewart's Theorem, prove that in any triangle the square of the measure of the internal bisector of any angle is equal to the product of the measures of the sides forming the bisected angle decreased by the product of the measures of the segments of the side to which this bisector is drawn.*

By Stewart's Theorem we obtain the following relationship:

$$c^2n + b^2m = a(t_a{}^2 + mn), \text{ or } t_a{}^2 + mn = \frac{c^2n + b^2m}{a},$$

as illustrated by Fig. S10-6.

But, $\frac{c}{b} = \frac{m}{n}$ (#47), therefore $cn = bm$.

Substituting in the above equation,

$$t_a{}^2 + mn = \frac{cbm + cbn}{m + n} = \frac{cb(m + n)}{m + n} = cb.$$

Hence, $t_a{}^2 = cb - mn$.

Challenge 1 *Can you also prove the theorem in Problem 10-6 without using Stewart's Theorem?*

As in Fig. S10-6, extend $\overline{AD}$, the bisector of $\angle BAC$, to meet the circumcircle of $\triangle ABC$ at E. Then draw $\overline{BE}$. Since $m\angle BAD = m\angle CAD$, and $m\angle E = m\angle C$ (#36), $\triangle ABE \sim \triangle ADC$, and $\frac{AC}{AD} = \frac{AE}{AB}$, or

$$(AC)(AB) = (AD)(AE) = (AD)(AD + DE)$$
$$= (AD)^2 + (AD)(DE). \quad \text{(I)}$$

However, $(AD)(DE) = (BD)(DC)$ (#52). (II)

Substituting (II) into (I), we obtain

$$(AD)^2 = (AC)(AB) - (BD)(DC),$$

or, using the letter designations in Fig. S10-6, $t_a^{\,2} = cb - mn$.

10-7 *The two shorter sides of a triangle measure 9 and 18. If the internal angle bisector drawn to the longest side measures 8, find the measure of the longest side of the triangle.*

Let $AB = 9$, $AC = 18$, and angle bisector $AD = 8$. (See Fig. S10-7.) Since $\frac{BD}{DC} = \frac{AB}{AC} = \frac{1}{2}$ (#47), we can let $BD = m = x$, so that $DC = n = 2x$. From the solution to Problem 10-6, we know that $t_a^{\,2} = bc - mn$, or $(AD)^2 = (AC)(AB) - (BD)(DC)$. Therefore, $(8)^2 = (18)(9) - 2x^2$, and $x = 7$. Thus, $BC = 3x = 21$.

Challenge *Find the measure of a side of a triangle if the other two sides and the bisector of the included angle have measures 12, 15, and 10, respectively.*

ANSWER: 18

S10-7

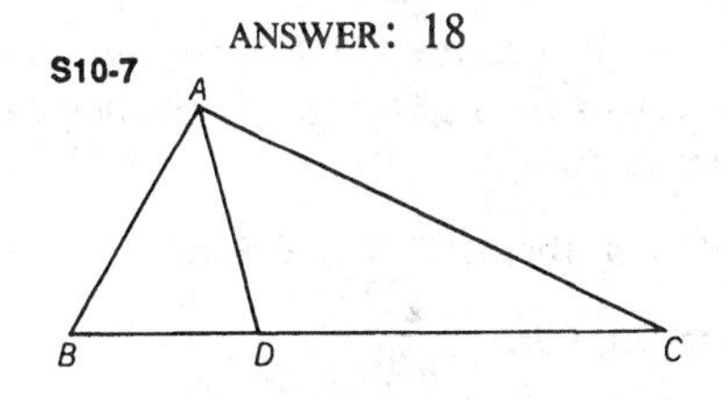

S10-8

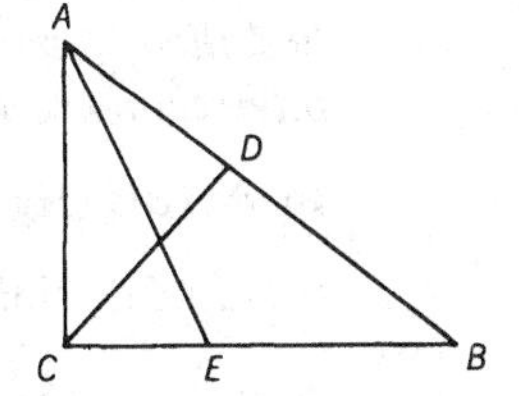

10-8 *In a right triangle, the bisector of the right angle divides the hypotenuse into segments that measure 3 and 4. Find the measure of the angle bisector of the larger acute angle of the right triangle.*

In right $\triangle ABC$, with right angle at C, and angle bisector $\overline{CD}$, $AD = 3$ while $DB = 4$. (See Fig. S10-8.)

Since $\frac{AC}{CB} = \frac{AD}{DB} = \frac{3}{4}$ (#47), $AC = 3x$, and $CB = 4x$.

By the Pythagorean Theorem, applied to $\triangle ABC$,

$$(3x)^2 + (4x)^2 = 7^2, \text{ and } x = \frac{7}{5}.$$

Thus, $AC = \frac{21}{5}$ and $CB = \frac{28}{5}$. Also, $\frac{AC}{AB} = \frac{CE}{EB}$ (#47).

Substituting, we get $\dfrac{\frac{21}{5}}{7} = \dfrac{CE}{\frac{28}{5} - CE}$. Thus $CE = \frac{21}{10}$ and $EB = \frac{7}{2}$.

The proof may be concluded using either one of the following methods.

METHOD I: From Solution 10-6, $(AE)^2 = (AC)(AB) - (CE)(EB)$. Substituting, we have $(AE)^2 = \left(\frac{21}{5}\right)(7) - \left(\frac{21}{10}\right)\left(\frac{7}{2}\right)$, and $AE = \dfrac{21\sqrt{5}}{10}$.

METHOD II: By the Pythagorean Theorem, applied to $\triangle ACE$,

$$(AE)^2 = (AC)^2 + (CE)^2;\text{ therefore, } AE = \frac{21\sqrt{5}}{10}.$$

S10-9

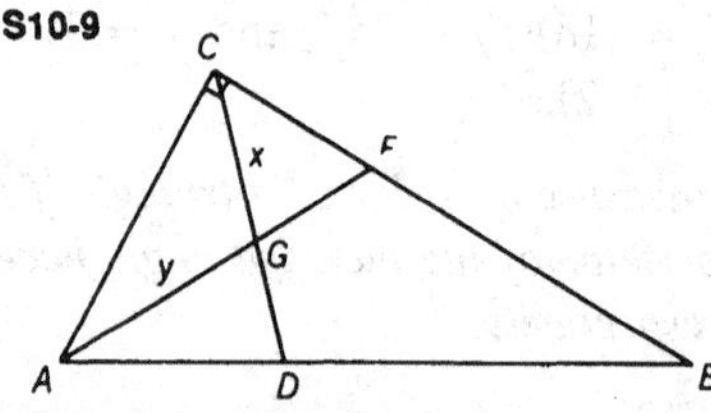

10-9 *In a 30–60–90 right triangle, if the measure of the hypotenuse is 4, find the distance from the vertex of the right angle to the point of intersection of the angle bisectors.*

In $\triangle ABC$ (Fig. S10-9), if $AB = 4$, then $AC = 2$ (#55c).

In $\triangle ACE$, since $m\angle CAE = 30$, $CE = \dfrac{2}{\sqrt{3}}$, (I)

and $AE = \dfrac{4}{\sqrt{3}}$. In $\triangle ACE$, $\dfrac{AC}{CE} = \dfrac{AG}{GE}$ (#47). (II)

If we let $AG = y$, then from equation (II), we find $GE = \dfrac{y}{\sqrt{3}}$.

Since $AG + GE = AE$, $y + \dfrac{y}{\sqrt{3}} = \dfrac{4}{\sqrt{3}}$, and $y = \dfrac{4}{1+\sqrt{3}} = 2\sqrt{3} - 2$. Thus, $AG = 2\sqrt{3} - 2$, (III)

and $GE = 2 - \dfrac{2\sqrt{3}}{3}$. (IV)

From Solution 10-6 we know that

$(CG)^2 = (AC)(CE) - (AG)(GE)$. (V)

Substituting (I), (III), and (IV) into (V), we get

$(CG)^2 = 8 - 4\sqrt{3}$.

Therefore, $CG = \sqrt{8 - 4\sqrt{3}} = \sqrt{6} - \sqrt{2}$.

HINTS

1-1 Express angles AFB, AEB, and ADB in terms of $\angle CAF$, $\angle CBF$, $\angle ABE$, and $\angle BAD$. Then apply Theorem #13.

1-2 Consider $\angle ADB$ as an exterior angle of $\triangle CDB$.

1-3 Examine the isosceles triangles.

1-4 METHOD I: Use Theorem #27 to show $\triangle FCA$ is isosceles.

METHOD II: Circumscribe a circle about $\triangle ABC$, extend $\overline{CE}$ to meet the circle at G. Then draw $\overline{GF}$.

1-5 To show $\overline{BP}$ is parallel $\overline{AE}$, use Theorem #7, after using Theorems #14 and #5. To show $\overline{BP}$ is perpendicular $\overline{AE}$, use Theorems #14 and #5 to prove that the bisector of $\angle A$ is also the bisector of the vertex angle of an isosceles triangle.

1-6 Extend $\overline{AM}$ through M to P so that $AM = MP$. Draw $\overline{BT}$; T is the midpoint of $\overline{AD}$. Then show that $\triangle TBP$ is isosceles. Use Theorems #21, #27, #12, and #8.

1-7 METHOD I: Draw a line through M parallel to $\overline{BC}$. Then use Theorems #27 and #8.

METHOD II: Extend $\overline{KM}$ to meet $\overline{CB}$ extended at G; then prove $\triangle KMC \cong \triangle GMC$.

1-8 Extend $\overline{CP}$ and $\overline{CQ}$ to meet $\overline{AB}$ at S and R, respectively. Prove that P and Q are the midpoints of $\overline{CS}$ and $\overline{CR}$, respectively; then use Theorem #26.

1-9 From E, the point of intersection of the diagonals of square $ABCD$, draw a line parallel to $\overline{BPQ}$. Use Theorems #25, #10, and #23.

1-10 METHOD I: Draw $\overline{AF} \perp \overline{DE}$, and draw $\overline{DG}$, where G is on $\overline{AF}$ and $m\angle FDG = 60$. Then show that $\overline{AF}$ is the perpendicular bisector of $\overline{DE}$. Apply Theorem #18.

METHOD II: Draw $\triangle AFD$ on side $\overline{AD}$ so that $m\angle FAD = m\angle FDA = 15$; then draw $\overline{FE}$. Now prove $m\angle EAB = 60$.

METHOD III: Draw equilateral $\triangle DFC$ externally on side $\overline{DC}$; then draw $\overline{EF}$. Show that $m\angle BAE = 60$.

METHOD IV: Extend $\overline{DE}$ and $\overline{CE}$ to meet $\overline{BC}$ and $\overline{AD}$ at K and H, respectively. Draw $\overline{AF}$ and $\overline{CG}$ perpendicular to $\overline{DK}$. Now prove $\overline{AF}$ is the perpendicular bisector of $\overline{DE}$.

1-11 Join E and F, and prove that $DGFE$ is an isosceles trapezoid.

1-12 Draw $\overline{CD}$, $\overline{CE}$, and the altitude from C to $\overline{AB}$; then prove triangles congruent.

1-13 Draw a line from one vertex (the side containing the given point) perpendicular to a diagonal of the rectangle; then draw a line from the given point perpendicular to the first line.

1-14 Prove various pairs of triangles congruent.

1-15 Use Theorems #26 and #10.

1-16 Draw a line through C and the midpoint of $\overline{AD}$; then prove that it is the perpendicular bisector of $\overline{TD}$.

1-17 Prove that the four given midpoints determine a parallelogram. Use Theorem #26.

1-18 Draw median $\overline{CGD}$. From D and E (the midpoint of $\overline{CG}$) draw perpendiculars to $\overline{XYZ}$. Show $\overline{QD}$ is the median of trapezoid $AXZB$. Then prove $QD = EP = \frac{1}{2}CY$.

1-19 Extend $\overline{BP}$ through P to E so that $BE = AQ$. Then draw $\overline{AE}$ and $\overline{BQ}$. Prove that $\overline{EMQ}$ is a diagonal of parallelogram $AEBQ$. Use Theorem #27.

1-20 Prove $\triangle AFE \cong \triangle BFC \cong \triangle DCE$.

1-21 (a) Prove four triangles congruent, thereby obtaining four equal sides; then prove one right angle.
(b) Prove that one diagonal of the square and one diagonal of the parallelogram share the same midpoint.

2-1 Consider $\triangle ADC$, then $\triangle ABC$. Apply Theorem #46.

2-2 METHOD I: Prove $\triangle BFC \sim \triangle PEB$; then manipulate the resulting proportions.

METHOD II: Draw a line from B perpendicular to $\overline{PD}$ at G. Then prove $\triangle GPB \cong \triangle EPB$.

2-3 Prove two pairs of triangles similar and equate ratios. Alternatively, extend the line joining the midpoints of the diagonals to meet one of the legs; then use Theorems #25 and #26.

2-4 Draw a line through D parallel to $\overline{BC}$ meeting $\overline{AE}$ at G. Obtain proportions from $\triangle ADG \sim \triangle ABE$ and $\triangle DGF \sim \triangle CEF$.

2-5 Draw a line through E parallel to $\overline{AD}$. Use this line with Theorems #25 and #26.

2-6 Prove $\triangle HEA \sim \triangle BEC$, and $\triangle BFA \sim \triangle GFC$; then equate ratios.

2-7 Extend $\overline{APM}$ to G so that $PM = MG$; also draw $\overline{BG}$ and $\overline{GC}$. Then use Theorem #46.

2-8 Show H is the midpoint of $\overline{AB}$. Then use Theorem #47 in $\triangle ABC$.

2-9 Prove $\triangle AFC \sim \triangle HGB$. Use proportions from these triangles, and also from $\triangle ABE \sim \triangle BHG$; apply Theorem #46.

2-10 Use proportions resulting from the following pairs of similar triangles:

$$\triangle AHE \sim \triangle ADM,\ \triangle AEF \sim \triangle AMC, \text{ and } \triangle BEG \sim \triangle BDC.$$

2-11 Prove $\triangle KAP \sim \triangle PAB$. Also consider $\angle PKA$ as an exterior angle of $\triangle KPB$ and $\triangle KPL$.

2-12 From points R and Q, draw perpendiculars to $\overline{AB}$. Prove various pairs of triangles similar.

2-13 Prove $\triangle ACZ \sim \triangle AYB$, and $\triangle BCZ \sim \triangle BXA$; then add the resulting proportions.

2-14 Draw lines through B and C, parallel to $\overline{AD}$, the angle bisector. Then apply the result of Problem 2-13.

2-15 Use the result of Problem 2-13.

2-16 Prove $\triangle FDG \sim \triangle ABG$, and $\triangle BGE \sim \triangle DGA$.

3-1 Apply the Pythagorean Theorem #55 in the following triangles: $\triangle ADC$, $\triangle EDC$, $\triangle ADB$, and $\triangle EDB$.

3-2 Use Theorem #29; then apply the Pythagorean Theorem to $\triangle DGB$, $\triangle EGA$, and $\triangle BGA$. (G is the centroid.)

3-3 Draw a line from C perpendicular to $\overline{HL}$. Then apply the Pythagorean Theorem to $\triangle ABC$ and $\triangle HGC$. Use Theorem #51.

3-4 Through the point in which the given line segment intersects the hypotenuse, draw a line parallel to either of the legs of the right triangle. Then apply Theorem #55.

3-5 METHOD I: Draw $\overline{AC}$ meeting $\overline{EF}$ at G; then apply the Pythagorean Theorem to $\triangle FBC$, $\triangle ABC$, and $\triangle EGC$.

METHOD II: Choose H on $\overline{EC}$ so that $EH = FB$; then draw $\overline{BH}$. Find BH.

3-6 Use the last two vectors (directed lines) and form a parallelogram with the extension of the first vector. Also drop a perpendicular to the extension of the first vector. Then use the Pythagorean Theorem. The Law of Cosines may also be used.

3-7 Draw the altitude to the side that measures 7. Then apply the Pythagorean Theorem to the two right triangles.

3-8 METHOD I: Construct $\triangle ABC$ so that $\overline{CG} \perp \overline{AB}$. (Why can this be done?) Then use Theorem #55.

METHOD II: Draw altitude $\overline{CJ}$. Apply the Pythagorean Theorem to $\triangle GJC$, $\triangle JEC$, and $\triangle JHC$.

3-9 Extend $\overline{BP}$ to meet $\overline{AD}$ at E; also draw a perpendicular from C to $\overline{AD}$. Use Theorems #51b and #46.

3-10 Use Theorems #55, #29, and #51b.

3-11 From the point of intersection of the angle bisectors, draw a line perpendicular to one of the legs of the right triangle. Then use Theorem #55.

3-12 Apply the Pythagorean Theorem to each of the six right triangles.

3-13 Use Theorems #41, and #29.

3-14 Draw a perpendicular from the centroid to one of the sides; then apply Theorem #55.

4-1 Use Theorem #34.

4-2 Draw $\overline{AO}$, $\overline{BC}$, and $\overline{OC}$. Prove $\triangle BEC \sim \triangle ABO$.

4-3 Draw $\overline{QA}$ and $\overline{QB}$; then prove $\triangle DAQ \sim \triangle CBQ$, and $\triangle QBE \sim \triangle QAC$. ($D$, C, and E are the feet of the perpendiculars on $\overleftrightarrow{PA}$, $\overleftrightarrow{AB}$, and $\overleftrightarrow{PB}$, respectively.)

4-4 Show that $\triangle GPB$ is isosceles.

4-5 Apply the Pythagorean Theorem to $\triangle DEB$, $\triangle DAB$, $\triangle AEC$, and $\triangle ABC$.

4-6 Extend $\overline{AO}$ to meet circle O at C; then draw $\overline{MA}$. Use Theorem #52 with chords $\overline{AOC}$ and $\overline{MPN}$.

4-7 Use Theorem #52 with chords $\overline{AB}$ and $\overline{CD}$.

4-8 Draw $\overline{BC}$ and $\overline{AD}$.

METHOD I: Show $\triangle CFD \sim \triangle DEA$, and $\triangle AEB \sim \triangle BFC$.

METHOD II: Use the Pythagorean Theorem in $\triangle AED$, $\triangle DFC$, $\triangle AEB$, and $\triangle BFC$.

4-9 From the center of the circle draw a perpendicular to the secant of measure 33. Then use Theorem #54.

4-10 Draw radii to points of contact; then draw $\overline{OB}$. Consider $\overline{OB}$ as an angle bisector in $\triangle ABC$. Use Theorem #47.

4-11 Draw $\overline{KO}$ and $\overline{LO}$. Show that $\angle KOL$ is a right angle.

4-12 Draw $\overline{DS}$ and $\overline{SJ}$. Use Theorems #51a and #52.

4-13 Draw $\overline{BD}$ and $\overline{CD}$. Apply Theorem #51b.

4-14 METHOD I: Draw $\overline{ED}$. Use Theorems #55c and 55d. Then prove $\triangle AEF \sim \triangle ABC$.

METHOD II: Use only similar triangles.

4-15 Use Theorems #18 and #55.

4-16 Prove $\triangle BEC \sim \triangle AED$, and $\triangle AEB \sim \triangle DEC$. E is the intersection of the diagonals.

4-17 Use Theorems #53, #50, #37, and #8.

4-18 Prove $\triangle DPB \sim \triangle BPC$, and $\triangle DAP \sim \triangle ACP$.

4-19 METHOD I: Draw diameter $\overline{BP}$ of the circumcircle. Draw $\overline{PT} \perp$ altitude $\overline{AD}$; draw $\overline{PA}$ and $\overline{CP}$. Prove $APCO$ is a parallelogram.

METHOD II: Let $AB = AC$. (Why is this permissible?) Then choose a point P so that $AP = BP$. Prove $\triangle ACD \sim \triangle BOD$.

4-20 Draw $\overline{PC}$, $\overline{ED}$, and $\overline{DC}$. Show that $\overline{PC}$ bisects $\angle BPA$.

4-21 Draw $\overline{DO}$ and $\overline{CDE}$ where E is on circle O. Use Theorems #30 and #52.

4-22 From O draw perpendiculars to $\overline{AB}$ and $\overline{CD}$; also draw $\overline{OD}$. Use Theorems #52 and #55.

4-23 For chords $\overline{AB}$ and $\overline{CD}$, draw $\overline{AD}$ and $\overline{CB}$. Also draw diameter $\overline{CF}$ and chord $\overline{BF}$. Use Theorem #55; also show that $AD = FB$.

4-24 Draw $\overline{MO}$, $\overline{NQ}$, and the common internal tangent. Show $MNQO$ is a parallelogram.

4-25 **(a)** Draw common internal tangent $\overline{AP}$. Use Theorem #53. Also prove $\triangle ADE \sim \triangle ABC$.

(b) METHOD I: Apply Theorem #15 in quadrilateral $ADPE$.

METHOD II: Show $\triangle ABC$ is a right triangle.

4-26 Draw $\overline{OA}$ and $\overline{O'B}$; then draw $\overline{AE} \perp \overline{OO'}$ and $\overline{BD} \perp \overleftrightarrow{OO'}$. Prove $ABO'O$ is a parallelogram.

4-27 Prove $\triangle AEO \sim \triangle AFC \sim \triangle ADO'$.

4-28 Extend the line of centers to the vertex of the square. Also draw a perpendicular from the center of each circle to a side of the square. Use Theorem #55a.

4-29 Apply the Pythagorean Theorem to $\triangle DEO$. E is the midpoint of $\overline{AO}$.

4-30 Find one-half the side of the square formed by joining the centers of the four smaller circles.

4-31 Draw radii to the points of contact. Use Theorem #55.

4-32 Use an indirect method. That is, assume the third common chord is not concurrent with the other two.

4-33 Show that the opposite angles are supplementary.

4-34 Show that quadrilateral $D'BB'D$ is cyclic.

4-35 Show that $\angle GFA \cong \angle DFB$ after proving $BDFO$ cyclic.

4-36 Show $\angle BRQ$ is supplementary to $\angle BCQ$.

4-37 Draw $\overline{DE}$. Show quadrilateral $DCEF$ is cyclic. Then find the measure of $\angle CED$.

4-38 Draw $\overline{AF}$. Show quadrilateral $AEFB$ is cyclic. What type of triangle is $\triangle ABE$?

4-39 Choose a point Q on $\overline{BP}$ such that $PQ = QC$. Prove $\triangle BQC \cong \triangle APC$.

4-40 METHOD I: Draw $\overline{BC}$, $\overline{OB}$, and $\overline{OC}$. Show quadrilateral $ABGC$ is cyclic, as is quadrilateral $ABOC$.

METHOD II: Draw $\overline{BG}$ and extend it to meet the circle at H. Draw $\overline{CH}$. Use Theorems #38, #18, and #30.

5-1 Draw $\overline{EC}$ and show that the area of $\triangle DEC$ is one-half the area of each of the parallelograms.

5-2 METHOD I: In $\triangle EDC$ draw altitude $\overline{EH}$. Use Theorems #28, #49, and #24.

METHOD II: Use the ratio between the areas of $\triangle EFG$ and $\triangle EDC$.

5-3 Compare the areas of the similar triangles.

5-4 Represent the area of each in terms of the radius of the circle.

5-5 Prove $\triangle ADC \sim \triangle AFO$.

5-6 METHOD I: Draw a line through D and perpendicular to $\overline{AB}$. Then draw $\overline{AQ}$ and $\overline{DQ}$. Use the Pythagorean Theorem in various right triangles.

METHOD II: Draw a line through P parallel to $\overline{BC}$ and meeting $\overline{AB}$ and $\overline{DC}$ (extended) at points H and F, respectively. Then draw a line from P perpendicular to $\overline{BC}$. Find the desired result by adding and subtracting areas.

5-7 Draw the altitude to the line which measures 14. Use similarity to obtain the desired result.

5-8 Use Formula #5b with each triangle containing $\angle A$.

5-9 Draw $\overline{DC}$. Find the ratio of the area of $\triangle DAE$ to the area of $\triangle ADC$.

5-10 Use Formula #5b with each triangle containing the angle between the specified sides.

5-11 METHOD I: From points C and D draw perpendiculars to $\overline{AB}$. Find the ratio between the areas of $\square AEDF$ and $\triangle ABC$.

METHOD II: Use similarity and Formula #5b for triangles containing $\angle A$.

5-12 Draw the line of centers O and Q. Then draw $\overline{NO}$, $\overline{NQ}$, $\overline{MO}$, and $\overline{MQ}$. Determine the type of triangle $\triangle KLN$ is.

5-13 Extend one of the medians one-third its length, through the side to which it is drawn; then join this external point with the two nearest vertices. Find the area of one-half the parallelogram.

5-14 Use Theorem #55e or Formula #5c to find the area of $\triangle ABC$. Thereafter, apply #29.

5-15 METHOD I: Draw the medians of the triangle. Use Theorems #26, #25, #29, and #55.

METHOD II: Use the result of Problem 5-14.

5-16 Draw a line through E parallel to $\overline{BD}$ meeting $\overline{AC}$ at G. Use Theorems #56 and #25.

5-17 Draw $\overline{EC}$. Compare the areas of triangles BEC and BAC. Then use Theorem #56 and its extension.

5-18 Through E, draw a line parallel to $\overleftrightarrow{AB}$ meeting $\overleftrightarrow{BC}$ and $\overleftrightarrow{AD}$ (extended) at points H and G, respectively. Then draw $\overline{AE}$ and $\overline{BE}$. Find the area of $\triangle AEB$.

5-19 Draw diagonal $\overline{AC}$. Use Theorem #29 in $\triangle ABC$. To obtain the desired result, subtract areas.

5-20 Draw $\overline{QB}$ and diagonal $\overline{BD}$. Consider each figure whose area equals one-half the area of parallelogram $ABCD$.

5-21 Draw $\overline{AR}$ and $\overline{AS}$. Express both areas in terms of RS, RT, and TS. Also use Theorems #32a, and #51a.

5-22 METHOD I: In equilateral $\triangle ABC$, draw a line through point P, the internal point, parallel to $\overline{BC}$ meeting $\overline{AB}$ and $\overline{AC}$ at E and F, respectively. From E draw $\overline{ET} \perp \overline{AC}$. Also draw $\overline{PH} \parallel \overline{AC}$ where H is on $\overline{AB}$. Show that the sum of the perpendiculars equals the altitude of the equilateral $\triangle ABC$, a constant for the triangle.

METHOD II: Draw $\overline{PA}$, $\overline{PB}$, and $\overline{PC}$; then add the areas of the three triangles APB, APC, and BPC. Show that the sum of the perpendiculars equals the altitude of equilateral $\triangle ABC$, a constant for the triangle.

6-1 Draw the radii of the inscribed circle to the points of tangency of the sides of the triangle. Also join the vertices to the center of the inscribed circle. Draw a line perpendicular at the incenter, to one of the lines drawn from the incenter to a vertex. Draw a line perpendicular to one of the sides at another vertex. Let the two perpendiculars meet. Extend the side to which the perpendicular was drawn through the point of intersection with the perpendicular so that the measure of the new line segment equals the semi-perimeter of the triangle.

6-2 Extend a pair of non-parallel opposite sides to form triangles with the other two sides. Apply Heron's Formula to the larger triangle. Then compare the latter area with the area of the quadrilateral.

6-3 **(a)** METHOD I: Use similar triangles to get $\frac{CN}{QM} = \frac{KN}{AM}$. Also prove $AS = AM$. Use Theorem #21-1 to prove rhombus.
METHOD II: Use similar triangles to show $\overline{AQ}$ is an angle bisector. Use #47 to show $\overline{SQ} \parallel \overline{AC}$, also show $AM = MQ$.

(b) Compare the areas of $\triangle BMQ$ and $\triangle AMQ$, also of $\triangle CSQ$ and $\triangle ASQ$.

6-4 Draw $\overline{AE}$ and $\overline{BF}$, where E and F are the points of tangency of the common external tangent with the two circles. Then draw $\overline{AN}$ (extended) and $\overline{BN}$. Use #47 twice to show that $\overline{CN}$ and $\overline{DN}$ bisect a pair of supplementary adjacent angles.

6-5 First find the area of the triangle by Heron's Formula (Formula #5c). Then consider the area of the triangle in terms of the triangles formed by joining P with the vertices. (Use Formula #5a). Do this for each of the four cases which must be considered.

6-6 METHOD I: In $\triangle ABC$, with angle bisectors $AE = BD$, draw $\angle DBF \cong \angle AEB$, $\overline{BF} \cong \overline{BE}$, $\overline{FG} \perp \overline{AC}$, $\overline{AH} \perp \overline{FH}$, where G and H lie on $\overline{AC}$ and $\overline{BF}$, respectively. Also draw $\overline{DF}$. Use congruent triangles to prove the base angles equal.

METHOD II: (indirect) In $\triangle ABC$, with angle bisectors $CE = BF$, draw $\overline{GF} \parallel \overline{EB}$ externally, and through E draw $\overline{GE} \parallel \overline{BF}$. Then draw $\overline{CG}$. Assume the base angles are not congruent.

METHOD III: (indirect) In $\triangle ABC$, with angle bisectors $\overline{BE} \cong \overline{DC}$, draw parallelogram $BDCH$; then draw $\overline{EH}$. Assume the base angles are not congruent. Use Theorem #42.

METHOD IV: (indirect) In $\triangle ABC$, with angle bisectors $\overline{BE}$ and $\overline{DC}$ of equal measure, draw $\angle FCD \cong \angle ABE$ where F is on $\overline{AB}$. Then choose a point G so that $BG = FC$. Draw $\overline{GH} \parallel \overline{FC}$, where H is on $\overline{BE}$. Prove $\triangle BGH \cong \triangle CFD$ and search for a contradiction. Assume $m\angle C > m\angle B$.

6-7 METHOD I: Draw $\overline{DH} \parallel \overline{AB}$ and $\overline{MN} \perp \overline{DH}$, where H is on the circle; also draw $\overline{MH}$, $\overline{QH}$, and $\overline{EH}$. Prove $\triangle MPD \cong \triangle MQH$.

METHOD II: Through P draw a line parallel to $\overline{CE}$, meeting $\overline{EF}$, extended through F, at K, and $\overline{CD}$ at L. Find the ratio $\frac{(MP)^2}{(MQ)^2}$.

METHOD III: Draw a line through E parallel to $\overline{AB}$, meeting the circle at G. Then draw $\overline{GP}$, $\overline{GM}$, and $\overline{GD}$. Prove $\triangle PMG \cong \triangle QME$.

METHOD IV: Draw the diameter through M and O. Reflect $\overline{DF}$ through this diameter; let $\overline{D'F'}$ be the image of $\overline{DF}$. Draw $\overline{CF'}$, $\overline{MF'}$, and $\overline{MD'}$. Also, let P' be the image of P. Prove that P' coincides with Q.

METHOD V: (Projective Geometry) Use harmonic pencil and range concepts.

6-8 METHOD I: Draw $\overline{DG} \parallel \overline{AB}$, where G is on $\overline{CB}$. Also draw $\overline{AG}$, meeting $\overline{DB}$ at F, and draw $\overline{FE}$. Prove that quadrilateral $DGEF$ is a kite (i.e. $GE = FE$ and $DG = DF$).

METHOD II: Draw $\overline{BF}$ so that $m\angle ABF = 20$ and F is on $\overline{AC}$. Then draw $\overline{FE}$. Prove $\triangle FEB$ equilateral, and $\triangle FDE$ isosceles.

METHOD III: Draw $\overline{DF} \parallel \overline{AB}$, where F is on $\overline{BC}$. Extend $\overline{BA}$ through A to G so that $AG = AC$. Then draw $\overline{CG}$. Use similarity and theorem #47 to prove that $\overline{DE}$ bisects $\angle FDB$.

METHOD IV: With B as center and $\overline{BD}$ as radius, draw a circle meeting $\overline{BA}$, extended, at F and $\overline{BC}$ at G. Then draw $\overline{FD}$ and $\overline{DG}$. Prove $\triangle FBD$ equilateral, and $\triangle DBG$ isosceles. Also prove $\triangle DCG \cong \triangle FDA$.

METHOD V: Using C as center, $\overline{AC}$ and $\overline{BC}$ as radii, and $\overline{AB}$ as a side, construct an 18-sided regular polygon.

METHOD VI: (Trigonometric Solution I) Use the law of sines in $\triangle AEC$ and $\triangle ABD$. Then prove $\triangle AEC \sim \triangle DEB$.

METHOD VII: (Trigonometric Solution II) Draw $\overline{AF} \parallel \overline{BC}$. Choose a point G on $\overline{AC}$ so that $AG = BE$. Extend $\overline{BG}$ to meet $\overline{AF}$ at H. Apply the law of sines to $\triangle ADB$ and $\triangle ABH$. Then prove $\triangle BDE \cong \triangle AHG$.

6-9 METHOD I: Rotate the given equilateral $\triangle ABC$ in its plane about point A through a counterclockwise angle of 60°. Let P' be the image of P. Find the area of quadrilateral $APCP'$ (when B is to the left of C), and the area of $\triangle BPC$.

METHOD II: Rotate each of the three triangles in the given equilateral triangle about a different vertex, so that there is now one new triangle on each side of the given equilateral triangle, thus forming a hexagon. Consider the area of the hexagon in parts, two different ways.

6-10 Rotate $\triangle DAP$ in its plane about point A through a counterclockwise angle of 90°. Express the area of $\triangle PP'B$ (P' is the image of P), in two different ways using Formula #5c, and Formula #5b. Investigate $\triangle PAP'$ and $\triangle APB$.

6-11 Prove a pair of overlapping triangles congruent.

Challenge 1 Draw two of the required lines. Draw the third line as two separate lines drawn from the point of intersection of the latter two lines, and going in opposite directions. Prove that these two smaller lines, in essence, combine to form the required third line.

Challenge 2 Use similarity to obtain three equal ratios. Each ratio is to contain one of the line segments proved congruent in Solution 6-11, while the measure of the other line segment in each ratio is a side of $\triangle KML$ where K, M, and L are the circumcenters.

6-12 METHOD I: Begin by fixing two angles of the given triangle to yield the desired equilateral triangle. Then prove a concurrency of the four lines at the vertex of the third angle of the given triangle.

METHOD II: This method begins like Method I. However, here we must prove that the lines formed by joining the third vertex of the given triangle to two of the closer vertices of the equilateral triangle are trisectors of the third angle (of the original triangle). In this proof an auxiliary circle is used.

6-13 Use similarity to prove that the orthocenter must lie on the line determined by the centroid and the circumcenter. The necessary constructions are a median, altitude, and perpendicular bisector of one side.

6-14 Draw the three common chords of pairs of circles. Show that the three quadrilaterals (in the given triangle) thus formed are each cyclic. (Note that there are two cases to be considered here.)

6-15 Draw the three common chords of pairs of circles. Use Theorems #30, #35, #36, and #48.

7-1 METHOD I: A line is drawn through A of cyclic quadrilateral $ABCD$, to meet $\overline{CD}$, extended, at P, so that $m\angle BAC = m\angle DAP$. Prove $\triangle BAC \sim \triangle DAP$, and $\triangle ABD \sim \triangle ACP$.

METHOD II: In quadrilateral $ABCD$, draw $\triangle DAP$ (internally) similar to $\triangle CAB$. Prove $\triangle BAP \sim \triangle CAD$. (The converse may be proved simultaneously.)

7-2 Draw $\overline{AF}$ and diagonal $\overline{AC}$. Use the Pythagorean Theorem; then apply Ptolemy's Theorem to quadrilateral $AFDC$.

7-3 Use the Pythagorean Theorem; then apply Ptolemy's Theorem to quadrilateral $AFBE$.

7-4 Draw $\overline{CP}$. Use the Pythagorean Theorem; then apply Ptolemy's Theorem to quadrilateral $BPQC$.

7-5 Draw $\overline{RQ}$, $\overline{QP}$, and $\overline{RP}$. Use similarity and Ptolemy's Theorem.

7-6 Prove that $ABCD$ is cyclic; then apply Ptolemy's Theorem.

7-7 Apply Ptolemy's Theorem to quadrilateral $ABPC$.

7-8 Apply Ptolemy's Theorem to quadrilateral $ABPC$.

7-9 Apply the result obtained in Problem 7-7 to $\triangle ABD$ and $\triangle ADC$.

7-10 Apply Ptolemy's Theorem to quadrilateral $ABPC$, and quadrilateral $BPCD$. Then apply the result obtained in Problem 7-7 to $\triangle BEC$.

7-11 Apply the result of Problem 7-8 to equilateral triangles AEC and BFD.

7-12 Consider $\overline{BD}$ in parts. Verify result with Ptolemy's Theorem.

7-13 Use the result of Problem 7-8.

7-14 Choose points P and Q on the circumcircle of quadrilateral $ABCD$ (on arc $\overparen{AD}$) so that $PA = DC$ and $QD = AB$. Apply Ptolemy's Theorem to quadrilaterals $ABCP$ and $BCDQ$.

7-15 On side $\overline{AB}$ of parallelogram $ABCD$ draw $\triangle AP'B \cong \triangle DPC$, externally. Also use Ptolemy's Theorem.

7-16 METHOD I: Draw the diameter from the vertex of the two given sides. Join the other extremity of the diameter with the remaining two vertices of the given triangle. Use Ptolemy's Theorem. (Note: There are two cases to be considered.)

METHOD II: Draw radii to the endpoints of the chord measuring 5. Then draw a line from the vertex of the two given sides perpendicular to the third side. Use Theorem #55c. Ptolemy's Theorem is not used in this method. (Note: There are two cases to be considered.)

8-1 METHOD I: Draw a line through C, parallel to $\overline{AB}$, meeting $\overline{PQR}$ at D. Prove that $\triangle DCR \sim \triangle QBR$, and $\triangle PDC \sim \triangle PQA$.

METHOD II: Draw $\overline{BM} \perp \overleftrightarrow{PR}$, $\overline{AN} \perp \overleftrightarrow{PR}$, and $\overline{CL} \perp \overleftrightarrow{PR}$, where M, N, and L are on $\overleftrightarrow{PQR}$. Prove that $\triangle BMQ \sim \triangle ANQ$, $\triangle LCP \sim \triangle NAP$, and $\triangle MRB \sim \triangle LRC$.

8-2 METHOD I: Compare the areas of the various triangles formed, which share the same altitude. (Note: There are two cases to be considered.)

METHOD II: Draw a line through A, parallel to $\overline{BC}$, meeting $\overline{CP}$ at S, and $\overline{BP}$ at R. Prove that $\triangle AMR \sim \triangle CMB$, $\triangle BNC \sim \triangle ANS$, $\triangle CLP \sim \triangle SAP$, and $\triangle BLP \sim \triangle RAP$. (Note: There are two cases to be considered.)

METHOD III: Draw a line through A and a line through C parallel to $\overline{BP}$, meeting $\overline{CP}$ and $\overline{AP}$ at S and R, respectively. Prove that $\triangle ASN \sim \triangle BPN$, and $\triangle BPL \sim \triangle CRL$; also use Theorem #49. (Note: There are two cases to be considered.)

METHOD IV: Consider $\overline{BPM}$ a transversal of $\triangle ACL$ and $\overline{CPN}$ a transversal of $\triangle ALB$. Then apply Menelaus' Theorem.

8-3 Apply Ceva's Theorem.

8-4 Use similarity, then Ceva's Theorem.

8-5 Use Theorem #47; then use Ceva's Theorem.

8-6 Use Theorem #47; then use Menelaus' Theorem.

8-7 Use Theorem #47; then use Menelaus' Theorem.

8-8 First use Ceva's Theorem to find BS; then use Menelaus' Theorem to find TB.

8-9 Use Menelaus' Theorem; then use Theorem #54.

8-10 Use both Ceva's and Menelaus' Theorems.

8-11 Consider $\overline{NGP}$ a transversal of $\triangle AKC$, and $\overline{GMP}$ a transversal of $\triangle AKB$. Then use Menelaus' Theorem.

8-12 Draw $\overline{AD} \perp \overline{BC}$, and $\overline{PE} \perp \overline{BC}$, where D and E lie on $\overline{BC}$. For both parts **(a)** and **(b)**, neither Ceva's Theorem nor Menelaus' Theorem is used. Set up proportions involving line segments and areas of triangles.

8-13 Extend $\overline{FE}$ to meet $\overleftrightarrow{CB}$ at P. Consider $\overline{AM}$ as a transversal of $\triangle PFC$ and $\triangle PEB$; then use Menelaus' Theorem.

8-14 Use one of the secondary results established in the solution of Problem 8-2, Method I. (See III, IV, and V.) Neither Ceva's Theorem nor Menelaus' Theorem is used.

8-15 Use Menelaus' Theorem and similarity.

8-16 Use Menelaus' Theorem, taking $\overline{KLP}$ and $\overline{MNP}$ as transversals of $\triangle ABC$ and $\triangle ADC$, respectively where P is the intersection of $\overline{AC}$ and $\overline{LN}$.

8-17 Use Theorems #36, #38, #48, and #53, followed by Menelaus' Theorem.

8-18 Taking $\overline{RSP}$ and $\overline{R'S'P'}$ as transversals of $\triangle ABC$, use Menelaus' Theorem. Also use Theorems #52 and #53.

8-19 Consider $\overline{RNH}$, $\overline{PLJ}$, and $\overline{MQI}$ transversals of $\triangle ABC$; use Menelaus' Theorem. Then use Ceva's Theorem.

8-20 Use Ceva's Theorem and Theorem #54.

8-21 Draw lines of centers and radii. Use Theorem #49 and Menelaus' Theorem.

8-22 Use Theorems #48, #46, and Menelaus' Theorem.

8-23 Use Menelaus' Theorem exclusively.

8-24 **(a)** Use Menelaus' Theorem and Theorem #34.
(b) Use Menelaus' Theorem, or use Desargues' Theorem (Problem 8-23).

8-25 Extend $\overline{DR}$ and $\overline{DQ}$ through R and Q to meet a line through C parallel to $\overline{AB}$, at points G and H, respectively. Use Theorem #48, Ceva's Theorem and Theorem #10. Also prove $\triangle GCD \cong \triangle HCD$.

8-26 METHOD I: Use the result of Problem 8-25, Theorem #47, and Menelaus' Theorem.

METHOD II: Use Desargues' Theorem (Problem 8-23).

8-27 Use Theorem #36a and the trigonometric form of Ceva's Theorem.

8-28 Use Theorems #18, #5, #46, and #47. Then use Menelaus' Theorem.

8-29 Consider transversals $\overline{BC}$, $\overline{AN}$, and $\overline{DE}$ of $\triangle XYZ$. Use Menelaus' Theorem.

8-30 Consider transversals $\overline{C''AB'}$, $\overline{A'B''C}$, $\overline{BA''C'}$ of $\triangle XYZ$. Use Menelaus' Theorem.

9-1 METHOD I: Prove quadrilaterals cyclic; then show that two angles are congruent, both sharing as a side the required line.

METHOD II: Prove quadrilaterals cyclic to show that two congruent angles are vertical angles (one of the lines forming these vertical angles is the required line).

METHOD III: Draw a line passing through a vertex of the triangle and parallel to a segment of the required line. Prove that the other segment of the required line is also parallel to the new line. Use Euclid's parallel postulate to obtain the desired conclusion.

9-2 Discover cyclic quadrilaterals to find congruent angles. Use Theorems #37, #36, and #8.

9-3 Prove X, Y, and Z collinear (the Simson Line); then prove $\triangle PAB \sim \triangle PXZ$.

9-4 Draw the Simson Lines of $\triangle ABC$ and $\triangle SCR$; then use the converse of Simson's Theorem.

9-5 Show that M is the point of intersection of the diagonals of a rectangle, hence the midpoint of $\overline{AP}$. Then use Theorem #31.

9-6 Draw various auxiliary lines, and use Simson's Theorem.

9-7 Use Simson's Theorem, and others to prove $\triangle LPM \sim \triangle KPN$.

9-8 Use the converse of Simson's Theorem, after showing that various Simson Lines coincide and share the same Simson point.

9-9 METHOD I: Extend an altitude to the circumcircle of the triangle. Join that point with the Simson point. Use Theorems #36, #8, #14, #5, #36a, #37, #7, and #18. Also use Simson's Theorem.

METHOD II: An isosceles (inscribed) trapezoid is drawn using one of the altitudes as part of one base. Other auxiliary lines are drawn. Use Theorems #9, #33, #21, #25, and Simson's Theorem.

9-10 Prove that each of the Simson Lines is parallel to a side of an inscribed angle. Various auxiliary lines are needed.

9-11 Use a secondary result obtained in the proof for Problem 9-10, line (III). Then show that the new angle is measured by arcs independent of point P.

9-12 Use the result of Solution 9-10, line (III).

10-1 Draw altitude $\overline{CE}$; then use the Pythagorean Theorem in various right triangles.

10-2 Apply Stewart's Theorem.

10-3 METHOD I: Use Stewart's Theorem.

METHOD II: Use Heron's Formula (Problem 6-1).

10-4 Apply Stewart's Theorem, using each of the interior lines separately. Also use the Pythagorean Theorem.

10-5 Use a secondary result obtained in the proof of Stewart's Theorem [See the solution to Problem 10-1, equations (II) and (IV).]

10-6 Apply Stewart's Theorem and Theorem #47.

10-7 Use the result obtained from Problem 10-6.

10-8 METHOD I: Use Theorems #47, and #55, and the result obtained from Problem 10-6.

METHOD II: Use Theorems #47 and #55.

10-9 Use Theorems #47 and #55, and the result obtained from Problem 10-6.

APPENDICES

APPENDIX I Selected Definitions, Postulates, and Theorems

1. If two angles are vertical angles then the two angles are congruent.
2. Two triangles are congruent if two sides and the included angle of the first triangle are congruent to the corresponding parts of the second triangle. (S.A.S.)
3. Two triangles are congruent if two angles and the included side of the first triangle are congruent to the corresponding parts of the second triangle. (A.S.A.)
4. Two triangles are congruent if the sides of the first triangle are congruent to the corresponding sides of the second triangle. (S.S.S.)
5. If a triangle has two congruent sides, then the triangle has two congruent angles opposite those sides. Also converse.
6. An equilateral triangle is equiangular. Also converse.
7. If a pair of corresponding angles formed by a transversal of two lines are congruent, then the two lines are parallel. Also converse.
8. If a pair of alternate interior angles formed by a transversal of two lines are congruent, then the lines are parallel. Also converse.
9. Two lines are parallel if they are perpendicular to the same line.
10. If a line is perpendicular to one of two parallel lines, then it is also perpendicular to the other.
11. If a pair of consecutive interior angles formed by a transversal of two lines are supplementary, then the lines are parallel. Also converse.
12. The measure of an exterior angle of a triangle equals the sum of the measures of the two non-adjacent interior angles.
13. The sum of the measures of the three angles of a triangle is 180, a constant.
14. The acute angles of a right triangle are complementary.

15 The sum of the measures of the four interior angles of a convex quadrilateral is 360, a constant.

16 Two triangles are congruent if two angles and a non-included side of the first triangle are congruent to the corresponding parts of the second triangle.

17 Two right triangles are congruent if the hypotenuse and a leg of one triangle are congruent to the corresponding parts of the other triangle.

18 Any point on the perpendicular bisector of a line segment is equidistant from the endpoints of the line segment. Two points equidistant from the endpoints of a line segment, determine the perpendicular bisector of the line segment.

19 Any point on the bisector of an angle is equidistant from the sides of the angle.

20 Parallel lines are everywhere equidistant.

21a The opposite sides of a parallelogram are parallel. Also converse.

21b The opposite sides of a parallelogram are congruent. Also converse.

21c The opposite angles of a parallelogram are congruent. Also converse.

21d Pairs of consecutive angles of a parallelogram are supplementary. Also converse.

21e A diagonal of a parallelogram divides the parallelogram into two congruent triangles.

21f The diagonals of a parallelogram bisect each other. Also converse.

21g A rectangle is a special parallelogram; therefore 21a through 21f hold true for the rectangle.

21h A rectangle is a parallelogram with congruent diagonals. Also converse.

21i A rectangle is a parallelogram with four congruent angles, right angles. Also converse.

21j A rhombus is a special parallelogram; therefore 21a through 21f hold true for the rhombus.

21k A rhombus is a parallelogram with perpendicular diagonals. Also converse.

21l A rhombus is a quadrilateral with four congruent sides. Also converse.

21m The diagonals of a rhombus bisect the angles of the rhombus.

21n A square has all the properties of both a rectangle and a rhombus; hence 21a through 21m hold true for a square.

22 A quadrilateral is a parallelogram if a pair of opposite sides are both congruent and parallel.

23 The base angles of an isosceles trapezoid are congruent. Also converse.

24 If a line segment is divided into congruent (or proportional) segments by three or more parallel lines, then any other transversal will similarly contain congruent (or proportional) segments determined by these parallel lines.

25 If a line contains the midpoint of one side of a triangle and is parallel to a second side of the triangle, then it will bisect the third side of the triangle.

26 The line segment whose endpoints are the midpoints of two sides of a triangle is parallel to the third side of the triangle and has a measure equal to one-half of the measure of the third side.

27 The measure of the median on the hypotenuse of a right triangle is one-half the measure of the hypotenuse.

28 The median of a trapezoid, the segment joining the midpoints of the non-parallel sides, is parallel to each of the parallel sides, and has a measure equal to one-half of the sum of their measures.

29 The three medians of a triangle meet in a point, the centroid, which is situated on each median so that the measure of the segment from the vertex to the centroid is two-thirds the measure of the median.

30 A line perpendicular to a chord of a circle and containing the center of the circle, bisects the chord and its major and minor arcs.

31 The perpendicular bisector of a chord of a circle contains the center of the circle.

32a If a line is tangent to a circle, it is perpendicular to a radius at the point of tangency.

32b A line perpendicular to a radius at a point on the circle is tangent to the circle at that point.

32c A line perpendicular to a tangent line at the point of tangency with a circle, contains the center of the circle.

32d The radius of a circle is only perpendicular to a tangent line at the point of tangency.

33 If a tangent line (or chord) is parallel to a secant (or chord) the arcs intercepted between these two lines are congruent.

34 Two tangent segments to a circle from an external point are congruent.

35 The measure of a central angle is equal to the measure of its intercepted arc.

36 The measure of an inscribed angle equals one-half the measure of its intercepted arc.

36a A quadrilateral is cyclic (i.e. may be inscribed in a circle) if one side subtends congruent angles at the two opposite vertices.

37 The opposite angles of a cyclic (inscribed) quadrilateral are supplementary. Also converse.

38 The measure of an angle whose vertex is on the circle and whose sides are formed by a chord and a tangent line, is equal to one-half the measure of the intercepted arc.

39 The measure of an angle formed by two chords intersecting inside the circle, is equal to half the sum of the measures of its intercepted arc and of the arc of its vertical angle.

40 The measure of an angle formed by two secants, or a secant and a tangent line, or two tangent lines intersecting outside the circle, equals one-half the difference of the measures of the intercepted arcs.

41 The sum of the measures of two sides of a non-degenerate triangle is greater than the measure of the third side of the triangle.

42 If the measures of two sides of a triangle are not equal, then the measures of the angles opposite these sides are also unequal, the angle with the greater measure being opposite the side with the greater measure. Also converse.

43 The measure of an exterior angle of a triangle is greater than the measure of either non-adjacent interior angle.

44 The circumcenter (the center of the circumscribed circle) of a triangle is determined by the common intersection of the perpendicular bisectors of the sides of the triangle.

45 The incenter (the center of the inscribed circle) of a triangle is determined by the common intersection of the interior angle bisectors of the triangle.

46 If a line is parallel to one side of a triangle it divides the other two sides of the triangle proportionally. Also converse.

47 The bisector of an angle of a triangle divides the opposite side into segments whose measures are proportional to the measures of the other two sides of the triangle. Also converse.

48 If two angles of one triangle are congruent to two corresponding angles of a second triangle, the triangles are similar. (A.A.)

49 If a line is parallel to one side of a triangle intersecting the other two sides, it determines (with segments of these two sides) a triangle similar to the original triangle.

50 Two triangles are similar if an angle of one triangle is congruent to an angle of the other triangle, and if the measures of the sides that include the angle are proportional.

51a The measure of the altitude on the hypotenuse of a right triangle is the mean proportional between the measures of the segments of the hypotenuse.

51b The measure of either leg of a right triangle is the mean proportional between the measure of the hypotenuse and the segment, of the hypotenuse, which shares one endpoint with the leg considered, and whose other endpoint is the foot of the altitude on the hypotenuse.

52 If two chords of a circle intersect, the product of the measures of the segments of one chord equals the product of the segments of the other chord.

53 If a tangent segment and a secant intersect outside the circle, the measure of the tangent segment is the mean proportional between the measure of the secant and the measure of its external segment.

54 If two secants intersect outside the circle, the product of the measures of one secant and its external segment equals the product of the measures of the other secant and its external segment.

55 (The Pythagorean Theorem) In a right triangle the sum of the squares of the measures of the legs equals the square of the measure of the hypotenuse. Also converse.

55a In an isosceles right triangle (45–45–90 triangle), the measure of the hypotenuse is equal to $\sqrt{2}$ times the measure of either leg.

55b In an isosceles right triangle (45–45–90 triangle), the measure of either leg equals one-half the measure of the hypotenuse times $\sqrt{2}$.

55c In a 30–60–90 triangle the measure of the side opposite the 30 angle is one-half the measure of the hypotenuse.

55d In a 30–60–90 triangle, the measure of the side opposite the 60 angle equals one-half the measure of the hypotenuse times $\sqrt{3}$.

55e In a triangle with sides of measures 13, 14, and 15, the altitude to the side of measure 14 has measure 12.

56 The median of a triangle divides the triangle into two triangles of equal area. An extension of this theorem follows. A line segment joining a vertex of a triangle with a point on the opposite side, divides the triangle into two triangles, the ratio of whose areas equals the ratio of the measures of the segments of this "opposite" side.

APPENDIX II Selected Formulas

1 The sum of the measures of the interior angles of an n-sided convex polygon $= (n - 2)180$.

2 The sum of the measures of the exterior angles of any convex polygon is constant, 360.

3 The area of a rectangle:
$K = bh$.

4a The area of a square:
$K = s^2$.

4b The area of a square:
$K = \frac{1}{2} d^2$.

5a The area of any triangle:
$K = \frac{1}{2} bh$, where b is the base and h is the altitude.

5b The area of any triangle:
$K = \frac{1}{2} ab \sin C$.

5c The area of any triangle:
$K = \sqrt{s(s - a)(s - b)(s - c)}$, where $s = \frac{1}{2}(a + b + c)$.

5d The area of a right triangle:
$K = \frac{1}{2} l_1 l_2$, where l is a leg.

5e The area of an equilateral triangle:
$K = \dfrac{s^2\sqrt{3}}{4}$, where s is any side.

5f The area of an equilateral triangle:
$K = \dfrac{h^2\sqrt{3}}{3}$, where h is the altitude.

6a The area of a parallelogram:
$K = bh$.

6b The area of a parallelogram:
$K = ab \sin C$.

7 The area of a rhombus:
$K = \frac{1}{2} d_1 d_2$.

8 The area of a trapezoid:
$K = \frac{1}{2} h(b_1 + b_2)$.

9 The area of a regular polygon:
$K = \frac{1}{2} ap$, where a is the apothem and p is the perimeter.

10 The area of a circle:
$K = \pi r^2 = \frac{\pi d^2}{4}$, where d is the diameter.

11 The area of a sector of a circle:
$K = \frac{n}{360} \pi r^2$, where n is the measure of the central angle.

12 The circumference of a circle:
$C = 2\pi r$.

13 The length of an arc of a circle:
$L = \frac{n}{360} 2\pi r$, where n is the measure of the central angle of the arc.

A CATALOG OF SELECTED

DOVER BOOKS

IN SCIENCE AND MATHEMATICS

Astronomy

CHARIOTS FOR APOLLO: The NASA History of Manned Lunar Spacecraft to 1969, Courtney G. Brooks, James M. Grimwood, and Loyd S. Swenson, Jr. This illustrated history by a trio of experts is the definitive reference on the Apollo spacecraft and lunar modules. It traces the vehicles' design, development, and operation in space. More than 100 photographs and illustrations. 576pp. 6 3/4 x 9 1/4. 0-486-46756-2

EXPLORING THE MOON THROUGH BINOCULARS AND SMALL TELESCOPES, Ernest H. Cherrington, Jr. Informative, profusely illustrated guide to locating and identifying craters, rills, seas, mountains, other lunar features. Newly revised and updated with special section of new photos. Over 100 photos and diagrams. 240pp. 8 1/4 x 11. 0-486-24491-1

WHERE NO MAN HAS GONE BEFORE: A History of NASA's Apollo Lunar Expeditions, William David Compton. Introduction by Paul Dickson. This official NASA history traces behind-the-scenes conflicts and cooperation between scientists and engineers. The first half concerns preparations for the Moon landings, and the second half documents the flights that followed Apollo 11. 1989 edition. 432pp. 7 x 10. 0-486-47888-2

APOLLO EXPEDITIONS TO THE MOON: The NASA History, Edited by Edgar M. Cortright. Official NASA publication marks the 40th anniversary of the first lunar landing and features essays by project participants recalling engineering and administrative challenges. Accessible, jargon-free accounts, highlighted by numerous illustrations. 336pp. 8 3/8 x 10 7/8. 0-486-47175-6

ON MARS: Exploration of the Red Planet, 1958-1978--The NASA History, Edward Clinton Ezell and Linda Neuman Ezell. NASA's official history chronicles the start of our explorations of our planetary neighbor. It recounts cooperation among government, industry, and academia, and it features dozens of photos from Viking cameras. 560pp. 6 3/4 x 9 1/4. 0-486-46757-0

ARISTARCHUS OF SAMOS: The Ancient Copernicus, Sir Thomas Heath. Heath's history of astronomy ranges from Homer and Hesiod to Aristarchus and includes quotes from numerous thinkers, compilers, and scholasticists from Thales and Anaximander through Pythagoras, Plato, Aristotle, and Heraclides. 34 figures. 448pp. 5 3/8 x 8 1/2. 0-486-43886-4

AN INTRODUCTION TO CELESTIAL MECHANICS, Forest Ray Moulton. Classic text still unsurpassed in presentation of fundamental principles. Covers rectilinear motion, central forces, problems of two and three bodies, much more. Includes over 200 problems, some with answers. 437pp. 5 3/8 x 8 1/2. 0-486-64687-4

BEYOND THE ATMOSPHERE: Early Years of Space Science, Homer E. Newell. This exciting survey is the work of a top NASA administrator who chronicles technological advances, the relationship of space science to general science, and the space program's social, political, and economic contexts. 528pp. 6 3/4 x 9 1/4. 0-486-47464-X

STAR LORE: Myths, Legends, and Facts, William Tyler Olcott. Captivating retellings of the origins and histories of ancient star groups include Pegasus, Ursa Major, Pleiades, signs of the zodiac, and other constellations. "Classic." – *Sky & Telescope.* 58 illustrations. 544pp. 5 3/8 x 8 1/2. 0-486-43581-4

A COMPLETE MANUAL OF AMATEUR ASTRONOMY: Tools and Techniques for Astronomical Observations, P. Clay Sherrod with Thomas L. Koed. Concise, highly readable book discusses the selection, set-up, and maintenance of a telescope; amateur studies of the sun; lunar topography and occultations; and more. 124 figures. 26 halftones. 37 tables. 335pp. 6 1/2 x 9 1/4. 0-486-42820-6

Chemistry

MOLECULAR COLLISION THEORY, M. S. Child. This high-level monograph offers an analytical treatment of classical scattering by a central force, quantum scattering by a central force, elastic scattering phase shifts, and semi-classical elastic scattering. 1974 edition. 310pp. 5 3/8 x 8 1/2. 0-486-69437-2

HANDBOOK OF COMPUTATIONAL QUANTUM CHEMISTRY, David B. Cook. This comprehensive text provides upper-level undergraduates and graduate students with an accessible introduction to the implementation of quantum ideas in molecular modeling, exploring practical applications alongside theoretical explanations. 1998 edition. 832pp. 5 3/8 x 8 1/2. 0-486-44307-8

RADIOACTIVE SUBSTANCES, Marie Curie. The celebrated scientist's thesis, which directly preceded her 1903 Nobel Prize, discusses establishing atomic character of radioactivity; extraction from pitchblende of polonium and radium; isolation of pure radium chloride; more. 96pp. 5 3/8 x 8 1/2. 0-486-42550-9

CHEMICAL MAGIC, Leonard A. Ford. Classic guide provides intriguing entertainment while elucidating sound scientific principles, with more than 100 unusual stunts: cold fire, dust explosions, a nylon rope trick, a disappearing beaker, much more. 128pp. 5 3/8 x 8 1/2. 0-486-67628-5

ALCHEMY, E. J. Holmyard. Classic study by noted authority covers 2,000 years of alchemical history: religious, mystical overtones; apparatus; signs, symbols, and secret terms; advent of scientific method, much more. Illustrated. 320pp. 5 3/8 x 8 1/2. 0-486-26298-7

CHEMICAL KINETICS AND REACTION DYNAMICS, Paul L. Houston. This text teaches the principles underlying modern chemical kinetics in a clear, direct fashion, using several examples to enhance basic understanding. Solutions to selected problems. 2001 edition. 352pp. 8 3/8 x 11. 0-486-45334-0

PROBLEMS AND SOLUTIONS IN QUANTUM CHEMISTRY AND PHYSICS, Charles S. Johnson and Lee G. Pedersen. Unusually varied problems, with detailed solutions, cover of quantum mechanics, wave mechanics, angular momentum, molecular spectroscopy, scattering theory, more. 280 problems, plus 139 supplementary exercises. 430pp. 6 1/2 x 9 1/4. 0-486-65236-X

ELEMENTS OF CHEMISTRY, Antoine Lavoisier. Monumental classic by the founder of modern chemistry features first explicit statement of law of conservation of matter in chemical change, and more. Facsimile reprint of original (1790) Kerr translation. 539pp. 5 3/8 x 8 1/2. 0-486-64624-6

MAGNETISM AND TRANSITION METAL COMPLEXES, F. E. Mabbs and D. J. Machin. A detailed view of the calculation methods involved in the magnetic properties of transition metal complexes, this volume offers sufficient background for original work in the field. 1973 edition. 240pp. 5 3/8 x 8 1/2. 0-486-46284-6

GENERAL CHEMISTRY, Linus Pauling. Revised third edition of classic first-year text by Nobel laureate. Atomic and molecular structure, quantum mechanics, statistical mechanics, thermodynamics correlated with descriptive chemistry. Problems. 992pp. 5 3/8 x 8 1/2. 0-486-65622-5

ELECTROLYTE SOLUTIONS: Second Revised Edition, R. A. Robinson and R. H. Stokes. Classic text deals primarily with measurement, interpretation of conductance, chemical potential, and diffusion in electrolyte solutions. Detailed theoretical interpretations, plus extensive tables of thermodynamic and transport properties. 1970 edition. 590pp. 5 3/8 x 8 1/2. 0-486-42225-9

Engineering

FUNDAMENTALS OF ASTRODYNAMICS, Roger R. Bate, Donald D. Mueller, and Jerry E. White. Teaching text developed by U.S. Air Force Academy develops the basic two-body and n-body equations of motion; orbit determination; classical orbital elements, coordinate transformations; differential correction; more. 1971 edition. 455pp. 5 3/8 x 8 1/2. 0-486-60061-0

INTRODUCTION TO CONTINUUM MECHANICS FOR ENGINEERS: Revised Edition, Ray M. Bowen. This self-contained text introduces classical continuum models within a modern framework. Its numerous exercises illustrate the governing principles, linearizations, and other approximations that constitute classical continuum models. 2007 edition. 320pp. 6 1/8 x 9 1/4. 0-486-47460-7

ENGINEERING MECHANICS FOR STRUCTURES, Louis L. Bucciarelli. This text explores the mechanics of solids and statics as well as the strength of materials and elasticity theory. Its many design exercises encourage creative initiative and systems thinking. 2009 edition. 320pp. 6 1/8 x 9 1/4. 0-486-46855-0

FEEDBACK CONTROL THEORY, John C. Doyle, Bruce A. Francis and Allen R. Tannenbaum. This excellent introduction to feedback control system design offers a theoretical approach that captures the essential issues and can be applied to a wide range of practical problems. 1992 edition. 224pp. 6 1/2 x 9 1/4. 0-486-46933-6

THE FORCES OF MATTER, Michael Faraday. These lectures by a famous inventor offer an easy-to-understand introduction to the interactions of the universe's physical forces. Six essays explore gravitation, cohesion, chemical affinity, heat, magnetism, and electricity. 1993 edition. 96pp. 5 3/8 x 8 1/2. 0-486-47482-8

DYNAMICS, Lawrence E. Goodman and William H. Warner. Beginning engineering text introduces calculus of vectors, particle motion, dynamics of particle systems and plane rigid bodies, technical applications in plane motions, and more. Exercises and answers in every chapter. 619pp. 5 3/8 x 8 1/2. 0-486-42006-X

ADAPTIVE FILTERING PREDICTION AND CONTROL, Graham C. Goodwin and Kwai Sang Sin. This unified survey focuses on linear discrete-time systems and explores natural extensions to nonlinear systems. It emphasizes discrete-time systems, summarizing theoretical and practical aspects of a large class of adaptive algorithms. 1984 edition. 560pp. 6 1/2 x 9 1/4. 0-486-46932-8

INDUCTANCE CALCULATIONS, Frederick W. Grover. This authoritative reference enables the design of virtually every type of inductor. It features a single simple formula for each type of inductor, together with tables containing essential numerical factors. 1946 edition. 304pp. 5 3/8 x 8 1/2. 0-486-47440-2

THERMODYNAMICS: Foundations and Applications, Elias P. Gyftopoulos and Gian Paolo Beretta. Designed by two MIT professors, this authoritative text discusses basic concepts and applications in detail, emphasizing generality, definitions, and logical consistency. More than 300 solved problems cover realistic energy systems and processes. 800pp. 6 1/8 x 9 1/4. 0-486-43932-1

THE FINITE ELEMENT METHOD: Linear Static and Dynamic Finite Element Analysis, Thomas J. R. Hughes. Text for students without in-depth mathematical training, this text includes a comprehensive presentation and analysis of algorithms of time-dependent phenomena plus beam, plate, and shell theories. Solution guide available upon request. 672pp. 6 1/2 x 9 1/4. 0-486-41181-8

Mathematics-Bestsellers

HANDBOOK OF MATHEMATICAL FUNCTIONS: with Formulas, Graphs, and Mathematical Tables, Edited by Milton Abramowitz and Irene A. Stegun. A classic resource for working with special functions, standard trig, and exponential logarithmic definitions and extensions, it features 29 sets of tables, some to as high as 20 places. 1046pp. 8 x 10 1/2. 0-486-61272-4

ABSTRACT AND CONCRETE CATEGORIES: The Joy of Cats, Jiri Adamek, Horst Herrlich, and George E. Strecker. This up-to-date introductory treatment employs category theory to explore the theory of structures. Its unique approach stresses concrete categories and presents a systematic view of factorization structures. Numerous examples. 1990 edition, updated 2004. 528pp. 6 1/8 x 9 1/4. 0-486-46934-4

MATHEMATICS: Its Content, Methods and Meaning, A. D. Aleksandrov, A. N. Kolmogorov, and M. A. Lavrent'ev. Major survey offers comprehensive, coherent discussions of analytic geometry, algebra, differential equations, calculus of variations, functions of a complex variable, prime numbers, linear and non-Euclidean geometry, topology, functional analysis, more. 1963 edition. 1120pp. 5 3/8 x 8 1/2. 0-486-40916-3

INTRODUCTION TO VECTORS AND TENSORS: Second Edition--Two Volumes Bound as One, Ray M. Bowen and C.-C. Wang. Convenient single-volume compilation of two texts offers both introduction and in-depth survey. Geared toward engineering and science students rather than mathematicians, it focuses on physics and engineering applications. 1976 edition. 560pp. 6 1/2 x 9 1/4. 0-486-46914-X

AN INTRODUCTION TO ORTHOGONAL POLYNOMIALS, Theodore S. Chihara. Concise introduction covers general elementary theory, including the representation theorem and distribution functions, continued fractions and chain sequences, the recurrence formula, special functions, and some specific systems. 1978 edition. 272pp. 5 3/8 x 8 1/2. 0-486-47929-3

ADVANCED MATHEMATICS FOR ENGINEERS AND SCIENTISTS, Paul DuChateau. This primary text and supplemental reference focuses on linear algebra, calculus, and ordinary differential equations. Additional topics include partial differential equations and approximation methods. Includes solved problems. 1992 edition. 400pp. 7 1/2 x 9 1/4. 0-486-47930-7

PARTIAL DIFFERENTIAL EQUATIONS FOR SCIENTISTS AND ENGINEERS, Stanley J. Farlow. Practical text shows how to formulate and solve partial differential equations. Coverage of diffusion-type problems, hyperbolic-type problems, elliptic-type problems, numerical and approximate methods. Solution guide available upon request. 1982 edition. 414pp. 6 1/8 x 9 1/4. 0-486-67620-X

VARIATIONAL PRINCIPLES AND FREE-BOUNDARY PROBLEMS, Avner Friedman. Advanced graduate-level text examines variational methods in partial differential equations and illustrates their applications to free-boundary problems. Features detailed statements of standard theory of elliptic and parabolic operators. 1982 edition. 720pp. 6 1/8 x 9 1/4. 0-486-47853-X

LINEAR ANALYSIS AND REPRESENTATION THEORY, Steven A. Gaal. Unified treatment covers topics from the theory of operators and operator algebras on Hilbert spaces; integration and representation theory for topological groups; and the theory of Lie algebras, Lie groups, and transform groups. 1973 edition. 704pp. 6 1/8 x 9 1/4. 0-486-47851-3

Mathematics–Logic and Problem Solving

PERPLEXING PUZZLES AND TANTALIZING TEASERS, Martin Gardner. Ninety-three riddles, mazes, illusions, tricky questions, word and picture puzzles, and other challenges offer hours of entertainment for youngsters. Filled with rib-tickling drawings. Solutions. 224pp. 5 3/8 x 8 1/2. 0-486-25637-5

MY BEST MATHEMATICAL AND LOGIC PUZZLES, Martin Gardner. The noted expert selects 70 of his favorite "short" puzzles. Includes The Returning Explorer, The Mutilated Chessboard, Scrambled Box Tops, and dozens more. Complete solutions included. 96pp. 5 3/8 x 8 1/2. 0-486-28152-3

THE LADY OR THE TIGER?: and Other Logic Puzzles, Raymond M. Smullyan. Created by a renowned puzzle master, these whimsically themed challenges involve paradoxes about probability, time, and change; metapuzzles; and self-referentiality. Nineteen chapters advance in difficulty from relatively simple to highly complex. 1982 edition. 240pp. 5 3/8 x 8 1/2. 0-486-47027-X

SATAN, CANTOR AND INFINITY: Mind-Boggling Puzzles, Raymond M. Smullyan. A renowned mathematician tells stories of knights and knaves in an entertaining look at the logical precepts behind infinity, probability, time, and change. Requires a strong background in mathematics. Complete solutions. 288pp. 5 3/8 x 8 1/2. 0-486-47036-9

THE RED BOOK OF MATHEMATICAL PROBLEMS, Kenneth S. Williams and Kenneth Hardy. Handy compilation of 100 practice problems, hints and solutions indispensable for students preparing for the William Lowell Putnam and other mathematical competitions. Preface to the First Edition. Sources. 1988 edition. 192pp. 5 3/8 x 8 1/2. 0-486-69415-1

KING ARTHUR IN SEARCH OF HIS DOG AND OTHER CURIOUS PUZZLES, Raymond M. Smullyan. This fanciful, original collection for readers of all ages features arithmetic puzzles, logic problems related to crime detection, and logic and arithmetic puzzles involving King Arthur and his Dogs of the Round Table. 160pp. 5 3/8 x 8 1/2. 0-486-47435-6

UNDECIDABLE THEORIES: Studies in Logic and the Foundation of Mathematics, Alfred Tarski in collaboration with Andrzej Mostowski and Raphael M. Robinson. This well-known book by the famed logician consists of three treatises: "A General Method in Proofs of Undecidability," "Undecidability and Essential Undecidability in Mathematics," and "Undecidability of the Elementary Theory of Groups." 1953 edition. 112pp. 5 3/8 x 8 1/2. 0-486-47703-7

LOGIC FOR MATHEMATICIANS, J. Barkley Rosser. Examination of essential topics and theorems assumes no background in logic. "Undoubtedly a major addition to the literature of mathematical logic." – *Bulletin of the American Mathematical Society.* 1978 edition. 592pp. 6 1/8 x 9 1/4. 0-486-46898-4

INTRODUCTION TO PROOF IN ABSTRACT MATHEMATICS, Andrew Wohlgemuth. This undergraduate text teaches students what constitutes an acceptable proof, and it develops their ability to do proofs of routine problems as well as those requiring creative insights. 1990 edition. 384pp. 6 1/2 x 9 1/4. 0-486-47854-8

FIRST COURSE IN MATHEMATICAL LOGIC, Patrick Suppes and Shirley Hill. Rigorous introduction is simple enough in presentation and context for wide range of students. Symbolizing sentences; logical inference; truth and validity; truth tables; terms, predicates, universal quantifiers; universal specification and laws of identity; more. 288pp. 5 3/8 x 8 1/2. 0-486-42259-3

Mathematics–Algebra and Calculus

VECTOR CALCULUS, Peter Baxandall and Hans Liebeck. This introductory text offers a rigorous, comprehensive treatment. Classical theorems of vector calculus are amply illustrated with figures, worked examples, physical applications, and exercises with hints and answers. 1986 edition. 560pp. 5 3/8 x 8 1/2. 0-486-46620-5

ADVANCED CALCULUS: An Introduction to Classical Analysis, Louis Brand. A course in analysis that focuses on the functions of a real variable, this text introduces the basic concepts in their simplest setting and illustrates its teachings with numerous examples, theorems, and proofs. 1955 edition. 592pp. 5 3/8 x 8 1/2. 0-486-44548-8

ADVANCED CALCULUS, Avner Friedman. Intended for students who have already completed a one-year course in elementary calculus, this two-part treatment advances from functions of one variable to those of several variables. Solutions. 1971 edition. 432pp. 5 3/8 x 8 1/2. 0-486-45795-8

METHODS OF MATHEMATICS APPLIED TO CALCULUS, PROBABILITY, AND STATISTICS, Richard W. Hamming. This 4-part treatment begins with algebra and analytic geometry and proceeds to an exploration of the calculus of algebraic functions and transcendental functions and applications. 1985 edition. Includes 310 figures and 18 tables. 880pp. 6 1/2 x 9 1/4. 0-486-43945-3

BASIC ALGEBRA I: Second Edition, Nathan Jacobson. A classic text and standard reference for a generation, this volume covers all undergraduate algebra topics, including groups, rings, modules, Galois theory, polynomials, linear algebra, and associative algebra. 1985 edition. 528pp. 6 1/8 x 9 1/4. 0-486-47189-6

BASIC ALGEBRA II: Second Edition, Nathan Jacobson. This classic text and standard reference comprises all subjects of a first-year graduate-level course, including in-depth coverage of groups and polynomials and extensive use of categories and functors. 1989 edition. 704pp. 6 1/8 x 9 1/4. 0-486-47187-X

CALCULUS: An Intuitive and Physical Approach (Second Edition), Morris Kline. Application-oriented introduction relates the subject as closely as possible to science with explorations of the derivative; differentiation and integration of the powers of x; theorems on differentiation, antidifferentiation; the chain rule; trigonometric functions; more. Examples. 1967 edition. 960pp. 6 1/2 x 9 1/4. 0-486-40453-6

ABSTRACT ALGEBRA AND SOLUTION BY RADICALS, John E. Maxfield and Margaret W. Maxfield. Accessible advanced undergraduate-level text starts with groups, rings, fields, and polynomials and advances to Galois theory, radicals and roots of unity, and solution by radicals. Numerous examples, illustrations, exercises, appendixes. 1971 edition. 224pp. 6 1/8 x 9 1/4. 0-486-47723-1

AN INTRODUCTION TO THE THEORY OF LINEAR SPACES, Georgi E. Shilov. Translated by Richard A. Silverman. Introductory treatment offers a clear exposition of algebra, geometry, and analysis as parts of an integrated whole rather than separate subjects. Numerous examples illustrate many different fields, and problems include hints or answers. 1961 edition. 320pp. 5 3/8 x 8 1/2. 0-486-63070-6

LINEAR ALGEBRA, Georgi E. Shilov. Covers determinants, linear spaces, systems of linear equations, linear functions of a vector argument, coordinate transformations, the canonical form of the matrix of a linear operator, bilinear and quadratic forms, and more. 387pp. 5 3/8 x 8 1/2. 0-486-63518-X

Mathematics-Probability and Statistics

Mathematics–Geometry and Topology

PROBLEMS AND SOLUTIONS IN EUCLIDEAN GEOMETRY, M. N. Aref and William Wernick. Based on classical principles, this book is intended for a second course in Euclidean geometry and can be used as a refresher. More than 200 problems include hints and solutions. 1968 edition. 272pp. 5 3/8 x 8 1/2. 0-486-47720-7

TOPOLOGY OF 3-MANIFOLDS AND RELATED TOPICS, Edited by M. K. Fort, Jr. With a New Introduction by Daniel Silver. Summaries and full reports from a 1961 conference discuss decompositions and subsets of 3-space; n-manifolds; knot theory; the Poincaré conjecture; and periodic maps and isotopies. Familiarity with algebraic topology required. 1962 edition. 272pp. 6 1/8 x 9 1/4. 0-486-47753-3

POINT SET TOPOLOGY, Steven A. Gaal. Suitable for a complete course in topology, this text also functions as a self-contained treatment for independent study. Additional enrichment materials make it equally valuable as a reference. 1964 edition. 336pp. 5 3/8 x 8 1/2. 0-486-47222-1

INVITATION TO GEOMETRY, Z. A. Melzak. Intended for students of many different backgrounds with only a modest knowledge of mathematics, this text features self-contained chapters that can be adapted to several types of geometry courses. 1983 edition. 240pp. 5 3/8 x 8 1/2. 0-486-46626-4

TOPOLOGY AND GEOMETRY FOR PHYSICISTS, Charles Nash and Siddhartha Sen. Written by physicists for physics students, this text assumes no detailed background in topology or geometry. Topics include differential forms, homotopy, homology, cohomology, fiber bundles, connection and covariant derivatives, and Morse theory. 1983 edition. 320pp. 5 3/8 x 8 1/2. 0-486-47852-1

BEYOND GEOMETRY: Classic Papers from Riemann to Einstein, Edited with an Introduction and Notes by Peter Pesic. This is the only English-language collection of these 8 accessible essays. They trace seminal ideas about the foundations of geometry that led to Einstein's general theory of relativity. 224pp. 6 1/8 x 9 1/4. 0-486-45350-2

GEOMETRY FROM EUCLID TO KNOTS, Saul Stahl. This text provides a historical perspective on plane geometry and covers non-neutral Euclidean geometry, circles and regular polygons, projective geometry, symmetries, inversions, informal topology, and more. Includes 1,000 practice problems. Solutions available. 2003 edition. 480pp. 6 1/8 x 9 1/4. 0-486-47459-3

TOPOLOGICAL VECTOR SPACES, DISTRIBUTIONS AND KERNELS, François Trèves. Extending beyond the boundaries of Hilbert and Banach space theory, this text focuses on key aspects of functional analysis, particularly in regard to solving partial differential equations. 1967 edition. 592pp. 5 3/8 x 8 1/2. 0-486-45352-9

INTRODUCTION TO PROJECTIVE GEOMETRY, C. R. Wylie, Jr. This introductory volume offers strong reinforcement for its teachings, with detailed examples and numerous theorems, proofs, and exercises, plus complete answers to all odd-numbered end-of-chapter problems. 1970 edition. 576pp. 6 1/8 x 9 1/4. 0-486-46895-X

FOUNDATIONS OF GEOMETRY, C. R. Wylie, Jr. Geared toward students preparing to teach high school mathematics, this text explores the principles of Euclidean and non-Euclidean geometry and covers both generalities and specifics of the axiomatic method. 1964 edition. 352pp. 6 x 9. 0-486-47214-0

Mathematics-History

THE WORKS OF ARCHIMEDES, Archimedes. Translated by Sir Thomas Heath. Complete works of ancient geometer feature such topics as the famous problems of the ratio of the areas of a cylinder and an inscribed sphere; the properties of conoids, spheroids, and spirals; more. 326pp. 5 3/8 x 8 1/2. 0-486-42084-1

THE HISTORICAL ROOTS OF ELEMENTARY MATHEMATICS, Lucas N. H. Bunt, Phillip S. Jones, and Jack D. Bedient. Exciting, hands-on approach to understanding fundamental underpinnings of modern arithmetic, algebra, geometry and number systems examines their origins in early Egyptian, Babylonian, and Greek sources. 336pp. 5 3/8 x 8 1/2. 0-486-25563-8

THE THIRTEEN BOOKS OF EUCLID'S ELEMENTS, Euclid. Contains complete English text of all 13 books of the Elements plus critical apparatus analyzing each definition, postulate, and proposition in great detail. Covers textual and linguistic matters; mathematical analyses of Euclid's ideas; classical, medieval, Renaissance and modern commentators; refutations, supports, extrapolations, reinterpretations and historical notes. 995 figures. Total of 1,425pp. All books 5 3/8 x 8 1/2.
Vol. I: 443pp. 0-486-60088-2
Vol. II: 464pp. 0-486-60089-0
Vol. III: 546pp. 0-486-60090-4

A HISTORY OF GREEK MATHEMATICS, Sir Thomas Heath. This authoritative two-volume set that covers the essentials of mathematics and features every landmark innovation and every important figure, including Euclid, Apollonius, and others. 5 3/8 x 8 1/2. Vol. I: 461pp. 0-486-24073-8
Vol. II: 597pp. 0-486-24074-6

A MANUAL OF GREEK MATHEMATICS, Sir Thomas L. Heath. This concise but thorough history encompasses the enduring contributions of the ancient Greek mathematicians whose works form the basis of most modern mathematics. Discusses Pythagorean arithmetic, Plato, Euclid, more. 1931 edition. 576pp. 5 3/8 x 8 1/2. 0-486-43231-9

CHINESE MATHEMATICS IN THE THIRTEENTH CENTURY, Ulrich Libbrecht. An exploration of the 13th-century mathematician Ch'in, this fascinating book combines what is known of the mathematician's life with a history of his only extant work, the Shu-shu chiu-chang. 1973 edition. 592pp. 5 3/8 x 8 1/2. 0-486-44619-0

PHILOSOPHY OF MATHEMATICS AND DEDUCTIVE STRUCTURE IN EUCLID'S ELEMENTS, Ian Mueller. This text provides an understanding of the classical Greek conception of mathematics as expressed in Euclid's Elements. It focuses on philosophical, foundational, and logical questions and features helpful appendixes. 400pp. 6 1/2 x 9 1/4. 0-486-45300-6

BEYOND GEOMETRY: Classic Papers from Riemann to Einstein, Edited with an Introduction and Notes by Peter Pesic. This is the only English-language collection of these 8 accessible essays. They trace seminal ideas about the foundations of geometry that led to Einstein's general theory of relativity. 224pp. 6 1/8 x 9 1/4. 0-486-45350-2

HISTORY OF MATHEMATICS, David E. Smith. Two-volume history – from Egyptian papyri and medieval maps to modern graphs and diagrams. Non-technical chronological survey with thousands of biographical notes, critical evaluations, and contemporary opinions on over 1,100 mathematicians. 5 3/8 x 8 1/2.
Vol. I: 618pp. 0-486-20429-4
Vol. II: 736pp. 0-486-20430-8

Physics

THEORETICAL NUCLEAR PHYSICS, John M. Blatt and Victor F. Weisskopf. An uncommonly clear and cogent investigation and correlation of key aspects of theoretical nuclear physics by leading experts: the nucleus, nuclear forces, nuclear spectroscopy, two-, three- and four-body problems, nuclear reactions, beta-decay and nuclear shell structure. 896pp. 5 3/8 x 8 1/2. 0-486-66827-4

QUANTUM THEORY, David Bohm. This advanced undergraduate-level text presents the quantum theory in terms of qualitative and imaginative concepts, followed by specific applications worked out in mathematical detail. 655pp. 5 3/8 x 8 1/2. 0-486-65969-0

ATOMIC PHYSICS AND HUMAN KNOWLEDGE, Niels Bohr. Articles and speeches by the Nobel Prize–winning physicist, dating from 1934 to 1958, offer philosophical explorations of the relevance of atomic physics to many areas of human endeavor. 1961 edition. 112pp. 5 3/8 x 8 1/2. 0-486-47928-5

COSMOLOGY, Hermann Bondi. A co-developer of the steady-state theory explores his conception of the expanding universe. This historic book was among the first to present cosmology as a separate branch of physics. 1961 edition. 192pp. 5 3/8 x 8 1/2. 0-486-47483-6

LECTURES ON QUANTUM MECHANICS, Paul A. M. Dirac. Four concise, brilliant lectures on mathematical methods in quantum mechanics from Nobel Prize-winning quantum pioneer build on idea of visualizing quantum theory through the use of classical mechanics. 96pp. 5 3/8 x 8 1/2. 0-486-41713-1

THE PRINCIPLE OF RELATIVITY, Albert Einstein and Frances A. Davis. Eleven papers that forged the general and special theories of relativity include seven papers by Einstein, two by Lorentz, and one each by Minkowski and Weyl. 1923 edition. 240pp. 5 3/8 x 8 1/2. 0-486-60081-5

PHYSICS OF WAVES, William C. Elmore and Mark A. Heald. Ideal as a classroom text or for individual study, this unique one-volume overview of classical wave theory covers wave phenomena of acoustics, optics, electromagnetic radiations, and more. 477pp. 5 3/8 x 8 1/2. 0-486-64926-1

THERMODYNAMICS, Enrico Fermi. In this classic of modern science, the Nobel Laureate presents a clear treatment of systems, the First and Second Laws of Thermodynamics, entropy, thermodynamic potentials, and much more. Calculus required. 160pp. 5 3/8 x 8 1/2. 0-486-60361-X

QUANTUM THEORY OF MANY-PARTICLE SYSTEMS, Alexander L. Fetter and John Dirk Walecka. Self-contained treatment of nonrelativistic many-particle systems discusses both formalism and applications in terms of ground-state (zero-temperature) formalism, finite-temperature formalism, canonical transformations, and applications to physical systems. 1971 edition. 640pp. 5 3/8 x 8 1/2. 0-486-42827-3

QUANTUM MECHANICS AND PATH INTEGRALS: Emended Edition, Richard P. Feynman and Albert R. Hibbs. Emended by Daniel F. Styer. The Nobel Prize–winning physicist presents unique insights into his theory and its applications. Feynman starts with fundamentals and advances to the perturbation method, quantum electrodynamics, and statistical mechanics. 1965 edition, emended in 2005. 384pp. 6 1/8 x 9 1/4. 0-486-47722-3

Physics

INTRODUCTION TO MODERN OPTICS, Grant R. Fowles. A complete basic undergraduate course in modern optics for students in physics, technology, and engineering. The first half deals with classical physical optics; the second, quantum nature of light. Solutions. 336pp. 5 3/8 x 8 1/2. 0-486-65957-7

THE QUANTUM THEORY OF RADIATION: Third Edition, W. Heitler. The first comprehensive treatment of quantum physics in any language, this classic introduction to basic theory remains highly recommended and widely used, both as a text and as a reference. 1954 edition. 464pp. 5 3/8 x 8 1/2. 0-486-64558-4

QUANTUM FIELD THEORY, Claude Itzykson and Jean-Bernard Zuber. This comprehensive text begins with the standard quantization of electrodynamics and perturbative renormalization, advancing to functional methods, relativistic bound states, broken symmetries, nonabelian gauge fields, and asymptotic behavior. 1980 edition. 752pp. 6 1/2 x 9 1/4. 0-486-44568-2

FOUNDATIONS OF POTENTIAL THERY, Oliver D. Kellogg. Introduction to fundamentals of potential functions covers the force of gravity, fields of force, potentials, harmonic functions, electric images and Green's function, sequences of harmonic functions, fundamental existence theorems, and much more. 400pp. 5 3/8 x 8 1/2. 0-486-60144-7

FUNDAMENTALS OF MATHEMATICAL PHYSICS, Edgar A. Kraut. Indispensable for students of modern physics, this text provides the necessary background in mathematics to study the concepts of electromagnetic theory and quantum mechanics. 1967 edition. 480pp. 6 1/2 x 9 1/4. 0-486-45809-1

GEOMETRY AND LIGHT: The Science of Invisibility, Ulf Leonhardt and Thomas Philbin. Suitable for advanced undergraduate and graduate students of engineering, physics, and mathematics and scientific researchers of all types, this is the first authoritative text on invisibility and the science behind it. More than 100 full-color illustrations, plus exercises with solutions. 2010 edition. 288pp. 7 x 9 1/4. 0-486-47693-6

QUANTUM MECHANICS: New Approaches to Selected Topics, Harry J. Lipkin. Acclaimed as "excellent" (*Nature*) and "very original and refreshing" (*Physics Today*), these studies examine the Mössbauer effect, many-body quantum mechanics, scattering theory, Feynman diagrams, and relativistic quantum mechanics. 1973 edition. 480pp. 5 3/8 x 8 1/2. 0-486-45893-8

THEORY OF HEAT, James Clerk Maxwell. This classic sets forth the fundamentals of thermodynamics and kinetic theory simply enough to be understood by beginners, yet with enough subtlety to appeal to more advanced readers, too. 352pp. 5 3/8 x 8 1/2. 0-486-41735-2

QUANTUM MECHANICS, Albert Messiah. Subjects include formalism and its interpretation, analysis of simple systems, symmetries and invariance, methods of approximation, elements of relativistic quantum mechanics, much more. "Strongly recommended." – *American Journal of Physics.* 1152pp. 5 3/8 x 8 1/2. 0-486-40924-4

RELATIVISTIC QUANTUM FIELDS, Charles Nash. This graduate-level text contains techniques for performing calculations in quantum field theory. It focuses chiefly on the dimensional method and the renormalization group methods. Additional topics include functional integration and differentiation. 1978 edition. 240pp. 5 3/8 x 8 1/2. 0-486-47752-5

Physics

MATHEMATICAL TOOLS FOR PHYSICS, James Nearing. Encouraging students' development of intuition, this original work begins with a review of basic mathematics and advances to infinite series, complex algebra, differential equations, Fourier series, and more. 2010 edition. 496pp. 6 1/8 x 9 1/4. 0-486-48212-X

TREATISE ON THERMODYNAMICS, Max Planck. Great classic, still one of the best introductions to thermodynamics. Fundamentals, first and second principles of thermodynamics, applications to special states of equilibrium, more. Numerous worked examples. 1917 edition. 297pp. 5 3/8 x 8. 0-486-66371-X

AN INTRODUCTION TO RELATIVISTIC QUANTUM FIELD THEORY, Silvan S. Schweber. Complete, systematic, and self-contained, this text introduces modern quantum field theory. "Combines thorough knowledge with a high degree of didactic ability and a delightful style." – *Mathematical Reviews.* 1961 edition. 928pp. 5 3/8 x 8 1/2. 0-486-44228-4

THE ELECTROMAGNETIC FIELD, Albert Shadowitz. Comprehensive undergraduate text covers basics of electric and magnetic fields, building up to electromagnetic theory. Related topics include relativity theory. Over 900 problems, some with solutions. 1975 edition. 768pp. 5 5/8 x 8 1/4. 0-486-65660-8

THE PRINCIPLES OF STATISTICAL MECHANICS, Richard C. Tolman. Definitive treatise offers a concise exposition of classical statistical mechanics and a thorough elucidation of quantum statistical mechanics, plus applications of statistical mechanics to thermodynamic behavior. 1930 edition. 704pp. 5 5/8 x 8 1/4. 0-486-63896-0

INTRODUCTION TO THE PHYSICS OF FLUIDS AND SOLIDS, James S. Trefil. This interesting, informative survey by a well-known science author ranges from classical physics and geophysical topics, from the rings of Saturn and the rotation of the galaxy to underground nuclear tests. 1975 edition. 320pp. 5 3/8 x 8 1/2. 0-486-47437-2

STATISTICAL PHYSICS, Gregory H. Wannier. Classic text combines thermodynamics, statistical mechanics, and kinetic theory in one unified presentation. Topics include equilibrium statistics of special systems, kinetic theory, transport coefficients, and fluctuations. Problems with solutions. 1966 edition. 532pp. 5 3/8 x 8 1/2. 0-486-65401-X

SPACE, TIME, MATTER, Hermann Weyl. Excellent introduction probes deeply into Euclidean space, Riemann's space, Einstein's general relativity, gravitational waves and energy, and laws of conservation. "A classic of physics." – *British Journal for Philosophy and Science.* 330pp. 5 3/8 x 8 1/2. 0-486-60267-2

RANDOM VIBRATIONS: Theory and Practice, Paul H. Wirsching, Thomas L. Paez and Keith Ortiz. Comprehensive text and reference covers topics in probability, statistics, and random processes, plus methods for analyzing and controlling random vibrations. Suitable for graduate students and mechanical, structural, and aerospace engineers. 1995 edition. 464pp. 5 3/8 x 8 1/2. 0-486-45015-5

PHYSICS OF SHOCK WAVES AND HIGH-TEMPERATURE HYDRO DYNAMIC PHENOMENA, Ya B. Zel'dovich and Yu P. Raizer. Physical, chemical processes in gases at high temperatures are focus of outstanding text, which combines material from gas dynamics, shock-wave theory, thermodynamics and statistical physics, other fields. 284 illustrations. 1966–1967 edition. 944pp. 6 1/8 x 9 1/4. 0-486-42002-7